RIEMANNIAN SUBMERSIONS

AND RELATED TOPICS

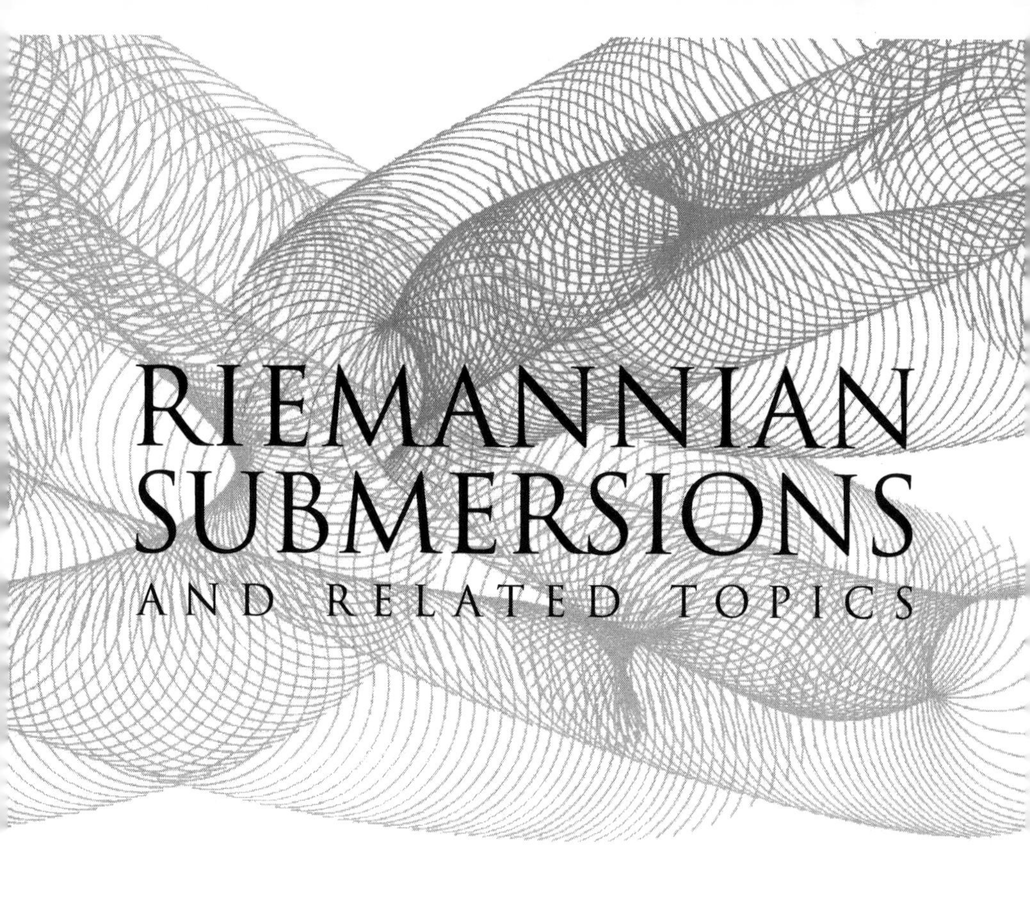

RIEMANNIAN SUBMERSIONS
AND RELATED TOPICS

Maria Falcitelli
University of Bari, Italy

Stere Ianus
University of Bucharest, Romania

Anna Maria Pastore
University of Bari, Italy

NEW JERSEY • LONDON • SINGAPORE • BEIJING • SHANGHAI • HONG KONG • TAIPEI • CHENNAI

Published by

World Scientific Publishing Co. Pte. Ltd.
5 Toh Tuck Link, Singapore 596224
USA office: 27 Warren Street, Suite 401-402, Hackensack, NJ 07601
UK office: 57 Shelton Street, Covent Garden, London WC2H 9HE

Library of Congress Cataloging-in-Publication Data
Falcitelli, Maria, 1954–
Riemannian submersions and related topics / Maria Falcitelli and Anna Maria Pastore, Stere Ianus.
p. cm.
Includes bibliographical references and index.
ISBN-13 978-981-238-896-4 (alk. paper)
ISBN-10 981-238-896-6 (alk. paper)
1. Riemannian submersions. I. Pastore, Anna Maria, 1945– . II. Ianus, Stere. III. Title.

QA649.F35 2004
516.3'6--dc22

2004047857

British Library Cataloguing-in-Publication Data
A catalogue record for this book is available from the British Library.

Printed in Singapore

To our parents

Preface

Immersions and submersions, which are special tools in Differential Geometry, also play a fundamental role in Riemannian Geometry, expecially when the involved manifolds carry an additional structure (of contact, Hermitian, quaternionic type, *etc.*). Even if submersions are, in a certain sense, a counterpart of immersions, the corresponding theories are quite different, also from a historical point of view.

The theory of isometric immersions, started with the work of Gauss on surfaces in the Euclidean 3-space, is classical and widely explained in many books, whereas the theory of Riemannian submersion goes back to four decades ago, when B. O'Neill and A. Gray, independently, formulated the basis of such theory, which has hugely been developed in the last two decades. Nowadays several works are still in progress. For instance, a new point of view on Riemannian submersions appears in a paper by H. Karcher in 1999.

Obviously the content of this book is not exhaustive, anyway, the results presented are enough to solve problems concerning many areas, like Theoretical Physics and the theory of Einstein, EinsteinWeyl spaces. This theory falls into a more general context, extensively treated in the Besse's book *Einstein Manifolds* and, more recently, in *Surveys in differential geometry: essays on Einstein manifolds*, edited by C. Le Brun and M. Wang.

At present, there is no work entirely devoted to a systematic exposition of the basic tools on the theory of Riemannian submersions and its developments. This is the aim of the present book, which is directed to graduate students as well as to researchers interested in Differential Geometry and Theoretical Physics. For this reason, many examples are explained in details and calculations, straightforward for specialists, are not omitted. Examples appear in each chapter. They often involve well-known definitions and concepts, which are presented in the subsequent chapters and can be found through the index. Anyway, the reader should be familiar with the basic concepts of Lie groups, principal bundles and Riemannian Geometry. We refer to the books of Kobayashi and Nomizu [172], Helgason [139], Yano and Kon [350] for a wide treatment of these subjects.

Concerning the content of each chapter, which we are going to outline, we only remark that Chap. 8 is rather independent of the others, with the exception of Chap. 1.

In Chap. 1, where the basic tools of the mentioned theory are given, the main properties of the invariant tensors introduced by B. O'Neill are stated, together with explicit formulas relating the curvatures of the total space, the fibres and the base space. We also describe classical examples, like Hopf fibrations and generalized Hopf fibrations.

Chapter 2 essentially concerns with Riemannian submersions having totally geodesic fibres. In particular we state the classification theorem, due to R. Escobales and A. Ranjan, on the Riemannian submersions with totally geodesic fibres and the standard m-sphere as total space. For any $n \geq 1$, the Hopf projections provide examples of the above submersions and the representation theory of Clifford algebras is applied to obtain the uniqueness result, up to equivalence. Then, combining a theorem of R. Escobales with a result of J. Ucci, we classify the Riemannian submersions from $\mathbf{P_{2n+1}}(\mathbf{C})$ onto $\mathbf{P_n}(\mathbf{Q})$ with complete, complex and totally geodesic fibres.

Almost Hermitian submersions are investigated into details in Chap. 3. In particular, we discuss the transference of geometric properties from the total space to the fibres and to the base space. We also describe some results, due to D. L. Johnson, in order to illustrate the relationship between the existence of Kähler submersions and of holomorphic connections on principal bundles over Kähler manifolds. Furthermore, we examine almost Hermitian submersions having integrable horizontal distribution, as it happens for Kähler, almost Kähler and nearly Kähler submersions. Considering locally conformal Kähler submersions, we explain the results of J. C. Marrero and J. Rocha and give a large class of examples, involving generalized Hopf manifolds, also known as Vaisman manifolds. Finally, we discuss almost complex conformal submersions.

Riemannian submersions between Riemannian manifolds equipped with an additional structure of (almost) contact type, firstly studied by B. Watson and D. Chinea, independently, are the subject of Chap. 4. Riemannian submersions from an almost contact metric manifold with an f-structure onto an almost quaternionic manifold are also considered. In particular, we show how 3-Sasakian manifolds can be related with quaternionic Kähler or hyperKähler manifolds. Submersions from a metric f-manifold onto an almost Hermitian manifold are also studied. Finally, we consider the concept of hyper f-structure due to G. Hernandez and examine into details Riemannian submersions with totally geodesic fibres from a $\mathcal{PS}$-manifold to a hyperKähler one.

Chapter 5 deals with the problem of determining Einstein metrics and EinsteinWeyl structures on manifolds which can be considered as the total space or the base space of a submersion. In particular, we explain the method used by W. Ziller for describing the invariant Einstein metrics on the spheres and on the projective spaces. Following the papers of M. Wang, W. Ziller and Y. Sakane, we also discuss the existence of Einstein metrics on the total space of a principal torus bundle over the product of KählerEinstein manifolds and, more generally, over a **C**-space. Conditions on the total space of a principal bundle, due to H. Pedersen and A. Swann, are also stated. We point out the interrelation between Riemannian submersions and Hermitian or almost contact EinsteinWeyl structures. Since in the compact case Hermitian EinsteinWeyl spaces are just Kähler Einstein or Vaisman manifolds, this subject is closely related to Chap. 3. Following I. Vaisman, H. Pedersen, Y. Poon and A. Swann, we prove the existence of a Kähler structure on the leaves space of a compact Vaisman manifold by the vertical foliation. Finally, we state recent results of F. Narita relating Sasakian structures with Einstein–Weyl structures on manifolds which are the total space of a submersion.

Using the concept of CR-submanifold of an almost Hermitian manifold, due to A. Bejancu, in Chap. 6 we consider CR-submersions and present a result of S. Kobayashi and some generalizations. Then, we show how the theory of Riemannian submersions allows to link submanifolds of Sasakian manifolds to submanifolds of Kähler ones; particular emphasis on the results of H. Reckziegel is given. Finally, using the concept of reflection with respect to a submanifold introduced by B. Y. Chen and L. Vanhecke, we prove recent theorems of F. Narita concerning submersions such that the reflections with respect to the fibres are isometries.

Chapter 7 deals with semi-Riemannian submersions. Lorentzian submersions with totally geodesic fibres are studied. In particular, we prove a theorem of M. Magid stating that, if the total space is the anti-de Sitter space, then the base is holomorphically isometric to the complex hyperbolic space. We also present the classification of semi-Riemannian submersions with totally geodesic fibres from a real or complex pseudo-hyperbolic space onto a Riemannian manifold. These results, due to G. Baditoiu and the second author, are closely linked to the subject of Chap. 2. Finally, we investigate semi-Riemannian submersions with totally umbilical or minimal fibres.

Applications of Riemannian submersions in Physics is the tool of the last chapter. In the theory of Kaluza–Klein type, one starts with the hypothesis that the space-time has $(4+m)$-dimensions. Next, because of some dynamical mechanism, one supposes that the ground state of this system is partially compactified, *i.e.* of type $M^4 \times M^m$, where M^4 denotes a 4-dimensional Minkowski space and M^m is a compact m-dimensional space. An interesting mechanism for space-time compactification is proposed in the form of a nonlinear sigma model. The general solutions of this model can be expressed in terms of harmonic maps satisfying the Einstein equations. A very general class of solutions is given by Riemannian submersions from the extra-dimensional space onto the space in which the scalar fields of the nonlinear sigma model take values.

Acknowledgments.
We want to express our gratitude to the people who helped us with the manuscript. In particular, we are in debt to Professor W. Ziller for his precious suggestions and to Professor M. Visinescu for helpful discussions on Chapter 8. We also thank Professors D. Blair, R. Escobales for their useful remarks and Professor L. Ornea for discussions on the subject.

Finally, we wish to thank the editorial board of World Scientific Publishing Co., in particular Ms. E.H. Chionh.

Maria Falcitelli
Stere Ianus
Anna Maria Pastore

Contents

Chapter 1

Riemannian Submersions

The aim of this chapter is to describe the basic tools of the theory of Riemannian submersions. Such a theory, hugely developed in the last three decades, arises from the papers of R. Hermann ([140]), B. O'Neill ([226]) and A. Gray ([126]). The basic terminology is the subject of Sec. 1.1.

In Sec. 1.2 we describe in details the standard Sasakian structure on the sphere $\mathbf{S^{2n+1}}$ and the Hopf fibration from $\mathbf{S^{2n+1}}$ onto the complex projective space $\mathbf{P_n(C)}$ equipped with the Fubini–Study metric.

Section 1.3 is devoted to state the main properties of the invariant tensors T, A introduced by B. O'Neill. Explicit formulas involving T, A and relating the Riemannian curvatures of the total and the base spaces are given. The generalized Hopf submersion from $\mathbf{S^{4n+3}}$ onto the quaternionic projective space $\mathbf{P_n(Q)}$ is described in Sec. 1.4; the relevance of this example is shown in Chap. 2. Other classical examples are given. In the last part of this section, a concise exposition of the theory developed by F. Tricerri and L. Vanhecke ([295]) on Riemannian homogeneous spaces is given. We explain how the existence of a suitable homogeneous structure on a Riemannian manifold (M, g) allows to consider M as the total space of a Riemannian submersion.

In Sec. 1.5, we characterize the geodesics on the total space of a Riemannian submersion by means of equations involving the invariants T, A. This allows to prove the B. O'Neill theorem, ([227]), stating that a Riemannian submersion is determined by its invariants and its differential map at one point.

Finally, in Sec. 1.6, we study the Clairaut submersions, proving the result of R. L. Bishop ([30]), which characterizes such submersions as those with totally umbilical fibres and prescribed mean curvature vector. Examples of Clairaut submersions are also given.

1.1 Riemannian Submersions

Let (M, g) and (B, g') be C^∞-Riemannian manifolds of dimension m and n, respectively. A surjective C^∞-map $\pi : M \to B$ is a *C^∞-submersion* if it has maximal rank at any point of M. The implicit function theorem states that the fibre over any $x \in B, \pi^{-1}(x)$, is a closed r-dimensional submanifold of $M, r = m-n$. Putting $\mathcal{V}_p = \ker \pi_{*p}$, for any $p \in M$, we obtain an integrable distribution $\mathcal{V}$ which corresponds to the foliation of M determined by the fibres of π, since each $\mathcal{V}_p$ coincides with the tangent space of $\pi^{-1}(x)$ at p, $\pi(p) = x$. Indeed, given a vector v in $T_p\pi^{-1}(x)$, let $c : [0,1] \to \pi^{-1}(x)$ be a curve such that $c(0) = p$, $\dot{c}(0) = v$. Since $(\pi \circ c)(t) = x$, $t \in [0,1]$, one has $\pi_*(\dot{c}(0)) = (\pi \circ c)_*(\frac{d}{dt}) = 0$, that is $v = \dot{c}(0) \in \mathcal{V}_p$. Therefore, $T_p\pi^{-1}(x)$ turns out to be an r-dimensional subspace of $\mathcal{V}_p$, and the equality of the dimensions implies $\mathcal{V}_p = T_p\pi^{-1}(x)$.

Each $\mathcal{V}_p$ is called the *vertical space* at p, $\mathcal{V}$ is the *vertical distribution*, the sections of $\mathcal{V}$ are the so-called *vertical vector fields* and determine a Lie subalgebra $\mathcal{X}^v(M)$ of $\mathcal{X}(M)$.

Let $\mathcal{H}$ be the complementary distribution of $\mathcal{V}$ determined by the Riemannian metric g. So, at any $p \in M$, one has the orthogonal decomposition $T_p(M) = \mathcal{V}_p \oplus \mathcal{H}_p$; $\mathcal{H}_p$ is called the *horizontal space* at p. The sections of the horizontal distribution $\mathcal{H}$ are the *horizontal vector fields*; they set up a subspace $\mathcal{X}^h(M)$ of $\mathcal{X}(M)$. For any $E \in \mathcal{X}(M)$, vE and hE denote the vertical and the horizontal components of E, respectively.

Moreover, $\mathcal{X}^c(M)$ stands for the space of the projectable vector fields on M, that is each element of $\mathcal{X}^c(M)$ is a vector field on M which is π-related to a vector field on B. It is easy to prove that $\mathcal{X}^c(M)$ is a Lie subalgebra of $\mathcal{X}(M)$.

Furthermore, one can consider the exact sequence:

$$0 \to \mathcal{X}^v(M) \overset{\alpha}{\to} \mathcal{X}^c(M) \overset{\beta}{\to} \mathcal{X}(B) \to 0 \tag{1.1}$$

where α and β are canonical maps.

The space $\mathcal{X}^b(M) = \mathcal{X}^c(M) \cap \mathcal{X}^h(M)$ is the space of the basic vector fields; it is isomorphic to $\mathcal{X}(B)$ via β. Given $X' \in \mathcal{X}(B)$, the basic vector field π-related to X' is named the *horizontal lift* of X'. Note that the basic vector fields locally span the horizontal distribution.

For a map $\pi : M \to B$, we denote by $d\pi$ or π_* the tangent map of π. Then, when X is a vector tangent to M at a point p, then $\pi_* X$ means $(\pi_*)_p X$. For a given projectable vector field X, sometimes we will write

$\pi_* X$ to denote the vector field X' to which X is π-related. Otherwise, any formula containing $\pi_* X$ has to be valued pointwise.

Definition 1.1 A C^∞-submersion $\pi : (M, g) \to (B, g')$ is called a *Riemannian submersion* if, at each point p of M, π_{*p} preserves the length of the horizontal vectors.

Since π is a submersion, π_{*p} determines a linear isomorphism between $\mathcal{H}_p$ and $T_{\pi(p)}B$, so that one has:

$$g_p(u, v) = g'_{\pi(p)}(\pi_{*p}u, \pi_{*p}v) \ , \quad u, v \in \mathcal{H}_p \ , \ p \in M \ , \tag{1.2}$$

and π_{*p} acts on $\mathcal{H}_p$ as a linear isometry.

Proposition 1.1 *Let $\pi : (M, g) \to (B, g')$ be a Riemannian submersion, and denote by ∇ and ∇' the Levi-Civita connections of M and B, respectively. If X, Y are basic vector fields, π-related to X', Y', one has:*

i) $g(X, Y) = g'(X', Y') \circ \pi$;
ii) $h[X, Y]$ *is the basic vector field π-related to* $[X', Y']$;
iii) $h(\nabla_X Y)$ *is the basic vector field π-related to* $\nabla'_{X'} Y'$;
iv) *for any vertical vector field V, $[X, V]$ is vertical.*

Proof. Property i) is an immediate consequence of (1.2). Moreover, for any $X, Y \in \mathcal{X}^b(M)$, π-related to X', Y', one has that $[X, Y]$ is π-related to $[X', Y']$, and ii) follows. We recall the formula which defines the Levi-Civita connection:

$$\begin{aligned} 2g(\nabla_X Y, Z) = {} & X(g(Y, Z)) + Y(g(X, Z)) - Z(g(X, Y)) \\ & + g([X, Y], Z) + g([Z, X], Y) - g([Y, Z], X) \ , \end{aligned} \tag{1.3}$$

for any $X, Y, Z \in \mathcal{X}(M)$. Then, considering X, Y, Z as the horizontal lifts of the vector fields X', Y', Z', one has: $X(g(Y, Z)) = X'(g'(Y', Z')) \circ \pi$, $g([X, Y], Z) = g'([X', Y'], Z') \circ \pi$ and combining with (1.3), one obtains:

$$g'(\pi_*(h\nabla_X Y), Z') \circ \pi = g(h\nabla_X Y, Z) = g'(\nabla'_{X'} Y', Z') \circ \pi \ .$$

Then, iii) follows, since π is surjective and Z' is arbitrarily chosen. Finally, for any $V \in \mathcal{X}^v(M)$ and $X \in \mathcal{X}^b(M)$, π-related to X', $[X, V]$ is π-related to $[X', 0]$, and iv) follows.

1.2 The Hopf Fibration

Let $i : (\mathbf{S^{2n+1}}, g) \to (\mathbf{R^{2n+2}}, <\,,\,>)$ be the standard isometric embedding of the unit sphere $\mathbf{S^{2n+1}}$ into $\mathbf{R^{2n+2}}$, equipped with the canonical metric. Any vector field X on $\mathbf{S^{2n+1}}$ is identified with $i_* X$.

We consider the so-called *Sasakian structure* on $\mathbf{S^{2n+1}}$ induced by the standard complex structure J on $\mathbf{C^{n+1}}$, identified with $\mathbf{R^{2n+2}}$, which is described as follows. Let N be the unit outward vector field normal to $\mathbf{S^{2n+1}}$. Putting:

$$JN = -\xi \,, \tag{1.4}$$

ξ turns out to be a global vector field tangent to $\mathbf{S^{2n+1}}$. Moreover, for any vector field X on $\mathbf{S^{2n+1}}$, the decomposition of JX in the tangent and normal components determines a $(1,1)$-tensor field φ and a 1-form η on $\mathbf{S^{2n+1}}$, such that:

$$JX = \varphi(X) + \eta(X)N \,. \tag{1.5}$$

Proposition 1.2 $(\mathbf{S^{2n+1}}, \varphi, \xi, \eta, g)$ *is an almost contact metric manifold.*

Proof. According to (1.4) and (1.5), for any $X \in \mathcal{X}(\mathbf{S^{2n+1}})$, one has:

$$X = -J(\varphi(X)) + \eta(X)\xi = -\varphi^2(X) - \eta(\varphi(X))N + \eta(X)\xi \,.$$

Then, separating the tangent and normal parts, one obtains:

$$\varphi^2 = -I + \eta \otimes \xi \,, \tag{1.6}$$

$$\eta \circ \varphi = 0 \,, \tag{1.7}$$

where I denotes the identity transformation of the tangent bundle of M. Moreover, (1.4) and (1.5) imply $N = \varphi(\xi) + \eta(\xi)N$, that is:

$$\varphi(\xi) = 0, \;\; \eta(\xi) = 1 \,. \tag{1.8}$$

Equations (1.6), (1.8) mean that (φ, ξ, η) is an almost contact structure on $\mathbf{S^{2n+1}}$ (see Sec. 4.1). Finally, since $<\,,\,>$ is J-Hermitian, using (1.5), one has:

$$g(\varphi(X), \varphi(Y)) = g(X,Y) - \eta(X)\eta(Y), \;\; X, Y \in \mathcal{X}(\mathbf{S^{2n+1}}) \,. \tag{1.9}$$

Thus, the almost contact structure is compatible with g.

Remark 1.1 Formulas (1.8) and (1.9) entail:

$$g(X,\xi) = \eta(X), \quad X \in \mathcal{X}(\mathbf{S^{2n+1}}) , \tag{1.10}$$

that is ξ is the dual vector field of η with respect to g. Moreover, (1.6), (1.7) and (1.9) also imply $g(X,\varphi Y) = -g(\varphi X, Y)$. Putting

$$\Phi(X,Y) = g(X,\varphi Y) , \tag{1.11}$$

one obtains the fundamental 2-form of the almost contact structure.

Proposition 1.3 *$(\mathbf{S^{2n+1}}, \varphi, \xi, \eta, g)$ is a Sasakian manifold.*

Proof. As in Sec. 4.1, a Sasakian manifold is defined as a normal, contact metric manifold (normal, almost contact metric manifold with the 1-form η of maximal rank). Thus, firstly we need to show that $\eta \wedge (d\eta)^n \neq 0$ everywhere. To this aim, using a result due to Blair ([33]), we prove the equivalent condition $d\eta = \Phi$. It is well known that, in this case, the Gauss and Weingarten equations for a hypersurface ([172] Vol. II) reduce to:

$$D_X Y = \nabla_X Y - g(X,Y)N ; \tag{1.12}$$

$$D_X N = X , \tag{1.13}$$

where D denotes the Levi-Civita connection on $\mathbf{R^{2n+2}}$. Applying (1.5), (1.12), (1.13), (1.10) one has:

$$(D_X J)Y = (\nabla_X \varphi)Y + g(\varphi(X),Y)N - g(X,Y)\xi + g(\nabla_X \xi, Y)N + \eta(Y)X ,$$

and, since $(\mathbf{R^{2n+2}}, J, <\, ,\, >)$ is a Kähler manifold, this relation gives:

$$(\nabla_X \varphi)Y = g(X,Y)\xi - \eta(Y)X , \quad \nabla_X \xi = -\varphi(X) , \quad \nabla_\xi \xi = 0 . \tag{1.14}$$

The skew-symmetry of Φ and the second relation above imply that ξ is a Killing vector field and, combining with (1.14), one has:

$$2d\eta(X,Y) = g(\nabla_X \xi, Y) - g(\nabla_Y \xi, X) = -2g(\varphi X, Y) = 2\Phi(X,Y) .$$

So, $d\eta = \Phi$ and (φ, ξ, η, g) is a contact metric structure. Finally, (1.14) implies the vanishing of the tensor field $[\varphi,\varphi] + 2d\eta \otimes \xi$, (*i.e.* the normality condition), where $[\varphi,\varphi]$ is the Nijenhuis tensor of φ, defined by

$$[\varphi,\varphi](X,Y) = [\varphi X, \varphi Y] - \varphi([\varphi X, Y]) - \varphi([X, \varphi Y]) + \varphi^2([X,Y]) ,$$

for any vector fields X, Y.

As proved in [33; 34], the first relation in (1.14) characterizes the Sasakian manifolds among the almost contact metric manifolds.

Theorem 1.1 (Hopf fibration). *The manifold* $\mathbf{S^{2n+1}}$ *is the total space of a principal fibre bundle over the* n*-dimensional complex projective space* $\mathbf{P_n(C)}$ *with structure group* $\mathbf{S^1}$*. The projection* $\pi : \mathbf{S^{2n+1}} \to \mathbf{P_n(C)}$ *is a Riemannian submersion with respect to the canonical metric* g *on* $\mathbf{S^{2n+1}}$ *and to a Kähler metric on* $\mathbf{P_n(C)}$ *with constant holomorphic sectional curvature* $c = 4$.

Proof. The first part of the statement is proved in [172], (Vol. II, Chap. IX, ex. 2.4). It can also be obtained applying a result due to Ogiue ([224]). The quotient space of $\mathbf{S^{2n+1}}$ with respect to the $\mathbf{S^1}$-action is identified with $\mathbf{P_n(C)}$. Let $\pi : \mathbf{S^{2n+1}} \to \mathbf{P_n(C)}$ be the projection. The vector field ξ defined in (1.4) is the fundamental vector field corresponding to the generator of the Lie algebra $\mathbf{R}$ of $\mathbf{S^1}$. Therefore, ξ is invariant with respect to the $\mathbf{S^1}$-action and, for any $p \in \mathbf{S^{2n+1}}$, ξ_p spans the vertical subspace at p. Moreover, since ξ is a Killing vector field, g is invariant under the $\mathbf{S^1}$-action. This property, combined with the condition $\eta(\xi) = 1$, implies that η can be regarded as the 1-form of a connection Γ on $\mathbf{S^{2n+1}}$. Obviously, π has maximal rank at any point, so π is a submersion whose horizontal distribution coincides with the Γ-horizontal distribution. Formula (1.14) also implies the vanishing of the Lie derivative $L_\xi\varphi$, that is the invariance of φ with respect to the $\mathbf{S^1}$-action. Denoting by R_a, $a \in \mathbf{S^1}$, the right translation on $\mathbf{S^{2n+1}}$, and by X^* the horizontal lift of $X' \in \mathcal{X}(\mathbf{P_n(C)})$, for any $p \in \mathbf{S^{2n+1}}$, one has:

$$\pi_{*pa}(\varphi_{pa}(X^*_{pa})) = \pi_{*pa}((R_{a*})_p(\varphi_p X^*_p)) = \pi_{*p}(\varphi_p X^*_p) \ ,$$

ensuring that φX^* is projectable. This allows to define a $(1,1)$-tensor field J on $\mathbf{P_n(C)}$ putting

$$JX' = \pi_*(\varphi X^*) \ , \tag{1.15}$$

i.e. JX' is the vector field to which φX^* is π-related. Thus, J turns out to be an almost complex structure on $\mathbf{P_n(C)}$, since applying (1.15) and (1.6), one has: $J^2X' = -X'$. Since $[\xi, X^*]$ is a vertical vector field, the symmetry of ∇, combined with (1.14), gives:

$$g([\xi, X^*], \xi) = \xi(g(X^*, \xi)) - g(X^*, \nabla_\xi \xi) + g(\varphi X^*, \xi) = 0 \ ,$$

and $[\xi, X^*]$ vanishes. Then, since ξ is Killing, we get $0 = (L_\xi g)(X^*, Y^*) =$

$\xi(g(X^*, Y^*))$, so we can consider the Riemannian metric G on $\mathbf{P_n}(\mathbf{C})$ defined by:

$$G(X', Y') \circ \pi = g(X^*, Y^*), \quad X', Y' \in \mathcal{X}(\mathbf{P_n}(\mathbf{C})), \tag{1.16}$$

which makes π a Riemannian submersion. Applying (1.16), (1.15), (1.9), one has:

$$G(JX', JY') \circ \pi = g(\varphi X^*, \varphi Y^*) = g(X^*, Y^*) = G(X', Y') \circ \pi ,$$

hence $(\mathbf{P_n}(\mathbf{C}), J, G)$ is an almost Hermitian manifold, whose 2-form Ω satisfies $\Omega(X', Y') \circ \pi = \Phi(X^*, Y^*)$. It is easy to verify that the Nijenhuis tensor of J is given by $N_J(X', Y') = \pi_*(([\varphi, \varphi] + 2d\eta \otimes \xi)(X^*, Y^*))$. So, $(\mathbf{P_n}(\mathbf{C}), J, G)$ turns out to be a Kähler manifold, since Φ is closed and the contact structure is normal. Finally, to compute the Riemannian curvature R' of $(\mathbf{P_n}(\mathbf{C}), G)$, by a routine calculation (see also Proposition 5.1 in [224]), one has:

$$\nabla_{X^*} Y^* = (\nabla'_{X'} Y')^* + \frac{1}{2}\eta[X^*, Y^*]\xi , \tag{1.17}$$

where ∇' is the Levi-Civita connection of G. Now, using $[\xi, X^*] = 0$ and (1.14) we get:

$$\nabla_\xi X^* = -\varphi X^* . \tag{1.18}$$

Formulas (1.17), (1.14) and $\Phi(X^*, Y^*) = d\eta(X^*, Y^*) = -\frac{1}{2}\eta[X^*, Y^*]$ entail:

$$\begin{aligned} G(\nabla'_{X'} \nabla'_{Y'} Z', W') \circ \pi &= g(\nabla_{X^*} \nabla_{Y^*} Z^*, W^*) - \tfrac{1}{2}\eta[Y^*, Z^*] g(\nabla_{X^*}\xi, W^*) \\ &= g(\nabla_{X^*} \nabla_{Y^*} Z^*, W^*) + \Phi(Y^*, Z^*)\Phi(X^*, W^*) ; \end{aligned}$$

$$\begin{aligned} G(\nabla'_{[X', Y']} Z', W') \circ \pi &= g(\nabla_{[X^*, Y^*]} Z^*, W^*) + 2\Phi(X^*, Y^*) g(\nabla_\xi Z^*, W^*) \\ &= g(\nabla_{[X^*, Y^*]} Z^*, W^*) + 2\Phi(X^*, Y^*)\Phi(Z^*, W^*) . \end{aligned}$$

Therefore, since g has constant sectional curvature $K = 1$, one has:

$$\begin{aligned} G(R'(X', Y')Z', W') = {} & G(Y', Z')G(X', W') - G(X', Z')G(Y', W') \\ & + G(Y', JZ')G(X', JW') - G(X', JZ')G(Y', JW') \\ & - 2G(X', JY')G(Z', JW') . \end{aligned}$$

Hence $(\mathbf{P_n}(\mathbf{C}), J, G)$ has constant holomorphic sectional curvature $c = 4$ ([172], Vol. II, p. 167).

Remark 1.2 A detailed description of the Hopf mapping from the point of view of complex manifolds is given in [169]. Namely, the standard Hermitian inner product $< , >$ on $\mathbf{C^{n+1}}$, defined by $< z, w >= \sum_{\alpha=0}^{n} z^{\alpha}\overline{w}^{\alpha}$, induces the Riemannian metric g^* on $(\mathbf{C^{n+1}})^*$ such that

$$< z, z > g_z^*(X^*, Y^*) = \mathrm{Re} < X^*, Y^* >,\ z \in (\mathbf{C^{n+1}})^* ,$$

for any $X^*, Y^* \in T_z(\mathbf{C^{n+1}})^*$. Then, at any point $z \in (\mathbf{C^{n+1}})^*$, one can consider the orthogonal decomposition

$$T_z(\mathbf{C^{n+1}})^* = T_z^v(\mathbf{C^{n+1}})^* \oplus T_z^h(\mathbf{C^{n+1}})^* ,$$

where $T_z^v(\mathbf{C^{n+1}})^*$ is the complex linear space spanned by z and $T_z^h(\mathbf{C^{n+1}})^*$ its $<,>$-orthogonal complement. Let $F : (\mathbf{C^{n+1}})^* \to \mathbf{P_n}(\mathbf{C})$ be the map whose value at any z is the linear space spanned by z, regarded as a point of $\mathbf{P_n}(\mathbf{C})$. Considering on $\mathbf{P_n}(\mathbf{C})$ the canonical structure of complex manifold, it is easy to prove that F is a submersion whose vertical and horizontal spaces at any point z coincide with $T_z^v(\mathbf{C^{n+1}})^*$ and $T_z^h(\mathbf{C^{n+1}})^*$, respectively. The metric on $\mathbf{P_n}(\mathbf{C})$ which makes F a Riemannian submersion is the Fubini–Study metric of constant holomorphic sectional curvature $c = 4$. Moreover, the unit sphere $\mathbf{S^{2n+1}} = \{z \in (\mathbf{C^{n+1}})^* \mid < z, z >= 1\}$, with the metric induced by g^*, is embedded in $(\mathbf{C^{n+1}})^*$ as a totally geodesic submanifold with constant sectional curvature 1. The restriction of F to $\mathbf{S^{2n+1}}$ is then a Riemannian submersion onto $\mathbf{P_n}(\mathbf{C})$, also known as the *Hopf submersion.*

Remark 1.3 More generally, let $\mathbf{S^{2n+1}}(1, k)$ be the Sasakian space form given by the sphere $\mathbf{S^{2n+1}}$ endowed with a Sasakian structure (φ, ξ, η, g), g being the metric of constant φ-sectional curvature $k > -3$, as described in [34]. This manifold belongs, for $c = 1$, to the class of c-Sasakian manifolds of constant φ-sectional curvature $k > -3c^2$ described in Example 3.2. Let $\pi : (\mathbf{S^{2n+1}}(1, k), g) \to (\mathbf{P_n}(\mathbf{C})(k + 3), G)$ be the Hopf fibration, where G is the Fubini–Study metric with holomorphic sectional curvature $k + 3$. For any real number $a \neq 0$ the metric $g_a = \pi^*G + a^2\eta \otimes \eta$ considered in [219] makes $\pi : (\mathbf{S^{2n+1}}(1, k), g_a) \to (\mathbf{P_n}(\mathbf{C})(k+3), G)$ a Riemannian submersion.

1.3 Fundamental Tensors and Fundamental Equations

A Riemannian submersion $\pi : (M, g) \to (B, g')$ determines two $(1, 2)$-tensor fields T, A on M. As in [226; 26], they are called the *fundamental tensor fields* or the *invariants* of π and are defined by means of the vertical and horizontal projections $v : \mathcal{X}(M) \to \mathcal{X}^v(M)$, $h : \mathcal{X}(M) \to \mathcal{X}^h(M)$, according to the formulas:

$$\begin{aligned} T(E,F) &= T_E F = h\nabla_{vE} vF + v\nabla_{vE} hF \ , \\ A(E,F) &= A_E F = v\nabla_{hE} hF + h\nabla_{hE} vF \ , \end{aligned} \tag{1.19}$$

for any $E,\ F \in \mathcal{X}(M)$. Here ∇ denotes the Levi-Civita connection of (M, g). It is easy to prove that T and A are, respectively, vertical and horizontal tensor fields, that is:

$$T_E F = T_{vE} F, \quad A_E F = A_{hE} F, \quad E,\ F \in \mathcal{X}(M) \ .$$

Using (1.19) and $\nabla g = 0$ we can easily obtain the following result.

Lemma 1.1 *For any $E, F, G \in \mathcal{X}(M)$ one has:*

$$g(T_E F, G) + g(T_E G, F) = 0, \quad g(A_E F, G) + g(A_E G, F) = 0 \ .$$

Let p be a point of M. Since for any $u, w \in T_pM$, $T_E F(p)$ and $A_E F(p)$ do not depend on the choice of the vector fields E, F on M such that $E(p) = u$, $F(p) = w$, it makes sense to consider the linear operators T_u, A_u on T_pM defined by:

$$T_u w = (T_E F)(p) \ , \quad A_u w = (A_E F)(p) \ , \tag{1.20}$$

where $E, F \in \mathcal{X}(M)$ and $E(p) = u, F(p) = w$.

Proposition 1.4 *For any $p \in M, u \in T_pM$, T_u and A_u are skew-symmetric operators on (T_pM, g_p) reversing the horizontal and the vertical subspaces at p.*

Proof. The skew-symmetry property follows from Lemma 1.1. Now, we only prove the relations:

$$T_u(\mathcal{V}_p) \subseteq \mathcal{H}_p \ , \quad T_u(\mathcal{H}_p) \subseteq \mathcal{V}_p \ , \quad u \in T_pM \ ,$$

since the inclusions corresponding to A_u are achieved with the same technique. Given $u \in T_pM$, let E be a vector field such that $E(p) = u$. For any $w \in \mathcal{V}_p$, one can consider a vertical vector field F such that $F(p) = w$. Then, (1.19) implies: $T_u w = (h\nabla_{vE} F)(p) \in \mathcal{H}_p$. Analogously, since any $w \in \mathcal{H}_p$ can be considered as the value at p of a horizontal vector field F, one has: $T_u w = (v\nabla_{vE} F)(p)$, so $T_u w$ is vertical.

Proposition 1.5 *The tensor fields T, A satisfy:*

$$\begin{aligned} & T_U W = T_W U \,, & & U, W \in \mathcal{X}^v(M) \,; \\ & A_X Y = -A_Y X = \tfrac{1}{2} v[X,Y] \,, & & X, Y \in \mathcal{X}^h(M) \,. \end{aligned} \tag{1.21}$$

Proof. The first formula is a consequence of the symmetry of ∇ and of the integrability of the vertical distribution. Now, we prove the vanishing of $A_X X$, for any $X \in \mathcal{X}^h(M)$. We may assume X basic. Then, for any $W \in \mathcal{X}^v(M)$, using iv) in Proposition 1.1, we have:

$$g(A_X X, W) = g(\nabla_X X, W) = -g(X, \nabla_W X) = -\frac{1}{2} W(g(X,X)) = 0 \,,$$

since $g(X,X)$ is constant on each fibre. So, $A_X X$ turns out to be a horizontal vector field; it must vanish, since $A_X X = v(\nabla_X X)$. Thus, A is alternating on the horizontal vector fields. Finally, if $X, Y \in \mathcal{X}^h(M)$, one has:

$$v[X,Y] = v(\nabla_X Y - \nabla_Y X) = A_X Y - A_Y X = 2A_X Y \,.$$

Remark 1.4 **(The geometric meaning of the invariants).**
From the second formula in (1.21) we have that the restriction of A to $\mathcal{X}^h(M) \times \mathcal{X}^h(M)$ measures the integrability of the horizontal distribution. Indeed, (1.21), Lemma 1.1 and the condition $A_U = 0$, $U \in \mathcal{X}^v(M)$, imply that $A = 0$ if and only if $\mathcal{H}$ is integrable. The tensor field A is named the *integrability tensor of* π. Let $\overline{\nabla}$ be the *Schouten connection* associated with the mutually orthogonal distributions $\mathcal{V}$ and $\mathcal{H}$ ([145]), *i.e.*:

$$\overline{\nabla}_E F = v(\nabla_E vF) + h(\nabla_E hF) \quad E, F \in \mathcal{X}(M) \,. \tag{1.22}$$

Then, via (1.19), one has:

$$\nabla_E F = \overline{\nabla}_E F + T_E F + A_E F \,. \tag{1.23}$$

It is easy to prove that $\overline{\nabla} g = 0$. Moreover, on any fibre $\pi^{-1}(x)$, $\overline{\nabla}$ coincides with the Levi-Civita connection with respect to the metric induced by g. In fact, for any $U, V \in \mathcal{X}^v(M)$, one has:

$$\overline{\nabla}_U W = v(\nabla_U W) \,, \quad T_U W = h(\nabla_U W) \,, \quad A_U W = 0 \,,$$

and (1.23) entails:

$$\nabla_U W = \overline{\nabla}_U W + T_U W, \quad U, W \in \mathcal{X}^v(M) \,. \tag{1.24}$$

So, the restriction of T to $\mathcal{X}^v(M) \times \mathcal{X}^v(M)$ acts as the second fundamental form of any fibre. In particular, the vanishing of T means that any fibre of

π is a totally geodesic submanifold of M. The converse statement is also true; it follows from Lemma 1.1 and the condition: $T_X = 0,\ X \in \mathcal{X}^h(M)$.

Remark 1.5 The following formulas are also an immediate consequence of (1.19):

$$\begin{aligned} \nabla_U X &= T_U X + h(\nabla_U X)\ ; \\ \nabla_X U &= v(\nabla_X U) + A_X U\ ; \\ \nabla_X Y &= A_X Y + h(\nabla_X Y)\ , \end{aligned} \tag{1.25}$$

for any $X, Y \in \mathcal{X}^h(M), U \in \mathcal{X}^v(M)$. Moreover, if X is basic then $h(\nabla_U X) = h(\nabla_X U) = A_X U$, $[X, U]$ being vertical.

As for the properties of the tensor fields ∇T, ∇A, Lemma 1.1 implies the skew-symmetry (with respect to g) of the $(1,1)$-tensor fields $(\nabla_E A)_F$ and $(\nabla_E T)_F$, respectively defined by:

$$(\nabla_E A)_F H = (\nabla_E A)(F, H),\ (\nabla_E T)_F H = (\nabla_E T)(F, H)\ .$$

Moreover, the following blocks of formulas are proved in [226], [126] and also appear in [26]:

$$\begin{aligned}
&\text{(a)}\ g((\nabla_E T)_U V, X) = g((\nabla_E T)_V U, X)\ , \\
&\qquad g((\nabla_E A)_X Y, U) = -g((\nabla_E A)_Y X, U)\ , \\[1ex]
&\text{(b)}\ (\nabla_X T)_Y = -T_{A_X Y}\ ,\ (\nabla_U T)_X = -T_{T_U X}\ , \\
&\qquad (\nabla_U A)_V = -A_{T_U V}\ ,\ (\nabla_X A)_U = -A_{A_X U}\ , \\[1ex]
&\text{(c)}\ g((\nabla_X T)_U V, W) = g(A_X V, T_U W) - g(A_X W, T_U V)\ , \\
&\qquad g((\nabla_U A)_X V, W) = g(T_U V, A_X W) - g(T_U W, A_X V)\ , \\[1ex]
&\text{(d)}\ g((\nabla_X T)_U Y, Z) = g(A_X Y, T_U Z) - g(A_X Z, T_U Y)\ , \\
&\qquad g((\nabla_X A)_Y U, V) = g(A_X U, A_Y V) - g(A_X V, A_Y U)\ , \\[1ex]
&\text{(e)}\ g((\nabla_U T)_V X, Y) = g(T_U X, T_V Y) - g(T_V X, T_U Y)\ , \\
&\qquad g((\nabla_X A)_Y Z, Z') = g(A_X Z, A_Y Z') - g(A_X Z', A_Y Z)\ , \\[1ex]
&\text{(f)}\ \sigma_{X,Y,Z}\, g((\nabla_X A)_Y Z, U) = \sigma_{X,Y,Z}\, g(A_X Y, T_U Z)\ , \\[1ex]
&\text{(g)}\ g((\nabla_U A)_X Y, V) + g((\nabla_V A)_X Y, U) \\
&\qquad = g((\nabla_Y T)_U V, X) - g((\nabla_X T)_U V, Y)\ ,
\end{aligned} \tag{1.26}$$

for any $E \in \mathcal{X}(M)$, $X, Y, Z, Z' \in \mathcal{X}^h(M)$, $U, V, W \in \mathcal{X}^v(M)$, σ denoting the cyclic sum.

Proposition 1.6 *Given a Riemannian submersion, one has:*

i) *if A is parallel, then A vanishes;*
ii) *if T is parallel, then T vanishes.*

Proof. We only prove i), the proof of ii) being analogous. Indeed, for any $X \in \mathcal{X}^h(M), W \in \mathcal{X}^v(M)$, via (1.26)(b), (1.21) and Lemma 1.1, one has:

$$g((\nabla_X A)_W X, W) = g(A_X(A_X W), W) = -g(A_X W, A_X W) .$$

Hence, if A is parallel, A_X vanishes on the vertical distribution and Lemma 1.1 implies $A_X = 0$. Then, A vanishes, since it is a horizontal tensor field.

Combining with Remark 1.4, the Riemannian submersions with parallel integrability tensor are characterized as those whose horizontal distribution is integrable, and the submersions with parallel tensor field T are the ones with totally geodesic fibres.

The tensor fields A, T and their covariant derivatives play a fundamental role in expressing the *Riemannian curvature* R of (M, g). According to [172], for any $E, F, G, H \in \mathcal{X}(M)$, we put:

$$R(E, F, G, H) = g(R(G, H, F), E) = g(([\nabla_G, \nabla_H] - \nabla_{[G,H]})F, E) .$$

As in [226], R^* denotes the (1,3)-tensor field on $\mathcal{X}^h(M)$ with values in $\mathcal{X}^h(M)$ which associates to any $X, Y, Z \in \mathcal{X}^h(M)$ and $p \in M$ the horizontal lift $R^*(X, Y, Z)_p$ of $R'_{\pi(p)}(\pi_{*p}(X_p), \pi_{*p}(Y_p), \pi_{*p}(Z_p))$, R' being the Riemannian curvature of (B, g'). Briefly, we write:

$$\pi_*(R^*(X, Y, Z)) = R'(\pi_* X, \pi_* Y, \pi_* Z) .$$

For any quadruplet of horizontal vector fields, we also put:

$$R^*(X, Y, Z, H) = g(R^*(Z, H, Y), X) = R'(\pi_* X, \pi_* Y, \pi_* Z, \pi_* H) \circ \pi .$$

Finally, $\hat{R}$ stands for the Riemannian curvature of any fibre $(\pi^{-1}(x), \hat{g}_x)$. Then, the corresponding Gauss and Codazzi equations lead to:

$$\begin{aligned} R(U, V, F, W) &= \hat{R}(U, V, F, W) + g(T_U W, T_V F) - g(T_V W, T_U F) , \\ R(U, V, W, X) &= g((\nabla_V T)(U, W), X) - g((\nabla_U T)(V, W), X) , \end{aligned} \tag{1.27}$$

for any $U, V, W, F \in \mathcal{X}^v(M), X \in \mathcal{X}^h(M)$. Moreover, the following formulas are stated in [226; 126]; a detailed proof is also given in [145]:

$$\begin{aligned} R(X,Y,Z,V) &= g((\nabla_Z A)(X,Y),V) + g(A_X Y, T_V Z) \\ &\quad - g(A_Y Z, T_V X) - g(A_Z X, T_V Y)\,, \\ R(X,Y,Z,H) &= R^*(X,Y,Z,H) - 2g(A_X Y, A_Z H) \\ &\quad + g(A_Y Z, A_X H) - g(A_X Z, A_Y H)\,, \end{aligned} \tag{1.28}$$

$$\begin{aligned} R(X,Y,V,W) &= g((\nabla_V A)(X,Y),W) - g((\nabla_W A)(X,Y),V) \\ &\quad + g(A_X V, A_Y W) - g(A_X W, A_Y V) \\ &\quad - g(T_V X, T_W Y) + g(T_W X, T_V Y)\,, \\ R(X,V,Y,W) &= g((\nabla_X T)(V,W),Y) + g((\nabla_V A)(X,Y),W) \\ &\quad - g(T_V X, T_W Y) + g(A_X V, A_Y W)\,, \end{aligned} \tag{1.29}$$

for any $X, Y, Z, H \in \mathcal{X}^h(M), V, W \in \mathcal{X}^v(M)$.

Remark 1.6 The fibres of the Hopf submersion $\pi : \mathbf{S^{2n+1}} \to \mathbf{P_n(C)}$ are totally geodesic submanifolds of $\mathbf{S^{2n+1}}$, since they are circles. Indeed, comparing (1.14) with (1.24), one obtains: $T(\xi,\xi) = 0$, which is equivalent to the vanishing of T. Moreover, the comparison of (1.17) with the last relation in (1.25) implies:

$$v[X^*,Y^*] = 2A_{X^*}Y^* = \eta[X^*,Y^*]\xi, \quad X^*, Y^* \in \mathcal{X}^b(\mathbf{S^{2n+1}})\,.$$

Then, for any $X, Y \in \mathcal{X}^h(\mathbf{S^{2n+1}})$, one has:

$$A_X Y = \frac{1}{2}\eta[X,Y]\xi = g(\varphi X, Y)\xi = < JX, Y > \xi\,,$$

and the horizontal distribution is not integrable.

Finally, we remark that the last part of Theorem 1.1 can be proved applying the second formula in (1.28).

Moreover, the submersion $\pi : (\mathbf{S^{2n+1}}(1,k), g_a) \to (P_n(\mathbf{C})(k+3), G)$, considered in Remark 1.3 has totally geodesic fibres. The corresponding invariant A acts on the horizontal distribution as: $A_X Y = -a g_a(X, \varphi Y)\tilde{\xi}$, where $a\tilde{\xi} = \xi$.

Equations (1.27), (1.28) and (1.29) allow to express the sectional curvatures $K(\alpha)$, α denoting a 2-plane in T_pM, $p \in M$. More precisely, if $\{U, V\}$ is an orthonormal basis of the vertical 2-plane α, one has:

$$K(\alpha) = \hat{K}(\alpha) + ||T_U V||^2 - g(T_U U, T_V V)\,, \tag{1.30}$$

$\hat{K}(\alpha)$ denoting the sectional curvature in the fibre through p. If $\{X, Y\}$ is an orthonormal basis of the horizontal 2-plane α, and $K'(\alpha')$ denotes the sectional curvature in (B, g') of the plane α' spanned by $\{\pi_* X, \pi_* Y\}$, then

$$K(\alpha) = K'(\alpha') - 3\|A_X Y\|^2 . \tag{1.31}$$

Finally, if $X \in \mathcal{H}_p$, $V \in \mathcal{V}_p$ are unit vectors spanning α, (1.26) and (1.28) imply:

$$K(\alpha) = g((\nabla_X T)(V, V), X) - \|T_V X\|^2 + \|A_X V\|^2 . \tag{1.32}$$

Now, we are going to compute the *Ricci tensor* ρ of (M, g). We call π-*adapted* a local orthonormal frame $\{X_i, U_j\}_{1 \le i \le n, 1 \le j \le r}$ on M such that each X_i is horizontal and each U_j is vertical.

Lemma 1.2 *Let U, V be vertical and X, Y horizontal vector fields. With respect to a π-adapted frame $\{X_i, U_j\}_{1 \le i \le n, 1 \le j \le r}$ one has:*

$$\begin{aligned} \textstyle\sum_{i=1}^n g(T_U X_i, T_V X_i) &= \textstyle\sum_{j=1}^r g(T_U U_j, T_V U_j) , \\ \textstyle\sum_{i=1}^n g(A_X X_i, A_Y X_i) &= \textstyle\sum_{j=1}^r g(A_X U_j, A_Y U_j) , \\ \textstyle\sum_{i=1}^n g(A_X X_i, T_U X_i) &= \textstyle\sum_{j=1}^r g(A_X U_j, T_U U_j) . \end{aligned} \tag{1.33}$$

Proof. We only prove the last formula, the technique for the others being the same. Since for any $i \in \{1, \ldots, n\}$ and $U \in \mathcal{X}^v(M)$, $T_U X_i$ is vertical, applying Lemma 1.1, we can write $T_U X_i = -\sum_{j=1}^r g(T_U U_j, X_i) U_j$. So, given $X \in \mathcal{X}^h(M)$, we get:

$$\begin{aligned} \textstyle\sum_{i=1}^n g(A_X X_i, T_U X_i) &= -\textstyle\sum_i \sum_j g(T_U U_j, X_i) g(A_X X_i, U_j) \\ &= \textstyle\sum_j \sum_i g(T_U U_j, X_i) g(A_X U_j, X_i) \\ &= \textstyle\sum_j g(A_X U_j, T_U U_j) . \end{aligned}$$

Let N be the horizontal vector field on M locally defined by:

$$N = \sum_{j=1}^r T_{U_j} U_j , \tag{1.34}$$

where $\{U_j\}_{1 \le j \le r}$ is a local orthonormal frame of $\mathcal{V}$. According to (1.24), $N = rH$, where H is the mean curvature vector field of any fibre.

Lemma 1.3 *For any $E \in \mathcal{X}(M)$, $X \in \mathcal{X}^h(M)$, one has:*

$$g(\nabla_E N, X) = \sum_{j=1}^{r} g((\nabla_E T)(U_j, U_j), X) . \tag{1.35}$$

Proof. The statement follows applying the covariant derivative ∇_E to (1.34) and simply proving that:

$$\sum_{j=1}^{r} g(T(\nabla_E U_j, U_j) + T_{U_j}(\nabla_E U_j), X) = 0 ,$$

for any horizontal vector field X. Indeed, since T is vertical and Proposition 1.5 and Lemma 1.1 hold, for any $j \in \{1, .., r\}$ one has:

$$g(T(\nabla_E U_j, U_j), X) + g(T_{U_j}(\nabla_E U_j), X) = -2g(T_{U_j} X, \nabla_E U_j) .$$

Then, writing the vertical field $T_{U_j} X$ as $\sum_{h=1}^{r} g(T_{U_j} X, U_h) U_h$, one obtains:

$$\begin{aligned} &\textstyle\sum_{j=1}^{r} g(T(\nabla_E U_j, U_j) + T_{U_j}(\nabla_E U_j), X) \\ &= 2 \textstyle\sum_{j,h=1}^{r} g(T_{U_j} U_h, X) g(U_h, \nabla_E U_j) = 0 . \end{aligned}$$

Proposition 1.7 *With respect to a π-adapted frame $\{X_i, U_j\}$, for any $U, V \in \mathcal{X}^v(M)$ and $X, Y \in \mathcal{X}^b(M)$, π-related to X', Y', the Ricci tensor ρ satisfies:*

$$\begin{aligned} \rho(U, V) &= \hat{\rho}(U, V) - g(N, T_U V) \\ &\quad + \textstyle\sum_i \{g((\nabla_{X_i} T)(U, V), X_i) + g(A_{X_i} U, A_{X_i} V)\} , \\ \rho(X, Y) &= \rho'(X', Y') \circ \pi + \tfrac{1}{2}\{g(\nabla_X N, Y) + g(\nabla_Y N, X)\} \\ &\quad -2 \textstyle\sum_i g(A_X X_i, A_Y X_i) - \sum_j g(T_{U_j} X, T_{U_j} Y) , \\ \rho(U, X) &= g(\nabla_U N, X) - \textstyle\sum_j g((\nabla_{U_j} T)(U_j, U), X) \\ &\quad + \textstyle\sum_i \{g((\nabla_{X_i} A)(X_i, X), U) - 2g(A_X X_i, T_U X_i)\} , \end{aligned} \tag{1.36}$$

where ρ' and $\hat{\rho}$, respectively, denote the Ricci tensor of (B, g') and the Ricci tensor of any fibre.

Proof. The first formula is obtained applying (1.27), (1.29), (1.26) and Lemma 1.2. Let X be a basic vector field π-related to X'; using (1.28),

(1.29), (1.21), (1.26) and Lemmas 1.2, 1.3, one has:

$$\rho(X,X) = \textstyle\sum_i\{R^*(X,X_i,X,X_i) - 3\|A_X X_i\|^2\}$$
$$+\textstyle\sum_j\{g((\nabla_X T)(U_j,U_j),X) - \|T_{U_j}X\|^2 + \|A_X U_j\|^2\}$$
$$= \rho'(X',X')\circ\pi + g(\nabla_X N, X) - 2\textstyle\sum_i \|A_X X_i\|^2 - \sum_j \|T_{U_j}X\|^2 \,.$$

Thus, the second relation is a consequence of the symmetry of g, ρ, ρ'. Finally, if U is vertical and X horizontal, using again (1.21), (1.26) and Lemma 1.3, one has:

$$\rho(U,X) = \textstyle\sum_i R(X,X_i,U,X_i) - \sum_j R(U,U_j,U_j,X)$$
$$= \textstyle\sum_i\{g((\nabla_{X_i}A)(X_i,X),U) + g(A_{X_i}X, T_U X_i) - g(A_X X_i, T_U X_i)\}$$
$$-\textstyle\sum_j\{g((\nabla_{U_j}T)(U,U_j),X) - g((\nabla_U T)(U_j,U_j),X)\}$$
$$= \textstyle\sum_i\{g((\nabla_{X_i}A)(X_i,X),U) - 2g(A_X X_i, T_U X_i)\}$$
$$-\textstyle\sum_j g((\nabla_{U_j}T)(U_j,U),X) + g(\nabla_U N, X)\,.$$

To compute the *scalar curvature* τ of (M,g), which depends on the scalar curvatures τ' of (B,g') and $\hat{\tau}$ of any fibre, we consider a π-adapted frame $\{X_i, U_j\}$ and apply (1.36). Thus, we obtain:

$$\tau = \textstyle\sum_j \rho(U_j,U_j) + \sum_i \rho(X_i,X_i)$$
$$= \hat{\tau} + \tau'\circ\pi - \|N\|^2 + \textstyle\sum_{i,j}\{g((\nabla_{X_i}T)(U_j,U_j),X_i) + g(A_{X_i}U_j, A_{X_i}U_j)$$
$$-g(T_{U_j}X_i, T_{U_j}X_i)\} + \textstyle\sum_i g(\nabla_{X_i}N, X_i) - 2\sum_{i,j} g(A_{X_j}X_i, A_{X_j}X_i)\,.$$

Then, using Lemma 1.3, since $\|A\|^2 = \sum_{i,j} g(A_{X_i}U_j, A_{X_i}U_j)$ and $\|T\|^2 = \sum_{i,j} g(T_{U_j}X_i, T_{U_j}X_i)$, we have:

$$\tau = \tau'\circ\pi + \hat{\tau} - \|N\|^2 - \|A\|^2 - \|T\|^2 + 2\sum_i g(\nabla_{X_i}N, X_i)\,. \quad (1.37)$$

1.4 Other Examples

Example 1.1 (Covering Maps). We recall that a *covering* of a manifold B is a connected manifold M equipped with a surjective differentiable map $\pi : M \to B$ such that each $x \in B$ has an open connected neighborhood U with $\pi^{-1}(U) = \bigcup U_\alpha$, where any U_α is an open connected component diffeomorphic with U via the restriction of π to U_α. Then π, which is called a *covering map*, turns out to be a submersion with discrete fibres. Since the vertical distribution determined by π is trivial, the pullback $\pi^* g$ of any Riemannian metric g on B to M provides a Riemannian structure on M such that $\pi : (M, \pi^* g) \to (B, g)$ is a Riemannian submersion.

Example 1.2 (Warped products). On the product manifold $M = M_1 \times M_2$ of two Riemannian manifolds $(M_1, g_1), (M_2, g_2)$ one considers the Riemannian metric $g = g_1 + f g_2$, f denoting a positive function on M_1. The manifold (M, g), which is also denoted by $M_1 \times_f M_2$, is called the *warped product* of (M_1, g_1) and (M_2, g_2) by f. It is easy to prove that the first projection $\pi_1 : M_1 \times_f M_2 \to M_1$ is a Riemannian submersion whose vertical and horizontal spaces at any point $p = (p_1, p_2)$ are respectively identified with $T_{p_2} M_2$, $T_{p_1} M_1$. Since the horizontal distribution is integrable, the invariant A associated with π_1 vanishes. To compute the invariant T, for any $U, V \in \mathcal{X}^v(M), X \in \mathcal{X}^h(M)$, applying (1.19), one obtains:

$$2g(T_U V, X) = 2g(\nabla_U V, X) = -X(g(U, V)) = -df(X) g_2(U, V) .$$

Therefore, $T_U V = -\frac{1}{2f} g(U, V) \mathrm{grad} f$, and any fibre of π_1 (which is identified with M_2) turns out to be a totally umbilical submanifold of $M_1 \times_f M_2$ with mean curvature vector field $H = -\frac{1}{2f} \mathrm{grad} f$. Thus, if f is a non-constant function, the fibres of π_1 are not minimal submanifolds of $M_1 \times_f M_2$. Moreover, the second projection $\pi_2 : M_1 \times_f M_2 \to M_2$ is not a Riemannian submersion, unless f is the constant function of value 1. But $f = 1$ means that the warped product coincides with the Riemannian product and, in this case, π_1 and π_2 are both Riemannian submersions with totally geodesic fibres and integrable horizontal distribution.

Example 1.3 (The tangent bundle and the unit sphere bundle). Let $\pi : TM \to M$ be the tangent bundle over the n-dimensional manifold M. As in [347], $\{x^q, y^q\}_{1 \le q \le n}$ denote the coordinates induced on $\pi^{-1}(U)$ by a system of local coordinates $\{x^q\}_{1 \le q \le n}$ defined in an open set U of M. The vertical lift of $X \in \mathcal{X}(M)$ is the vector field X^V on TM locally defined by:

$$X^V{}_{|\pi^{-1}(U)} = (X^q \circ \pi)\frac{\partial}{\partial y^q} ,$$

where $X_{|U} = X^q \frac{\partial}{\partial x^q}$. Since $\frac{\partial}{\partial y^q} = \frac{\partial}{\partial x^q}^V$, $q \in \{1, \dots, n\}$, π is a submersion whose vertical distribution $\mathcal{V}$ is locally spanned by $\{\frac{\partial}{\partial x^q}^V\}$. We recall that any linear connection ∇ on M determines the so-called horizontal distribution $\mathcal{H}$ on TM; $\mathcal{H}$ and $\mathcal{V}$ are complementary distributions ([172]). For any $X \in \mathcal{X}(M)$ the horizontal lift of X with respect to ∇ is the vector field X^H on TM locally defined by:

$$X^H{}_{|\pi^{-1}(U)} = X^q(\frac{\partial}{\partial x^q} - \Gamma^p_{qj} y^j \frac{\partial}{\partial y^p}) ,$$

where Γ^p_{ij} are the components of ∇ in U. Moreover, $\mathcal{H}$ turns out to be the distribution on TM locally spanned by $\{\frac{\partial}{\partial x^q}^H\}_{1 \le q \le n}$.

Any Riemannian metric g on M determines the so-called *Sasakian metric* g^D on TM or the *diagonal lift of* g ([263; 347]), defined by:

$$\begin{aligned} g^D(X^H, Y^H) &= g(X,Y) \circ \pi , \\ g^D(X^V, Y^V) &= g(X,Y) \circ \pi , \\ g^D(X^H, Y^V) &= g^D(X^V, Y^H) = 0 , \end{aligned} \tag{1.38}$$

for any $X, Y \in \mathcal{X}(M)$, where the horizontal lifts are evaluated with respect to the Levi-Civita connection on (M, g). Since $\mathcal{V}$ and $\mathcal{H}$ are mutually orthogonal with respect to g^D, $\mathcal{H}$ coincides with the horizontal distribution of the submersion π and $\pi : (TM, g^D) \to (M, g)$ is a Riemannian submersion. Any fibre turns out to be a flat, totally geodesic submanifold of TM. Indeed, the invariant T vanishes, since the Levi-Civita connection $\tilde{\nabla}$ on (TM, g^D) satisfies: $\tilde{\nabla}_{X^V} Y^V = 0$, for any $X, Y \in \mathcal{X}(M)$. Since also $[X^V, Y^V] = 0$, the curvature $\tilde{R}$ of $\tilde{\nabla}$ satisfies: $\tilde{R}(X^V, Y^V, Z^V) = 0$, so the flatness of the fibres follows from (1.27). Let R be the Riemannian curvature of (M, g). For any $X, Y \in \mathcal{X}(M)$, $\gamma(R(X,Y))$ denotes the vertical vector field defined by:

$$\gamma(R(X,Y))_\xi = R(X,Y,Z)^V_\xi , \quad \xi \in T_x M ,$$

where Z is a vector field extending ξ. Then, according to the formula: $[X^H, Y^H] = [X,Y]^H - \gamma(R(X,Y))$, one gets: $2A_{X^H} Y^H = -\gamma(R(X,Y))$, and the horizontal distribution is integrable if and only if g is flat. A straightforward application of (1.28) and (1.29) gives the explicit expression of $\tilde{R}$ on any triplet of vector fields. Such formulas were firstly proved by O. Kowalski ([175]).

Another metric $\tilde{g}$ on TM which makes $\pi : (TM, \tilde{g}) \to (M, g)$ a Riemannian submersion with totally geodesic fibres is the *Cheeger–Gromoll metric* introduced in [58] and explicitly described in [210] and in [270]. It is given by:

$$\tilde{g}(X^H, Y^H) = g(X, Y) \circ \pi \,, \quad \tilde{g}(X^H, Y^V) = \tilde{g}(X^V, Y^H) = 0 \,,$$

$$\tilde{g}(X^V, Y^V)(P) = \tfrac{1}{1+r}(g_x(X_x, Y_x) + g_x(X_x, u) g_x(Y_x, u)) \,,$$

for any $P = (x, u) \in TM$, $X, Y \in \mathcal{X}(M)$, with $r = g_x(u, u)$. Even in this case $\mathcal{H}$ coincides with the horizontal distribution of π and the invariant $\tilde{A}$ acts as A on $\mathcal{H}$, while

$$\begin{aligned} \tilde{g}(\tilde{A}_{X^H} Z^V, Y^H)(P) &= \tfrac{1}{2(1+r)} g_x(R_x(X_x, Y_x, u), Z_x) \\ &= \tfrac{1}{1+r} g^D(A_{X^H} Z^V, Y^H)(P) \,. \end{aligned}$$

We remark that the two metrics $g^D, \tilde{g}$ have different geometric properties. For instance, taking $M = \mathbf{S^n}$ with the standard metric, then the Cheeger–Gromoll metric on $T\mathbf{S^n}$ has non-negative sectional curvatures ([210]), whereas this is not true for the Sasakian metric.

The *unit sphere bundle* is the hypersurface $T_1 M$ of TM locally defined by the equation $g_{ij} y^i y^j = 1$, *i.e.* $T_1 M = \{(x, u) \in TM \mid g_x(u, u) = 1\}$. The vector field N on $T_1 M$ such that $N_{|\pi^{-1}(U)} = y^i \frac{\partial}{\partial y^i}$ is the unit normal vector field to $T_1 M$. Let g' be the metric on $T_1 M$ induced by g^D; so the restriction $\pi' = \pi_{|T_1 M} : (T_1 M, g') \to (M, g)$ is a Riemannian submersion with fibres isometric with $\mathbf{S^{n-1}}$, whose vertical distribution is the orthogonal complement of N in $\mathcal{V}$.

Another example of Riemannian submersion can be obtained considering the Sasaki–Mok metric induced on the bundle of the linear frames over a Riemannian manifold (M, g) ([72]).

Example 1.4 (The generalized Hopf fibration). It is well known that for any positive integer n, the unit sphere $\mathbf{S^{4n+3}}$ can be regarded as the total space of a principal $\mathbf{S^3}$-bundle over the quaternionic projective space $\mathbf{P_n(Q)}$. More precisely, one can realize the sphere $\mathbf{S^{4n+3}}$ as the symmetric space $Sp(n+1)/Sp(n)$.

We recall that $Sp(n)$ is the quaternionic unitary group (symplectic group), *i.e.* the compact Lie group consisting of the linear transformations of the right-vector space $\mathbf{Q^n}$ which preserve the quaternionic form $\sum_{\alpha=1}^{n} d\bar{q}^\alpha dq^\alpha$. It is isomorphic with the subgroup of $Sp(n+1)$ consisting of the symplectic transformations of $\mathbf{Q^{n+1}}$ which fix the point $x_0 = (1, 0, \dots, 0)$.

Any symplectic transformation of $\mathbf{Q^{n+1}}$ naturally induces a transformation of $\mathbf{P_n}(\mathbf{Q})$ and it is easy to prove that $Sp(n+1)$ also acts transitively on $\mathbf{P_n}(\mathbf{Q})$. Moreover, the group $Sp(n) \times \mathbf{S^3}$ (isomorphic with $Sp(n) \times Sp(1)$) can be identified with the subgroup of the linear transformations of $\mathbf{P_n}(\mathbf{Q})$ fixing the element of $\mathbf{P_n}(\mathbf{Q})$ determined by x_0. Thus, $\mathbf{P_n}(\mathbf{Q})$ can be realized as the symmetric space $Sp(n+1)/(Sp(n) \times Sp(1))$ and the inclusion $Sp(n) \to Sp(n) \times Sp(1)$ induces a map $\pi : \mathbf{S^{4n+3}} \to \mathbf{P_n}(\mathbf{Q})$ acting as the projection of a principal fibre bundle with structure group $\mathbf{S^3}$ and fibre $Sp(1) \cong \mathbf{S^3}$ ([276]).

As a consequence of a result of Bérard-Bergery, ([19], or theorem 9.80 [26]), π is a Riemannian submersion with totally geodesic fibres from $(\mathbf{S^{4n+3}}, g)$ to $(\mathbf{P_n}(\mathbf{Q}), g')$, g denoting the metric of constant curvature 1 and g' the one with constant quaternionic sectional curvature 4. We call π the *generalized Hopf fibration* or the *natural fibration*. In [126] A. Gray explicitly computes the sectional curvatures of $(\mathbf{P_n}(\mathbf{Q}), g')$ by means of (1.30), (1.31), (1.32). To evaluate the invariant A, we regard $(\mathbf{S^{4n+3}}, g)$ as a Riemannian submanifold of the Euclidean space $\mathbf{R^{4n+4}}$. Let us consider the almost complex structures I, J, K on $\mathbf{R^{4n+4}}$, respectively induced by the right multiplication for $\mathbf{i}, \mathbf{j}, \mathbf{k}$ in $\mathbf{Q^{n+1}}$ and let N be the outward unit normal to $\mathbf{S^{4n+3}}$. Then IN, JN, KN are globally defined vector fields on $\mathbf{S^{4n+3}}$ spanning an integrable distribution, which coincides with the vertical distribution $\mathcal{V}$ of π ([86]). Thus, the projective space $\mathbf{P_n}(\mathbf{Q})$ can be identified as the leaves space $\mathbf{S^{4n+3}}/\mathcal{V}$. Let X, Y be horizontal vector fields. Since $D_X N = X$, D denoting the Levi-Civita connection of $\mathbf{R^{4n+4}}$, one has:

$$\begin{aligned} & g(A_X Y, IN) = -g(A_X IN, Y) = -g(IX, Y) \,, \\ & g(A_X Y, JN) = -g(JX, Y) \,, \quad g(A_X Y, KN) = -g(KX, Y) \,. \end{aligned}$$

Therefore, one obtains:

$$A_X Y = -g(IX, Y) IN - g(JX, Y) JN - g(KX, Y) KN \,.$$

Example 1.5 (Riemannian submersions and Riemannian Homogeneous manifolds). Another class of Riemannian submersions occurs on connected, simply connected, homogeneous Riemannian manifolds. A well known theorem of W. Ambrose and I. M. Singer ([5]) states that a connected, simply-connected and complete Riemannian manifold (M, g) is homogeneous, *i.e.* there exists a transitive and effective group G of isometries of M, if and only if there exists a $(1,2)$-tensor field T on M such that:

$$\text{(A-S)} \qquad \begin{array}{rl} \text{i)} & g(T_X Y, Z) + g(T_X Z, Y) = 0\,, \\ \text{ii)} & (\nabla_X R)_{YZ} = [T_X, R_{YZ}] - R_{T_X YZ} - R_{YT_X Z}\,, \\ \text{iii)} & (\nabla_X T)_Y = [T_X, T_Y] - T_{T_X Y}\,, \end{array}$$

for any vector fields X, Y, Z on M. Here ∇ denotes the Riemannian connection and R its curvature tensor field. In [295] F. Tricerri and L. Vanhecke define a *homogeneous (Riemannian) structure* on an n-dimensional Riemannian manifold (M, g) as a $(1,2)$-tensor field T which is a solution of the system (A-S). They also prove that the existence of such a structure on (M, g) implies only the local homogeneity of the manifold even if one assumes that the manifold is connected and complete.

Now, we briefly outline their classification. For a point $p \in M$, consider the vector space $\mathcal{T}(W)$, where $W = T_pM$, consisting of the (0,3)-tensors t having the same symmetries as a homogeneous structure, *i.e.* verifying $t(X,Y,Z) = -t(X,Z,Y)$ for any $X, Y, Z \in W$, where $t(X,Y,Z) = g(t(X,Y),Z)$. Under the natural action of $O(n)$, $n = dim\, M$, defined by

$$at(X,Y,Z) = t(a^{-1}X, a^{-1}Y, a^{-1}Z), \quad a \in O(n)\,,$$

$\mathcal{T}(W)$, with the scalar product

$$< t, t' >= \sum_{i,j,k=1}^{n} t(e_i, e_j, e_k) t'(e_i, e_j, e_k)$$

where $\{e_i\}$, $i \in \{1, \ldots, n\}$ is an orthonormal basis of W, decomposes into irreducible invariant subspaces. If $n \geq 3$ one obtains three invariant irreducible subspaces $\mathcal{T}_i(W)$ $1 \leq i \leq 3$, and eight invariant subspaces, while, if $n = 2$, then $\mathcal{T}(W)$ is irreducible and coincides with $\mathcal{T}_1(W)$. A homogeneous structure T on (M, g) belongs to the class $\mathcal{T}_i$ if for each $p \in M$, T_p belongs to the corresponding subspace $\mathcal{T}_i(W)$. When $n \geq 3$, this gives rise to the eight classes listed in Table 1.1, each of them is invariant under isomorphisms of homogeneous structures. We put $(n-1)\xi = \sum_{i=1}^{n} T(E_i, E_i)$, $c_{12}(T)(Z) = \sum_{i=1}^{n} g(T(E_i, E_i), Z)$, for any $Z \in \mathcal{X}(M)$, with $\{E_1, \ldots, E_n\}$ arbitrary local orthonormal vector fields.

Table 1.1 Classification of homogeneous Riemannian structures

Class	Algebraic defining condition
Symmetric spaces	$T = 0$
$\mathcal{T}_1$	$T(X,Y) = g(X,Y)\xi - g(\xi,Y)X$
$\mathcal{T}_2$	$c_{12}(T) = 0, \quad \sigma_{XYZ}\, g(T(X,Y),Z) = 0$
$\mathcal{T}_3$	$T(X,Y) + T(Y,X) = 0$
$\mathcal{T}_1 \oplus \mathcal{T}_2$	$\sigma_{XYZ}\, g(T(X,Y),Z) = 0$
$\mathcal{T}_1 \oplus \mathcal{T}_3$	$T(X,Y) + T(Y,X) = 2g(X,Y)\xi - g(X,\xi)Y - g(Y,\xi)X$
$\mathcal{T}_2 \oplus \mathcal{T}_3$	$c_{12}(T) = 0$
$\mathcal{T}_1 \oplus \mathcal{T}_2 \oplus \mathcal{T}_3$	no condition

Denoting by ω be the 1-form dual of ξ with respect to the metric g, called the fundamental 1-form of the homogeneous structure T, it is easy to verify that ω is a closed non-vanishing form if either T belongs to $\mathcal{T}_1$ or T is a proper structure belonging to $\mathcal{T}_1 \oplus \mathcal{T}_3$, *i.e.* $T \notin \mathcal{T}_1, T \notin \mathcal{T}_3$. Moreover, the manifolds admitting such a structure are locally isometric with the hyperbolic spaces $\mathbf{H^n}$ of dimension $n \geq 2$ and $n \geq 4$, respectively ([295; 238]). Note that $\mathbf{H^n}$ can be considered as a warped product, hence as the total space of a Riemannian submersion ([26]).

Therefore, it is meaningful to study the proper homogeneous structures T belonging to the class $\mathcal{T}_1 \oplus \mathcal{T}_2$ ($T \notin \mathcal{T}_1, T \notin \mathcal{T}_2$), with the additional condition $d\omega = 0$. Some examples in this sense, in dimension $n = 3, n = 4$, can be found in [295; 239].

Theorem 1.2 *Let (M,g) be a Riemannian manifold of dimension $n \geq 3$ equipped with a proper homogeneous structure $T \in \mathcal{T}_1 \oplus \mathcal{T}_2$ such that $d\omega = 0$. Then, the distribution $\mathcal{D}$ orthogonal to ξ is integrable and the integral manifolds are isoparametric hypersurfaces, of constant mean curvature $c = \|\xi\|$, carrying a homogeneous structure $\overline{T} \in \mathcal{T}_2$.*

Theorem 1.3 *Let (M,g) be a connected, simply connected Riemannian manifold equipped with a proper homogeneous structure $T \in \mathcal{T}_1 \oplus \mathcal{T}_2$ such that $d\omega = 0$. Then M is the total space of a Riemannian submersion.*

Proof. Since $\pi_1(M) = 0$ and $d\omega = 0$, there exists a function $f : M \to \mathbf{R}$ such that $df = \omega$ and $\xi = \mathrm{grad} f$. The previous theorem implies that f is an isoparametric function and the maximal integral manifolds of $\mathcal{D}$ are the level sets of f. Since ξ is never zero, the level hypersurfaces are regular, so there are no focal varieties of f. Again, since $\mathrm{grad} f$ never vanishes, for any $p \in M$ the tangent map f_{*p} is surjective and then f is a submersion. Its fibres are the level sets of f, so that the vertical distribution $\mathcal{V}$ coincides with $\mathcal{D}$, whereas the horizontal one $\mathcal{H}$ is spanned by ξ. Furthermore, each fibre is not totally geodesic since its mean curvature is a non-zero constant. Finally, considering $(\mathbf{R}, g_0)$ with $g_0 = c^{-2}dt^2$, for any $p \in M$, one has:

$$(g_0)_{f(p)}(f_{*p}\xi_p, f_{*p}\xi_p) = c^{-2}(\omega_p(\xi_p))^2 = g_p(\xi_p, \xi_p) .$$

So, f_{*p} induces an isometry from $\mathcal{H}_p$ onto $T_{f(p)}\mathbf{R}$ and $f : (M, g) \to (\mathbf{R}, g_0)$ is a Riemannian submersion.

Remark 1.7 The existence of the function f allows to consider the manifold (M, g) as a manifold foliated by the level sets of f. The obtained foliation of codimension 1 is a Riemannian foliation with bundle-like metric g ([288]). It is not harmonic since its leaves have non-zero constant mean curvature.

Remark 1.8 The invariant A vanishes and the other invariant, here denoted by Q to avoid confusion, is given by:

$$\begin{aligned}
&Q(X,Y) = c^{-2}g(T(X,Y),\xi)\xi\ , && X,Y \in \mathcal{X}^v(M)\ ,\\
&Q(X,\xi) = T(X,\xi) = \nabla_X\xi,\ \ Q(\xi,X) = 0\ , && X \in \mathcal{X}^v(M)\ ,\\
&Q(\xi,\xi) = 0\ . &&
\end{aligned}$$

Moreover, using the O'Neill formulas for the curvature tensor field, one obtains that the mixed sectional curvatures, *i.e.* the sectional curvatures of 2-planes in T_pM containing ξ_p, are non-positive and at least one of them is negative ([239]).

We recall that a Riemannian submersion is locally a warped product if and only if the invariant A vanishes, the mean curvature vector field H is basic and the trace-free component Q_0 of the invariant Q vanishes ([26]). Now, in the present case, $H = \xi$, so that the first two conditions are satisfied and the following results hold.

Theorem 1.4 *Let (M, g) be an n-dimensional connected, simply connected Riemannian manifold with a proper homogeneous structure T in $\mathcal{T}_1 \oplus \mathcal{T}_2$ having closed fundamental 1-form. If (M, g) is a warped product, then (M, g) is isometric to the hyperbolic space $\mathbf{H^n}$ of constant sectional curvature $k = -||\xi||^2$ and $n \geq 6$.*

Theorem 1.5 *The hyperbolic space $\mathbf{H^n}, n \geq 6$, admits a proper structure $T \in \mathcal{T}_1 \oplus \mathcal{T}_2$ with closed fundamental 1-form, if and only if the Euclidean space $\mathbf{R^{n-1}}$ admits a parallel, non-vanishing homogeneous structure $\overline{T} \in \mathcal{T}_2$.*

For the proofs, see [239]. Moreover, we point out that the existence of a homogeneous structure of class $\mathcal{T}_2$ on $\mathbf{R^{n-1}}, n \geq 6$ is still an open problem

([177]). Finally, analogous results are obtained in [240] for a Riemannian manifold carrying a proper homogeneous structure in $\mathcal{T}_1 \oplus \mathcal{T}_2 \oplus \mathcal{T}_3$ having closed fundamental 1-form.

Example 1.6 (Riemannian submersions and biquotients). In [84] J. Eschenburg constructs an interesting class of compact manifolds which are realized as biquotients, *i.e.* as the base space of a homogeneous principal bundle. More precisely, let G be a compact Lie group and K a closed subgroup of G. The map

$$((a,k),a') \in G \times K \times G \mapsto aa'k^{-1} \in G\ ,$$

defines a transitive action of $G_0 = G \times K$ on G and the isotropy group of the unit element is $\Delta K = \{(k,k) \mid k \in K\}$. Consider a bi-invariant metric $<,>_0$ on G and the induced metric on K. This gives a bi-invariant metric on $G \times K$, denoted again by $<,>_0$. Since ΔK acts isometrically and freely on G_0 by right translations, the orbit space $G_0/\Delta K$ inherits from G_0 a Riemannian metric which makes the projection $\pi : G_0 \to G_0/\Delta K$ a Riemannian submersion. One considers the map $F : G_0/\Delta K \to G$ such that $F(\Delta K(a,k)) = ak^{-1}$ for any $(a,k) \in G_0$, which acts as a diffeomorphism. Therefore G inherits from G_0 a Riemannian metric $<,>$ and the map $\phi : (G \times K, <,>_0) \to (G, <,>)$ defined by $\phi(a,k) = ak^{-1}$, $a \in G$, $k \in K$ is a Riemannian submersion. To determine the ϕ-horizontal space $\mathcal{H}_p$ at $p = (1,1) \in G \times K$, let $\mathfrak{g}$, $\mathfrak{k}$ be the Lie algebras of G, K, respectively. Consider the $<,>_0$-orthogonal splitting $\mathfrak{g} = \mathfrak{k} \oplus \mathfrak{p}$, and for any $X \in \mathfrak{g}$, put $X = X_{\mathfrak{k}} + X_{\mathfrak{p}}$, $X_{\mathfrak{k}}$,$X_{\mathfrak{p}}$ being the projections of X on $\mathfrak{k}$, $\mathfrak{p}$. A vector $(X,Y) \in \mathfrak{g} \oplus \mathfrak{k}$ is ϕ-horizontal if and only if for any $Z \in \mathfrak{k}$, $<(X,Y),(Z,Z)>_0 = 0$ or, equivalently, $Y = -X_{\mathfrak{k}}$. Then one obtains $\mathcal{H}_p = \{(X_1 + X_2, -X_2) \mid X_1 \in \mathfrak{p}, X_2 \in \mathfrak{k}\}$ and easily proves that the ϕ-horizontal lift of any vector $X \in \mathfrak{g}$ is $(X_{\mathfrak{p}} + \frac{1}{2}X_{\mathfrak{k}}, -\frac{1}{2}X_{\mathfrak{k}})$. This allows to compute the action of $<,>$ on $\mathfrak{g}$. In fact, given $X \in \mathfrak{g}$, its norm (with respect to $<,>$) is given by:

$$||X||^2 = ||X_{\mathfrak{p}} + \frac{1}{2}X_{\mathfrak{k}}||_0^2 + ||\frac{1}{2}X_{\mathfrak{k}}||_0^2 = ||X_{\mathfrak{p}}||^2 + \frac{1}{2}||X_{\mathfrak{k}}||_0^2\ .$$

Thus, in the new metric, all vectors in $\mathfrak{k}$ are shortened by the factor $2^{-1/2}$. Now, we consider a closed subgroup U of G_0, which acts isometrically on G by $((u_1,u_2),a) \mapsto u_1 a u_2^{-1}$. This action is free if and only if, for any $(u_1,u_2) \in U$, $(u_1,u_2) \neq (1,1)$, u_1 is not conjugate to u_2. In this case, G/U is a manifold and one can consider the Riemannian metric $<,>'$ induced by $<,>$ in such a way that the projection $p : (G, <,>) \to (G/U, <,>')$ is a Riemannian submersion. Equation (1.31) entails that $(G/U, <,>')$ has non-negative sectional curvatures.

This construction was firstly used by Gromoll and Meyer ([134]) to obtain a metric of non-negative sectional curvatures on an exotic 7-sphere. Further examples are explained in [84], like the spaces described by Aloff and Wallach ([4]), who consider a suitable flag manifold as the base space of infinitely many $\mathbf{S^1}$-bundles.

1.5 Geodesics and O'Neill Theorem

Given a Riemannian submersion $\pi : (M, g) \to (B, g')$, let $c : I \to M$ be a (differentiable) curve and $\gamma = \pi \circ c$ its projection on B, both assumed regular curves. For any $t \in I$, we put:

$$\dot{c}(t) = E(t) + W(t) \ , \quad E(t) \in \mathcal{H}_{c(t)} \ , \quad W(t) \in \mathcal{V}_{c(t)} \ . \tag{1.39}$$

In particular, c is called a *horizontal curve* if, for any t, $\dot{c}(t) \in \mathcal{H}_{c(t)}$. In any case, $E(t)$ is a basic vector field along c, since it is the horizontal lift of the vector field $\dot{\gamma}(t)$ tangent to γ. Let X, E, W be extensions of $\dot{c}(t), E(t), W(t)$ to M, with $W \in \mathcal{X}^v(M)$ and $E \in \mathcal{X}^b(M)$; indeed, one can consider E as the basic vector field π-related to an extension E' of $\dot{\gamma}$. Analogously, for any vector field $Y(t)$ along c, we put:

$$Y(t) = F(t) + U(t) \ , \quad F(t) \in \mathcal{H}_{c(t)}, \quad U(t) \in \mathcal{V}_{c(t)} \ , \tag{1.40}$$

and denote by Y, F, U extensions of $Y(t), F(t), U(t)$ with $F \in \mathcal{X}^b(M)$ and $U \in \mathcal{X}^v(M)$. Then $F' = \pi_* F$ is a vector field on B which extends the vector field $\pi_{*c(t)}(Y(t))$; according to iii) in Proposition 1.1 and applying (1.22), $\overline{\nabla}_E F = h(\nabla_E F)$ is the basic vector field π-related to $\nabla'_{E'} F'$. On the other hand, using (1.23), (1.39), (1.40) and the properties of A, T, one has:

$$\nabla_X Y = \overline{\nabla}_X U + \overline{\nabla}_E F + \overline{\nabla}_W F + T_W F + T_W U + A_E F + A_E U \ , \tag{1.41}$$

where $\overline{\nabla}_X U, T_W F, A_E F \in \mathcal{X}^v(M)$, while the other vector fields on the right are horizontal. Moreover, applying (1.22), the symmetry of ∇, iv) in Proposition 1.1 and (1.19), one obtains:

$$\overline{\nabla}_W F = h(\nabla_W F) = h(\nabla_F W + [W, F]) = A_F W \ ,$$

and, separating the horizontal and the vertical components in (1.41), one derives, along c:

$$\begin{aligned} v(\nabla_{\dot c}Y(t)) &= (\overline{\nabla}_{\dot c}U + T_W F + A_E F)(t)\ , \\ h(\nabla_{\dot c}Y(t)) &= (\overline{\nabla}_E F + A_F W + T_W U + A_E U)(t)\ . \end{aligned} \tag{1.42}$$

As an immediate application of this result, one obtains the characterizing equations of the geodesics on M.

Proposition 1.8 *If $c : I \to M$ is a regular curve and $E(t), W(t)$ denote the horizontal and vertical components of its tangent vector field, then c is a geodesic on M if and only if*

$$\begin{aligned} &(\overline{\nabla}_{\dot c}W + T_W E)(t) = 0\ , \\ &(\overline{\nabla}_E E + 2A_E W + T_W W)(t) = 0\ . \end{aligned} \tag{1.43}$$

Proposition 1.9 *Let $c : I \to M$ be a geodesic, with $E(t) = h(\dot c(t))$ and $W(t) = v(\dot c(t))$. Then, the curve $\gamma = \pi \circ c$ is a geodesic on B if and only if:*

$$(2A_E W + T_W W)(t) = 0\ , \quad t \in I\ . \tag{1.44}$$

Proof. We recall that $\overline{\nabla}_E E$ is the horizontal lift of $\nabla'_{\dot\gamma}\dot\gamma$. Since $\mathcal{X}^b(M)$ and $\mathcal{X}(B)$ are isomorphic, γ is a geodesic if and only if $(\overline{\nabla}_E E)(t) = 0$, $t \in I$. Thus, the statement is an immediate consequence of the second equation in (1.43).

Corollary 1.1 *The projection on B of a horizontal geodesic on M is a geodesic.*

Proposition 1.10 *Let $c : I \to M$ be a geodesic. If the tangent vector $\dot c(t_0)$ at a point $p_0 = c(t_0)$ is horizontal, then c is horizontal.*

Proof. Indeed, expressing $\dot c(t)$ as in (1.40), via (1.25), (1.24), (1.22) and (1.21), one has:

$$\begin{aligned} h(\nabla_{\dot c}W(t)) &= (A_E W + T_W W)(t)\ , \\ v(\nabla_{\dot c}W(t)) &= \overline{\nabla}_{\dot c}W(t) = -(T_W E)(t)\ , \end{aligned}$$

that is

$$\nabla_{\dot c}W(t) = (A_E W + T_W(W - E))(t)\ .$$

Since the zero vector field along c is a solution of this equation and the equation admits a unique solution, for any prescribed initial value, the assertion follows. Note that this proposition states that the horizontal distribution of π is totally geodesic.

Now, we are going to prove a fundamental theorem due to O' Neill ([227]), which states that a Riemannian submersion is determined by its invariants and its differential map at a point. First of all, we need to characterize the basic vector fields.

Lemma 1.4 *Let $\pi : (M, g) \to (B, g')$ be a Riemannian submersion with connected fibres. A horizontal vector field X is basic if and only if*

$$\nabla_V X = T_V X + A_X V \ , \quad V \in \mathcal{X}^v(M) \ . \tag{1.45}$$

In particular, if the fibres are totally geodesic, (1.45) *reduces to:*

$$\nabla_V X = A_X V \ , \quad V \in \mathcal{X}^v(M) \ .$$

Proof. In fact, if $X \in \mathcal{X}^b(M)$, for any vertical vector field $V, [X, V]$ is vertical and via (1.25), (1.19) one has:

$$\nabla_V X = T_V X + h(\nabla_V X) = T_V X + h(\nabla_X V) = T_V X + A_X V \ .$$

Vice versa, given a horizontal vector field X satisfying (1.45) and using Lemma 1.1 and (1.21), for any $V \in \mathcal{X}^v(M), Y \in \mathcal{X}^b(M)$, π-related to Y', one obtains:

$$\begin{aligned} V(g(X,Y)) &= g(\nabla_V X, Y) + g(X, \nabla_V Y) \\ &= g(T_V X + A_X V, Y) + g(X, T_V Y + A_Y V) \\ &= -g(A_X Y, V) - g(A_Y X, V) = 0 \ . \end{aligned}$$

Thus, $g(X, Y)$ is constant on any fibre. Then, if p, q belong to the same fibre *i.e.* $\pi(p) = \pi(q) = x$, we have that $g_p(X_p, Y_p) = g_q(X_q, Y_q)$ implies $g'_x(\pi_{*p} X_p, Y'_x) = g'_x(\pi_{*q} X_q, Y'_x)$. Since π has maximal rank at any point, this gives $\pi_{*p} X_p = \pi_{*q} X_q$, *i.e.* X is projectable and then basic, since it is horizontal.

Theorem 1.6 **(O'Neill).** *Let $\pi, \overline{\pi} : (M, g) \to (B, g')$ be Riemannian submersions, with M connected. If π and $\overline{\pi}$ have the same fundamental tensor fields and $\pi_{*p} = \overline{\pi}_{*p}$ at a point p of M, then $\pi = \overline{\pi}$.*

Proof. Let D denote the set of the points p of M such that $\pi_{*p} = \overline{\pi}_{*p}$ (thus, $\pi(p) = \overline{\pi}(p)$ for the points of D). Since D is closed and non-empty and M is connected, to obtain the statement we prove that each point in D admits a neighborhood where π and $\overline{\pi}$ coincide. Clearly, at each point p of D, the horizontal and vertical spaces with respect to $\pi, \overline{\pi}$ coincide; they are denoted by $\mathcal{H}_p$, $\mathcal{V}_p$. Let p be a point of D, and $\pi(p) = x_0$. Then, according to (1.24), a curve $\alpha : I \to \pi^{-1}(x_0)$, with $0 \in I$ and $\alpha(0) = p$, is a geodesic of $\pi^{-1}(x_0)$ if and only if

$$(\nabla_{\dot\alpha}\dot\alpha)(t) = (T_{\dot\alpha}\dot\alpha)(t)\ ,\quad t \in I\ ,\ \dot\alpha(0) \in \mathcal{V}_p\ . \tag{1.46}$$

Vice versa, any curve $\alpha : I \to \pi^{-1}(x_0)$ whose tangent vector field satisfies (1.46) is the geodesic on $\pi^{-1}(x_0)$ with initial condition $(p, \dot\alpha(0))$. Thus, the supports of the curves satisfying (1.46) set up a convex neighborhood F of p in $\pi^{-1}(x_0)$. Let X be a vector field defined on F with values in TM. Then X is basic if and only if

$$X_p \in \mathcal{H}_p,\ \ \nabla_V X = A_X V + T_V X\ ,\quad V \in \mathcal{X}(F)\ . \tag{1.47}$$

Indeed, since the vector fields tangent to F are vertical, Lemma 1.4 implies that X satisfies (1.47), when X is basic. Vice versa, assuming (1.47), we prove that X is horizontal and thus X turns out to be basic by virtue of Lemma 1.4. Indeed, since F is a convex neighborhood of p, given q in F and $W \in \mathcal{V}_q$ one can consider a geodesic $\beta : I \to \pi^{-1}(x_0)$ such that $\beta(0) = p$, $\beta(t_0) = q$, $\dot\beta(t_0) = W$. Then β satisfies (1.46) and, since $\nabla g = 0$, applying (1.47), one has: $\frac{d}{dt} g(X(t), \dot\beta(t)) = 0$, *i.e.* $g(X, \dot\beta)$ is constant along β. Thus $g_q(X_q, W) = g_p(X_p, \dot\beta(0)) = 0$, and so X_q is horizontal. Now, regarding F as a submanifold of M, we denote by $\nu(F)$ the normal bundle of F. Since the exponential map $\exp : \nu(F) \to M$ is regular at $(p, 0_p)$, there exists a neighborhood U of $(p, 0_p)$ diffeomorphic to a neighborhood $\tilde U$ of p via exp. Thus, any $m \in \tilde U$ can be expressed as $m = \exp_q X_q = \gamma(1)$, where $q \in F$ and γ is the geodesic of M with initial condition $(q, X_q) \in U$. Since $X_q \in \mathcal{H}_q$, γ is a horizontal geodesic and, according to Corollary 1.1, $\pi \circ \gamma$ is the geodesic on B with initial condition $(x_0, \pi_{*q}(X_q))$. Then, one has: $\pi(m) = \pi(\gamma(1)) = \exp_{x_0} \pi_{*q}(X_q)$. Let $\overline{X}$ be the π-basic vector field defined on F such that $\overline{X}(q) = X_q$. According to (1.47), $\overline{X}$ is basic with respect to $\overline{\pi}$, also. Moreover, $\pi(m) = \exp_{x_0} \pi_{*p}(\overline{X}(p))$. Analogously, one can write $\overline{\pi}(m) = \exp_{x_0} \overline{\pi}_{*p}(\overline{X}(p))$; then, since $p \in D$, one has $\overline{\pi}(m) = \pi(m)$. Thus, π and $\overline{\pi}$ coincide on $\tilde U$.

1.6 Clairaut Submersions

Let S be a revolution surface in $\mathbf{R}^3$ with rotation axis d. For any $p \in S$ we denote by $r(p)$ the distance from p to d. Given a geodesic $c : I \to S$ on S, let $w(t)$ be the angle between $c(t)$ and the meridian curve through $c(t), t \in I$.

A classical theorem due to Clairaut states that for any geodesic c on S the product $r(c(t))\sin w(t)$ is independent of t, *i.e.* $(r\circ c)\sin w$ is a constant function. In [30] Bishop introduces the notion of Clairaut submersion and characterizes such a submersion by means of a remarkable property of the fibres.

Definition 1.2 A Riemannian submersion $\pi:(M,g)\to(B,g')$ is called a *Clairaut submersion* if there exists a positive function r on M such that, for any geodesic c on M, the function $(r\circ c)\sin w$ is constant, where, for any t, $w(t)$ is the angle between $\dot{c}(t)$ and the horizontal space at $c(t)$.

Theorem 1.7 (Bishop). *Let $\pi:(M,g)\to(B,g')$ be a Riemannian submersion with connected fibres. Then π is a Clairaut submersion with $r=e^f$ if and only if each fibre is totally umbilical and has mean curvature vector field $H=-\mathrm{grad}f$.*

Proof. We begin showing that π is a Clairaut submersion with $r=e^f$ if and only if for any geodesic $c:I\to M$, one has, along c:

$$g(V(t),V(t))g(\dot{c}(t),(\mathrm{grad}f)_{c(t)})+g((T_VV)(t),E(t))=0\,, \qquad (1.48)$$

where $t\in I$, $E(t)=h(\dot{c}(t))$, $V(t)=v(\dot{c}(t))$ and, as usual, we denote by E, V horizontal and vertical extensions of the horizontal and vertical components of $\dot{c}(t)$, respectively. To this aim, let $c:I\to M$ be a geodesic and, according to the decomposition $\dot{c}(t)=E(t)+V(t)$, let $w(t)$ denote the angle in $[0,\pi]$ between $\dot{c}(t)$ and $E(t)$. If $\sin w(t)$ vanishes at a point t_0, then we have $V(t_0)=0$, the geodesic c turns out to be horizontal at $c(t_0)$ and, applying Proposition 1.10, c is horizontal. Thus, for any function r on M, $r(c(t))\sin w(t)$ identically vanishes and (1.48) is satisfied. Hence, we can only examine the case of non-horizontal geodesics, *i.e.* $\sin w(t)$ never vanishing. Putting $a=||\dot{c}(t)||^2$, which is constant since c is a geodesic, one obtains:

$$g_{c(t)}(E(t),E(t))=a\cos^2 w(t)\,,\quad g_{c(t)}(V(t),V(t))=a\sin^2 w(t)\,. \quad (1.49)$$

Thus, computing the derivative with respect to t of the second relation, one has, along c:

$\frac{d}{dt}g(V,V)=2a\sin w\cos w\frac{d}{dt}w\;;$

$\frac{d}{dt}g(V,V)=2g(\nabla_{\dot{c}}V,V)=2g(v(\nabla_{\dot{c}}V),V)=-2g(T_VE,V)=2g(T_VV,E)\;,$

i.e.

$$a \sin w(t) \cos w(t) \frac{d}{dt} w(t) = (g(T_V V, E))(c(t)) \ . \tag{1.50}$$

It follows that π is a Clairaut submersion with $r = e^f$ if and only if one has $\frac{d}{dt}(\exp(f \circ c) \sin w) = 0$. Multiplying with the nonzero factor $a \sin w(t)$, using (1.50) and (1.49), this is equivalent to:

$$g(V, V)(c(t)) \frac{d(f \circ c)}{dt}(t) + (g(T_V V, E))(c(t)) = 0 \ , \tag{1.51}$$

which can be rewritten as (1.48). Finally, we prove that (1.48) is equivalent to the total umbilicity of the fibres, with the prescribed vector field H. Indeed, assuming (1.48) and considering any geodesic c on M with initial vertical tangent vector, grad$f$ turns out to be horizontal. Thus, the function f is constant on any fibre, the fibres being connected. Now, for any X, W horizontal and vertical vector fields and for any $p \in M$, applying (1.48) to the unique geodesic having initial conditions $(p, X_p + W_p)$, we obtain:

$$g(W, W)\text{grad} f + T_W W = 0 \ ,$$

and, by polarization:

$$g(U, W)\text{grad} f + T_U W = 0 \ . \tag{1.52}$$

Since the restriction of T to $\mathcal{X}^v(M) \times \mathcal{X}^v(M)$ acts as the second fundamental form of any fibre, (1.52) means that any fibre is totally umbilical with mean curvature vector field $H = -\text{grad} f$. Vice versa, the hypothesis that the fibres are totally umbilical with mean curvature vector field $H = -\text{grad} f$ implies that grad$f$ is horizontal. Thus, for any geodesic c on M, putting $\dot{c}(t) = E(t) + V(t)$, one has:

$$\begin{aligned} g_{c(t)}(V(t), V(t)) g_{c(t)}(\dot{c}(t), \text{grad}_{c(t)} f) &= (g(V, V) g(E, \text{grad} f))(c(t)) \\ &= -(g(E, T_V V))(c(t)) \ , \end{aligned}$$

i.e. (1.48) holds.

Example 1.7 The fibres of the first projection $\pi_1 : M_1 \times_f M_2 \to M_1$ from the warped product of (M_1, g_1) and (M_2, g_2) by f are totally umbilical with mean curvature vector $H = -\text{grad}(\log f^{\frac{1}{2}})$ (see Example 1.2). Thus, if M_2 is connected, π_1 is a Clairaut submersion with $r = f^{\frac{1}{2}}$. This example falls within a wide class of Clairaut submersions considered in [214] and described as follows.

Example 1.8 Let $\pi : (M, g) \to (B, g')$ be a Riemannian submersion with connected, totally geodesic fibres and σ a function on B. The metric g_σ on M given by:

$$g_\sigma(X, Y) = g(X, Y)\ ,\ g_\sigma(X, U) = 0\ ,\ g_\sigma(U, V) = e^{-2\sigma\circ\pi} g(U, V)\ ,$$

for any $X, Y \in \mathcal{X}^h(M)$, $U, V \in \mathcal{X}^v(M)$, makes π a Clairaut submersion. Indeed, g_σ and g determine the same horizontal distribution and induce homothetic metrics on the fibres. For any $U, V \in \mathcal{X}^v(M)$, $X \in \mathcal{X}^h(M)$, the Levi-Civita connection ∇' of (M, g_σ) satisfies:

$$\begin{aligned} g_\sigma(\nabla'_U V, X) &= \tfrac{1}{2}(-X(g_\sigma(U, V)) + g_\sigma([X, U], V) - g_\sigma([V, X], U)) \\ &= d(\sigma \circ \pi)(X) g_\sigma(U, V) + e^{-2\sigma\circ\pi} g_\sigma(\nabla_U V, X)\ . \end{aligned}$$

Hence, the invariant tensor T' on (M, g_σ) is given by:

$$\begin{aligned} T'_U V = h(\nabla'_U V) &= g_\sigma(U, V)\mathrm{grad}(\sigma \circ \pi) + e^{-2\sigma\circ\pi} T_U V \\ &= g_\sigma(U, V)\mathrm{grad}(\sigma \circ \pi)\ . \end{aligned}$$

Thus, any fibre is a totally umbilical submanifold of (M, g_σ) with mean curvature vector field $H = \mathrm{grad}(\sigma \circ \pi)$, *i.e.* π is a Clairaut submersion with $r = e^{-\sigma\circ\pi}$.

Chapter 2

Submersions with Totally Geodesic Fibres

This chapter is essentially devoted to the classification of the Riemannian submersions with totally geodesic fibres and total space $\mathbf{S^m}$.

Firstly, in Sec. 2.1, we prove the Hermann theorem, ([140]), stating that a Riemannian submersion with totally geodesic fibres and complete total space acts as the projection of a bundle associated with a suitable principal fibre bundle.

Section 2.2 is devoted to the properties of Riemannian submersions with totally geodesic fibres, when the total space has non-positive, respectively non-negative, or possibly constant sectional curvatures. Some of these results are useful in proving the classification theorem stated by R. Escobales ([86]) and A. Ranjan ([246]) on the Riemannian submersions with totally geodesic fibres and the standard m-sphere as total space.

In Sec. 2.3 we prove in details the mentioned theorem, stating that, in the above hypotheses, the given submersion acts as the projection of only three possible fibre bundles. For any $n \geq 1$, the Hopf submersion and the generalized Hopf submersion considered in Chap. 1 provide examples of the above projections. Moreover, we explicitly describe an $\mathbf{S^7}$-fibration from $\mathbf{S^{15}}$ on the sphere $\mathbf{S^8}(\frac{1}{2})$.

The representation theory of Clifford algebras is applied in Sec. 2.4 to obtain the uniqueness result, *i.e.* to prove that, up to equivalence, the previous examples are the only Riemannian submersions with totally geodesic fibres and total space $\mathbf{S^m}$.

In the last Sec. 2.5, we discuss the minimality and the totally geodesicity of the submanifold obtained lifting a submanifold of the base space to the total space. Then, via the Hopf submersion, we construct a Riemannian submersion $\rho : \mathbf{P_{2n+1}(C)} \to \mathbf{P_n(Q)}$ with fibres isometric to $\mathbf{P_1(C)}$. Combining a theorem of R. Escobales ([87]) with a result of J. Ucci ([302]), up to equivalence, ρ turns out to be the only Riemannian submersion whose fibres are complete, complex and totally geodesic submanifolds of $\mathbf{P_{2n+1}(C)}$.

2.1 Riemannian Submersions with Complete Total Space

Let $\pi : (M, g) \to (B, g')$ be a Riemannian submersion and $\gamma : [a, b] \to B$ a curve. Horizontal lifts of γ, *i.e.* horizontal curves on M which project on γ, exist at least locally. In fact, given a curve $\gamma : [a, b] \to B$ with $\gamma(a) = x$, for any $p \in \pi^{-1}(x)$ there exist a chart (U, φ), $\varphi = \{x^1, \ldots, x^m\}$ of M centered at p and a chart (V, ψ), $\psi = \{y^1, \ldots, y^n\}$ of B centered at x, such that $\pi(U) = V$ and the local frame in U is given by $\{\frac{\partial}{\partial x^i}\}_{1 \le i \le m}$ where $\{\frac{\partial}{\partial x^i}\}_{n+1 \le i \le m}$ locally span the vertical distribution. For any $i \in \{1, \ldots, n\}$ we decompose $\frac{\partial}{\partial x^i}$ in its horizontal and vertical components writing:

$$\frac{\partial}{\partial x^i} = h(\frac{\partial}{\partial x^i}) + \sum_{k=n+1}^{m} a_i^k \frac{\partial}{\partial x^k} ,$$

where the a_i^k's are differentiable functions on U. Then, we choose $\delta > 0$ such that $\gamma([a, a + \delta]) \subset V$ and we look for a horizontal lift $\tilde{\gamma}_p : [a, a + \delta] \to M$ of $\gamma_{|[a,a+\delta]}$ starting at p. Denoting by $y^i(t)$, $i = 1, \ldots, n$ and by $x^i(t), i = 1, \ldots, m$ the components of γ in V and $\tilde{\gamma}_p$ in U, respectively, we need to impose the conditions:

$$x^i(t) = y^i(t) \;\; i = 1, \ldots, n ,$$

and to solve the system of linear differential equations:

$$\dot{x}^k(t) = -\sum_{i=1}^{n} \dot{x}^i(t) a_i^k(x^1(t), \ldots, x^m(t)) , \;\; k = n + 1, \ldots, m , \qquad (2.1)$$

with initial condition $x^k(a) = x^k(p)$, $k = n + 1, \ldots, m$. Since a solution of (2.1) is locally uniquely determined on the initial condition and smoothly depends on the data, we get the existence of a horizontal lift of γ starting at p and defined in a maximal interval $[a, \varepsilon[$, $a < \varepsilon < b$, where ε is chosen to satisfy the additional condition $\tilde{\gamma}_p([a, \varepsilon[) \subset U$.

The horizontal distribution $\mathcal{H}$ of π is called *Ehresmann-complete* if any curve $\gamma : [a, b] \to B$ can be globally horizontally lifted to M, starting at any point of $\pi^{-1}(\gamma(a))$.

Now, we are going to prove the result due to R. Hermann, stating that $\mathcal{H}$ is Ehresmann-complete, provided that (M, g) is complete. To this aim, it is enough to show that the maximal horizontal lift $\tilde{\gamma}_p$ of γ is prolongable at ε. Since $\tilde{\gamma}_p$ is horizontal, for any $t \in [a, \varepsilon[$, $\|\dot{\tilde{\gamma}}_p(t)\| = \|\dot{\gamma}(t)\|$, thus $\tilde{\gamma}_p$ and γ have the same arc-length parameter. So, the continuity of γ and the completeness of M imply the existence of the limit point $q = \lim_{t\to\varepsilon} \tilde{\gamma}_p(t) \in M$. Moreover, one has $\pi(q) = \lim_{t\to\varepsilon} \gamma(t) = \gamma(\varepsilon)$ and this allows to extend $\tilde{\gamma}_p$ at ε.

Fixed $\pi : (M, g) \to (B, g')$ with Ehresmann-complete horizontal distribution, any curve $\gamma : [a, b] \to B$ joining two points x and y determines the differentiable map $F_\gamma : \pi^{-1}(x) \to \pi^{-1}(y)$ such that:

$$F_\gamma(p) = \tilde{\gamma}_p(b), \quad p \in \pi^{-1}(x) \,. \tag{2.2}$$

Proposition 2.1 *Let $\pi : (M, g) \to (B, g')$ be a Riemannian submersion with (M, g) complete. For any curve γ on B joining two points x and y, the map F_γ is a diffeomorphism between the fibres $\pi^{-1}(x)$ and $\pi^{-1}(y)$. Moreover, if the fibres are totally geodesic, F_γ is an isometry.*

Proof. In fact, given $\gamma : [0, t_0] \to B$ with $\gamma(0) = x, \gamma(t_0) = y$, let c be the reparametrization of γ defined by:

$$c(t) = \gamma(t_0 - t), \quad t \in [0, t_0] \,. \tag{2.3}$$

The map F_c defined as in (2.2) acts as the inverse map of F_γ; thus F_γ is a diffeomorphism. Now, we observe that the fibres are complete manifolds, being closed submanifolds of M. Assuming that the invariant tensor T vanishes, we prove that F_γ preserves the distance between points. In fact, let p, q be two points in $\pi^{-1}(x)$ and $\alpha_0 : [0,1] \to \pi^{-1}(x)$ a minimizing geodesic, with $\alpha_0(0) = p, \alpha_0(1) = q$. The map $\alpha : [0,1] \times [0, t_0] \to M$ defined by:

$$\alpha(r, t) = \tilde{\gamma}_{\alpha_0(r)}(t) \,, \quad r \in [0,1],\, t \in [0, t_0] \,, \tag{2.4}$$

is a deformation of α_0 by horizontal curves covering γ. Namely, for any r, $\alpha(r, \cdot) = \tilde{\gamma}_{\alpha_0(r)}$ is a horizontal curve. Since $\pi(\alpha(r,t)) = \pi(\tilde{\gamma}_{\alpha_0(r)}(t)) = \gamma(t)$, for any $t \in [0, t_0]$, the curve $\alpha_t = \alpha(\cdot, t) : [0,1] \to \pi^{-1}(\gamma(t))$ has vertical tangent vector field, denoted by $\dot{\alpha}_t$. Moreover, if $t = 0$, one has: $\alpha(r, 0) = \tilde{\gamma}_{\alpha_0(r)}(0) = \alpha_0(r)$, *i.e.* α is a horizontal deformation of α_0. For any $t' \in [0, t_0]$, we denote by $E_{t'}$ the horizontal vector field such that $E_{t'}(r) = \frac{\partial \alpha}{\partial t}(r, t')$. According to (1.19), since $E_{t'}, \dot{\alpha}_{t'}$ are, respectively, horizontal and vertical vector fields, one gets: $v(\nabla_{\dot{\alpha}_{t'}} E_{t'}) = T_{v(\dot{\alpha}_{t'})} E_{t'} = 0$, and so

$$g(\dot{\alpha}_{t'}(r), \nabla_{\dot{\alpha}_{t'}(r)} E_{t'}) = 0, \quad r \in [0,1] \,. \tag{2.5}$$

This implies that the length function L of the family of curves $\{\alpha_t\}_{t\in[0,t_0]}$ is constant. Indeed, assuming that any curve α_t is arc-length parametrized, the derivative of the function $L(t) = \int_0^1 g(\dot{\alpha}_t(r), \dot{\alpha}_t(r))^{\frac{1}{2}} dr$ is:

$$L'(t) = \int_0^1 g(\dot{\alpha}_t(r), \dot{\alpha}_t(r))^{-\frac{1}{2}} g(\nabla_{\dot{\alpha}_t(r)} E_t, \dot{\alpha}_t(r)) dr \ .$$

Thus (2.5) implies $L'(t) = 0$, $t \in [0, t_0]$, *i.e.* the constancy of L. Therefore the curve α_0, which realizes the distance between the points p and q, has the same length as the curve α_{t_0}, joining $\alpha(0, t_0) = \tilde{\gamma}_p(t_0) = F_\gamma(p)$ and $\alpha(1, t_0) = \tilde{\gamma}_q(t_0) = F_\gamma(q)$. This proves that F_γ is distance-decreasing. Actually, F_γ preserves the distances, since the same argument applies to the map $F_\gamma^{-1} = F_c$, c being defined as in (2.3).

Corollary 2.1 *Let $\pi : (M, g) \to (B, g')$ be a Riemannian submersion with totally geodesic fibres. If M is complete, the flow of any basic vector field on M gives rise to an isometry between the fibres.*

Proof. Let X be a basic vector field and $\phi : I \times U \to M$, $I = (-\varepsilon, \varepsilon)$, $\varepsilon > 0$, the flow of X, *i.e.* $\{\phi_t = \phi(t, \cdot)\}_{t\in I}$ is the 1-parameter group of local transformations generated by X ([172]). We recall that, for any $p \in U$, the orbit of p is the integral curve of X starting at p, *i.e.* the curve $\alpha : I \to M$ such that $\alpha(t) = \phi_t(p), t \in I$. We denote by ψ the flow of $X' = \pi_* X$; then, one has:

$$\psi_t \circ \pi = \pi \circ \phi_t \ , \qquad t \in I \ . \tag{2.6}$$

In fact, given $p \in U$, one considers the projection of the orbit of p, *i.e.* the curve $\pi \circ \alpha : I \to B$, whose tangent vector field is given by: $\pi_{*\phi_t(p)}(X_{\phi_t(p)}) = X'_{\pi(\phi_t(p))}$. Since $\pi(\alpha(0)) = \pi(p), \pi \circ \alpha$ is the integral curve of X' starting at $\pi(p)$, thus $\pi \circ \alpha$ coincides with the orbit of $\pi(p)$, *i.e.* $\pi(\phi_t(p)) = \pi(\alpha(t)) = \psi_t(\pi(p)), t \in I$. Given x in $\pi(U)$, $t_0 \in \mathbf{R}$, $0 < t_0 < r$, let $\gamma : [0, t_0] \to B$ be the curve such that $\gamma(t) = \psi_t(x), t \in [0, t_0]$. Formula (2.6) implies that, for any $p \in \pi^{-1}(x)$, the restriction to $[0, t_0]$ of the orbit α starting at p is the horizontal lift of γ with initial point p, *i.e.* $F_\gamma(p) = \alpha(t_0) = \phi_{t_0}(p)$, where $F_\gamma : \pi^{-1}(x) \to \pi^{-1}(y)$, $y = \gamma(t_0)$ is defined as in (2.2). This means that F_γ coincides with the restriction of ϕ_{t_0} to the fibre $\pi^{-1}(x)$ and, via Proposition 2.1, ϕ_{t_0} is an isometry.

Proposition 2.2 *Let $\pi : (M, g) \to (B, g')$ be a Riemannian submersion with totally geodesic fibres. If X, Y are basic vector fields, the restriction of $A_X Y$ to any fibre is a Killing vector field.*

Proof. We prove that, given the basic vector fields X, Y, the Lie derivative $L_{A_X Y}(g)$ vanishes on the pairs of vertical vector fields. In fact, fixed $V, W \in \mathcal{X}^v(M)$, (1.25) and (1.21) and Lemma 1.4 imply:

$$\begin{aligned} g(\nabla_V(A_X Y), W) &= g((\nabla_V A)_X Y, W) + g(A_{A_X V} Y, W) + g(A_X(A_Y V), W) \\ &= g((\nabla_V A)_X Y, W) + g(A_Y W, A_X V) - g(A_Y V, A_X W) , \end{aligned}$$

and then, since $T = 0$ and (1.26, g) holds, one has:

$$\begin{aligned} (L_{A_X Y} g)(V, W) &= g(\nabla_V(A_X Y), W) + g(V, \nabla_W(A_X Y)) \\ &= g((\nabla_V A)_X Y, W) + g(V, (\nabla_W A)_X Y) = 0 . \end{aligned}$$

Now, we give a sketch of the proof of the Hermann theorem. For more details, see [140; 26].

Theorem 2.1 (Hermann). *Let $\pi : (M, g) \to (B, g')$ be a Riemannian submersion, with M connected and complete. Then B is complete. Moreover, if the fibres are totally geodesic, then π acts as the projection of a bundle associated with a principal fibre bundle with structure group the Lie group of isometries of the fibre.*

Proof. To prove the geodesic-completeness of B, one considers a geodesic $c : [a, b] \to B$. Putting $c(a) = x$, for any $p \in \pi^{-1}(x)$ the horizontal lift γ of c starting at p is a geodesic on M. Indeed, the basic vector field $\overline{\nabla}_{\dot\gamma}\dot\gamma = h(\nabla_{\dot\gamma}\dot\gamma)$ vanishes, since it is π-related to $\nabla_{\dot c}\dot c$. Since M is complete, γ can be extended to a (horizontal) geodesic α of M and Corollary 1.1 implies that α projects to a geodesic of B extending c. Now, assuming that the fibres are totally geodesic, fix a point x_0 in B and put $F = \pi^{-1}(x_0)$. Let G be the Lie group of the isometries of F and, for any x in B, G_x be the set of isometries from F on $\pi^{-1}(x)$. Putting $E' = \bigcup_{x \in B} G_x$, the family of maps $\{L_a\}_{a \in G}$, where $L_a(f) = f \circ a$, for any f in E', determines a free left action of G on E'. Moreover, one considers the map $\pi' : E' \to B$ such that $\pi'(G_x) = \{x\}$, for any $x \in B$. Now, E' is equipped with the differentiable structure obtained by considering an open covering $\{U_i\}_{i \in I}$ of B consisting of geodesically convex sets. Then, for any U_i, one picks a point x_i in U_i and a curve σ_i joining x_0 and x_i. Since U_i is geodesically convex, for any $x \in U_i$, via σ_i one can construct a curve $\sigma_{x,i}$ joining x_0 and x. According to Proposition 2.1, the map $F_{\sigma_{x,i}}$ turns out to be an isometry between $\pi^{-1}(x_0)$ and $\pi^{-1}(x)$. This allows to define, for any $U_i \cap U_j \neq \emptyset$, the maps $f_i : U_i \to E'$, $g_{ji} : U_i \cap U_j \to G$ such that $f_i(x) = F_{\sigma_{x,i}}$ and $g_{ji}(x) = F^{-1}_{\sigma_{x,i}} \circ F_{\sigma_{x,j}}$. It is easy to prove that any g_{ji} is smooth and, for any $x \in U_i \cap U_j$, $f_j(x) = g_{ji}(x) f_i(x)$. Then, the families $\{g_{ji}\}$ and $\{f_i\}$, respectively, set up a set of transition functions and a set of cross sections for a coordinate bundle isomorphic to $E'(B,G)$ and so E' is equipped with the differentiable structure determined via such isomorphism ([276]). It is also easy to prove that $\pi : M \to B$ acts as the projection of a bundle isomorphic to the bundle associated with $E'(B,G)$, with F as fibre.

The following results are proved in [313] (see also [26] 9.59).

Theorem 2.2 *Given a principal bundle $P(B,G)$, let $\pi : M \to B$ be the projection of the vector bundle associated with P with fibre F and group G. Let g' be a Riemannian metric on B and $\tilde{g}$ a G-invariant metric on F. If $\mathcal{H}$ is the horizontal distribution on M determined by a connection on P, then M admits a Riemannian metric g which makes $\pi : (M,g) \to (B,g')$ a Riemannian submersion with horizontal distribution $\mathcal{H}$ and totally geodesic fibres.*

Remark 2.1 Recently, in [95], Farafanova explicitly determined the metric g in the above theorem. The Sasakian metrics g^D on the tangent bundle TM and on the unit sphere bundle T_1M over a Riemannian manifold (M,g) (Example 1.3) are obtained as particular cases of this procedure, respectively considering the metric $\tilde{g}$ on the fibre as the canonical metric on $\mathbf{R^n}$ and on $\mathbf{S^{n-1}}$.

We recall that a map $f : X \to Y$ between connected Riemannian manifolds is called *totally geodesic* if for any geodesic γ on X, the curve $f \circ \gamma$ is a geodesic on Y.

Totally geodesic Riemannian submersions are considered in [313], where the following result is proved.

Theorem 2.3 (Vilms). *Let $\pi : (M,g) \to (B,g')$ be a totally geodesic Riemannian submersion, with* $\dim M > \dim B$. *If M is complete and simply connected, then M is a Riemannian product and π acts as the projection on one of the factors.*

2.2 Submersions with Totally Geodesic Fibres

Lemma 2.1 *Let $\pi : (M,g) \to (B,g')$ be a Riemannian submersion with totally geodesic fibres. Then, for any $X,Y,Z \in \mathcal{X}^h(M)$, $W \in \mathcal{X}^v(M)$, one* *has:*

$$\begin{aligned} &(A_X A_Y + A_Y A_X)(W) = -v(R(W,X,Y) + R(W,Y,X)) \ ; \\ &v((\nabla_X A)_Y Z) = -v(R(Y,Z,X)) \ ; \\ &h((\nabla_X A)_Y W) = h(R(X,W,Y)) \ . \end{aligned} \tag{2.7}$$

Proof. Since the invariant T vanishes, the last formula in (1.29) reduces to:

$$R(X,W,Y,V) = g((\nabla_W A)_X Y, V) + g(A_X W, A_Y V) \ ,$$

for any $X, Y \in \mathcal{X}^h(M)$, $V, W \in \mathcal{X}^v(M)$. Then, one has:

$$\begin{aligned} g(R(W,X,Y) + R(W,Y,X), V) &= g((\nabla_W A)_X Y + (\nabla_W A)_Y X, V) \\ &\quad + g(A_X W, A_Y V) + g(A_Y W, A_X V) \\ &= -g(A_Y A_X W + A_X A_Y W, V) \ . \end{aligned}$$

This proves the first relation; the second one follows immediately from (1.28) and also implies the last statement. In fact, since $(\nabla_X A)_Y$ is skew-symmetric, for any $Z \in \mathcal{X}^h(M)$, one has:

$$g((\nabla_X A)_Y W, Z) = -g((\nabla_X A)_Y Z, W) = g(R(X,W,Y), Z) \ .$$

Proposition 2.3 *Let $\pi : (M,g) \to (B,g')$ be a Riemannian submersion with totally geodesic fibres. If M has non-positive sectional curvatures, then the horizontal distribution is integrable and B has non-positive sectional curvatures. Moreover, if M has negative sectional curvatures, so has B and the fibres are discrete. In the additional hypothesis that M is complete and connected, π acts as a covering projection.*

Proof. Assuming that the integrability tensor A does not vanish, one can consider unit vectors X, V at a point p of M, X horizontal, V vertical, such that $A_X V \neq 0$. Since T vanishes, the sectional curvature of the 2-plane α spanned by X, V, given by $K(\alpha) = \|A_X V\|^2$, is strictly positive. Thus, when M has non-positive sectional curvatures, the invariant A vanishes and B also has non-positive sectional curvatures, according to (1.31). Let now assume that M has negative sectional curvatures. If a fibre $\pi^{-1}(x)$ is not discrete, given $p \in \pi^{-1}(x)$, the sectional curvature of any 2-plane in T_pM containing a nonzero vertical vector vanishes and this entails a contradiction. Moreover, since the vertical distribution is trivial, at any point p of M, π_{*p} is an injective isometry, *i.e.* π is an isometric immersion. When M is complete, a well-known theorem states that π acts as the projection of a covering space ([172] Vol. I).

Proposition 2.4 *Let $\pi : (M, g) \to (B, g')$ be a Riemannian submersion with totally geodesic r-dimensional fibres, $1 \leq r < \dim M$. If M has positive sectional curvatures, then one has:*

i) *for any (nonzero) horizontal vector field X, A_X induces an injective, not surjective, map from the vertical into the horizontal distribution;*
ii) $\dim M < 2 \dim B$ *;*
iii) *B has positive sectional curvatures.*

Proof. Let p be a point of M and X a nonzero horizontal vector at p. Since A_X interchanges the horizontal and the vertical spaces, its restriction to $\mathcal{V}_p$ can be considered as a linear map $A_X : \mathcal{V}_p \to \mathcal{H}_p$. Since $T = 0$, for any unit vector $V \in \mathcal{V}_p$ the sectional curvature of the 2-plane spanned by X, V, given by $||X||^{-2}||A_X V||^2$, is strictly positive, thus A_X is injective. Moreover, $g(A_X V, X) = -g(A_X X, V) = 0$, *i.e.* X is orthogonal to $A_X(\mathcal{V}_p)$. So, one has: $r = \dim \mathcal{V}_p < \dim \mathcal{H}_p = \dim B$, $\dim M = r + \dim B < 2 \dim B$. The last part of the statement follows from (1.31).

We recall that a Riemannian manifold is said to be *δ-pinched*, $0 < \delta < 1$, if the sectional curvature of any 2-plane α satisfies: $\delta \leq K(\alpha) \leq 1$.

Proposition 2.5 *Let $\pi : (M, g) \to (B, g')$ be a Riemannian submersion with totally geodesic fibres. If (M, g) is δ-pinched, then g' is homothetic with a $\frac{1}{4}\delta$-pinched Riemannian metric.*

Proof. Assuming that (M, g) is δ-pinched, we prove that:

$$\delta \leq K'(\alpha') \leq 4 \,, \tag{2.8}$$

for any 2-plane α' tangent to B. In fact, given x in B and a 2-plane α' in $T_x B$ we consider an orthonormal base $\{X, Y\}$ of a horizontal 2-plane α which projects on α' and denote by β the plane spanned by X and the vertical vector $V = ||A_Y X||^{-1} A_Y X$. Since $||A_X V||^2 = K(\beta) \leq 1$, one has:

$$||A_X Y|| = g(A_Y X, V) = g(A_X V, Y) \leq 1 \,,$$

hence $K'(\alpha') = K(\alpha) + 3||A_X Y||^2$ satisfies (2.8), and thus $(B, 4g')$ is a $\frac{1}{4}\delta$-pinched manifold.

Corollary 2.2 *Let $\pi : (\mathbf{S^n}, g) \to (B, g')$ be a Riemannian submersion with totally geodesic fibres, g denoting the canonical metric. Assuming that the dimension of the fibres ranges in $\{1, \ldots, n-1\}$, then B is $\frac{1}{4}$-pinched.*

Lemma 2.2 *Let $\pi : (M, g) \to (M', g')$ and $\tau : (M', g') \to (M'', g'')$ be Riemannian submersions, such that $\tau \circ \pi$ has totally geodesic fibres. Then τ also has totally geodesic fibres.*

Proof. We prove that the τ-vertical distribution is totally geodesic. To this aim, we consider τ-vertical vector fields X', Y' on M' and their π-horizontal lifts X, Y to M. Since $\tau \circ \pi$ has totally geodesic fibres and $(\tau \circ \pi)_* X = \tau_* X' = 0$, $(\tau \circ \pi)_* Y = 0$, the covariant derivative $\nabla_X Y$ with respect to the Levi-Civita connection on M is $(\tau \circ \pi)$-vertical, too. Moreover, applying Proposition 1.1, the π-horizontal component of $\nabla_X Y$ is π-related to $\nabla'_{X'} Y'$, ∇' being the Levi-Civita connection on M', hence $\tau_*(\nabla'_{X'} Y') = \tau_*(\pi_*(h\nabla_X Y)) = (\tau \circ \pi)_*(\nabla_X Y) = 0$, *i.e.* $\nabla'_{X'} Y'$ is vertical with respect to τ.

To state the main theorem of this section, we need to consider suitable connections on the horizontal distribution $\mathcal{H}$ of a Riemannian submersion. More precisely, we denote by ∇^h the connection on $\mathcal{H}$ induced by the Levi-Civita connection ∇, *i.e.*

$$\nabla^h_E X = h(\nabla_E X), \;\; E \in \mathcal{X}(M), \; X \in \mathcal{X}^h(M) \,. \tag{2.9}$$

Comparing with (1.22), one has: $\nabla^h_E X = \overline{\nabla}_E X$, thus ∇^h coincides with the connection on $\mathcal{H}$ induced by the Schouten connection. Moreover, the Levi-Civita connection ∇' on (B, g') canonically induces a connection ∇^* on $\mathcal{H}$, regarded as the pull-back of TB via π. The invariant A links ∇^h and ∇^*, according to:

$$\nabla^h_E X - \nabla^*_E X = A_X(vE), \;\; E \in \mathcal{X}(M), \; X \in \mathcal{X}^h(M) \,. \tag{2.10}$$

In fact, since for any $X \in \mathcal{X}^h(M)$, $V \in \mathcal{X}^v(M)$, $\nabla^*_V X = 0$ and $A_X V = h(\nabla_V X)$, then (2.10) holds when E is vertical. Now, we prove (2.10) when E is a basic vector field. We can also assume that $X \in \mathcal{X}^b(M)$. Then, $\nabla^*_E X$ is the horizontal lift of $\nabla'_{E'} X'$, with $\pi_* E = E', \pi_* X = X'$. On the other hand, by Proposition 1.1, $\nabla^h_E X = h(\nabla_E X)$ is the horizontal lift of $\nabla'_{E'} X'$, and then $\nabla^h_E X - \nabla^*_E X = 0$. Thus, (2.10) holds also in this case, since $A_X(vE)$ vanishes.

Theorem 2.4 *Let $\pi : (M, g) \to (B, g')$ be a Riemannian submersion with totally geodesic fibres and* $\dim M \geq 3$. *If (M, g) has constant sectional curvature, then (B, g') is a locally symmetric space.*

Proof. We can prove the condition $(\nabla'_{E'} R')(X', Y', Z') = 0$ considering at any $x \in B$ local vector fields E', X', Y', Z' defined in a neighborhood of x and such that $\nabla'_{E'}X' = \nabla'_{E'}Y' = \nabla'_{E'}Z' = 0$ at x. Let E, X, Y, Z denote the horizontal lifts of E', X', Y', Z', respectively. Since $\nabla^*_E(R^*(X,Y,Z))$ is the horizontal lift of $\nabla'_{E'}(R'(X',Y',Z'))$, which coincides with $(\nabla'_{E'}R')(X',Y',Z')$ at x, we have to prove that $\nabla^*_E(R^*(X,Y,Z))$ vanishes at any point $p \in \pi^{-1}(x)$. Equivalently, since E is horizontal and (2.10) holds, we have to prove:

$$(\nabla^h_E(R^*(X,Y,Z)))_p = 0\,. \tag{2.11}$$

Let c denote the sectional curvature of (M,g). Applying (1.28), one has:

$$R^*(X,Y,Z) = c(g(Y,Z)X - g(X,Z)Y) + 2A_Z A_X Y - A_X A_Y Z + A_Y A_X Z\,,$$

and then:

$$\nabla^h_E(R^*(X,Y,Z)) = 2\nabla^h_E(A_Z A_X Y) - \nabla^h_E(A_X A_Y Z) + \nabla^h_E(A_Y A_X Z)\,.$$

Each summand vanishes at p. Indeed, we start from:

$$\begin{aligned}\nabla^h_E(A_Z A_X Y) = h((\nabla_E A)_Z A_X Y &+ A_{\nabla_E Z}(A_X Y) \\ &+ A_Z(\nabla_E A_X Y))\,.\end{aligned} \tag{2.12}$$

Since $A_X Y$ is vertical, (2.7) implies:

$$\begin{aligned}h((\nabla_E A)_Z A_X Y) &= h(R(E, A_X Y, Z)) \\ &= c\,h(g(A_X Y, Z)E - g(E,Z)A_X Y) = 0\,.\end{aligned}$$

Moreover, since $h(\nabla_E Z)$ projects on $\nabla'_{E'}Z'$ and A is horizontal, one has $A_{\nabla_E Z} = 0$ at p. Finally, by definition,

$$h(A_Z(\nabla_E A_X Y)) = h(\nabla_Z(v(\nabla_E A_X Y)))\,,$$

and since $A_{\nabla_E X} = 0$ at p, one has at p:

$$v(\nabla_E(A_X Y)) = v((\nabla_E A)_X Y + A_X(\nabla_E Y))\,.$$

Moreover, since $\nabla_E Y$ is vertical, $A_X(\nabla_E Y)$ is horizontal and so its vertical component vanishes. Finally, (2.7) implies:

$$v((\nabla_E A)_X Y) = -v(R(X,Y,E)) = c\,v(g(X,E)Y - g(Y,E)X) = 0\,.$$

This proves $v(\nabla_E A_X Y) = 0$ at p and then $\nabla^h_E(A_Z A_X Y) = 0$ at p.

We end this section stating a theorem of J. C. Nash, which gives a condition for the existence of a metric with positive definite Ricci tensor on the total space of a principal bundle. Further results connected with this subject are in [221].

Theorem 2.5 *Let $\pi : P \to M$ be the projection of a principal bundle with compact, semisimple structural group G. If M is compact and admits a Riemannian metric g' with positive definite Ricci tensor, then P has a G-invariant metric g_0 with positive definite Ricci curvature. Moreover, $\pi : (P, g_0) \to (M, g')$ a Riemannian submersion.*

Proof. Let $\hat{g}$ be a bi-invariant metric on G and denote by $< , >$ the corresponding inner product on the Lie algebra $\mathfrak{g}$ of G. We also fix a principal connection θ on P. For any $t \in \mathbf{R}$, $t > 0$, we consider the metric g_t on P pointwise acting as

$$g_t(X,Y) = g'(\pi_* X, \pi_* Y) \circ \pi + t^2 < \theta(X), \theta(Y) > ,$$

for any vectors $X, Y \in T_pP$, $p \in P$, and we put $g = g_1$. It is easy to prove that, for each t, $\pi : (P, g_t) \to (M, g')$ is a Riemannian submersion with totally geodesic fibres isometric to $(G, \hat{g})$ and horizontal distribution $\mathcal{H}$ independent of t and determined by θ (see also Secs. 5.2, 5.4). Let R_t be the Riemannian curvature of (P, g_t) and put $R = R_1$. From (1.27), (1.28), (1.29) one easily obtains, for each X, Y horizontal vector fields and V, W vertical ones:

$$\begin{aligned}
&\text{i)}\ R_t(X,Y,X,Y) = R^*(X,Y,X,Y) - 3t^2\hat{g}(A_XY, A_XY)\ ;\\
&\text{ii)}\ R_t(X,V,X,V) = t^4 g(A_XV, A_XV)\ ;\\
&\text{iii)}\ R_t(X,V,W,V) = 0\ ;\\
&\text{iv)}\ R_t(X,Y,V,Y) = t^2 R(X,Y,V,Y)\ .
\end{aligned}$$

Moreover, considering a vertical vector V at a point $p \in P$, we denote by $\overline{V}$ the (unique) element of $\mathfrak{g}$ whose corresponding fundamental vector field coincides with V at p. Then, for any V, W in $\mathcal{V}_p$, we have:

$$\text{v)}\ R_t(V,W,V,W) = \tfrac{1}{4}t^2 < [\overline{V}, \overline{W}], [\overline{V}, \overline{W}] > .$$

Now, we compute the Ricci tensor ρ_t of (P, g_t) at any point p. We fix $X \in T_pP$, such that $g_t(X,X) = 1$. Then, there exist an orthonormal basis $\{e_1, \dots, e_r\}$ of $(\mathfrak{g}, < , >)$, an orthonormal basis $\{h_1, \dots, h_n\}$ of $\mathcal{H}_p$, real numbers a, b such that $a^2 + b^2 = 1$ and $X = \frac{a}{t} e_1^*(p) + b h_1$, where e_1^* is the fundamental vector field corresponding to e_1. Put $Y = \frac{b}{t} e_1^*(p) - a h_1$ and consider $\{X, Y, \frac{1}{t} e_2^*(p), \dots, \frac{1}{t} e_r^*(p), h_2, \dots, h_n\}$ as a g_t-orthonormal basis of T_pP, so that:

$$\begin{aligned}
\rho_t(X,X) = {} & R_t(X,Y,X,Y) + t^{-2} \textstyle\sum_{i=2}^r R_t(X, e_i^*(p), X, e_i^*(p)) \\
& + \textstyle\sum_{j=2}^n R_t(X, h_j, X, h_j)\ .
\end{aligned}$$

Using ii) we have

$$\begin{aligned} R_t(X,Y,X,Y) &= t^{-2}(a^2+b^2)^2 R_t(h_1, e_1^*(p), h_1, e_1^*(p)) \\ &= t^2 g(A_{h_1}(e_1^*(p)), A_{h_1}(e_1^*(p))) \geq 0 \,. \end{aligned}$$

Fixed $i \in \{2, \dots, r\}$ via v), ii), iii) we also get

$$\begin{aligned} R_t(X, e_i^*(p), X, e_i^*(p)) &= t^{-2}a^2 R_t(e_1^*(p), e_i^*(p), e_1^*(p), e_i^*(p)) \\ &\quad +2abt^{-1} R_t(h_1, e_i^*(p), e_1^*(p), e_i^*(p)) \\ &\quad +b^2 R_t(h_1, e_1^*(p), h_1, e_1^*(p)) \\ &= \tfrac{a^2}{4} < [e_1, e_i], [e_1, e_i] > \\ &\quad +b^2 t^4 g(A_{h_1}(e_i^*(p)), A_{h_1}(e_i^*(p))) \,. \end{aligned}$$

Finally, considering $j \in \{2, \dots, n\}$, we have

$$\begin{aligned} R_t(X, h_j, X, h_j) &= t^2 a^2 g(A_{h_j}(e_1^*(p)), A_{h_j}(e_1^*(p))) + 2abtR(e_1^*(p), h_j, h_1, h_j) \\ &\quad +b^2 R'(\pi_* h_1, \pi_* h_j, \pi_* h_1, \pi_* h_j) - 3b^2 t^2 g(A_{h_1} h_j, A_{h_1} h_j) \,. \end{aligned}$$

These relations imply:

$$\begin{aligned} \rho_t(X,X) &\geq \tfrac{1}{4} a^2 t^{-2} \textstyle\sum_{i=2}^r < [e_1, e_i], [e_1, e_i] > + b^2 \rho'(\pi_* h_1, \pi_* h_1) \\ &\quad + bt(2a \textstyle\sum_{j=2}^n R(e_1^*(p), h_j, h_1, h_j) - 3bt \textstyle\sum_{j=2}^n g(A_{h_1} h_j, A_{h_1} h_j)) \,, \end{aligned}$$

where ρ' is the Ricci tensor of (M, g'). By the hypotheses, there exists $H_0 \in \mathbf{R}$ such that

$$\rho'(X', X') \geq H_0 g'(X', X') \geq 0 \,,$$

for any X' tangent to M. Moreover, the Ricci tensor of $(G, \hat{g})$ satisfies

$$\hat{\rho}(e_1^*(p), e_1^*(p)) = \frac{1}{4} \sum_{i=2}^{r} < [e_1, e_i], [e_1, e_i] > \geq C_0 \,,$$

where C_0 is a positive constant such that $\hat{\rho}(U,U) \geq C_0 \hat{g}(U,U)$, for any vertical vector U. Therefore $\rho_t(X,X) \geq a^2 t^{-2} C_0 + b^2 H_0 + o(t)$. So, the metric $g_0 = g_{t_0}$ corresponding to a small enough value t_0 satisfies the statement.

2.3 Riemannian Submersions from the Spheres

In [86] R. Escobales deals with the problem of classifying, up to equivalence, the Riemannian submersions with connected and totally geodesic fibres from the m-sphere $\mathbf{S^m}$ of radius 1, equipped with the canonical metric. Such results are improved by A. Ranjan ([246]), who also synthetically proves the classification theorem. The Hermann theorem (Theorem 2.1) allows to consider any Riemannian submersion from $\mathbf{S^m}$ with totally geodesic fibres as the projection of a fibre bundle. Moreover, we recall the notion of equivalence between submersions ([86]).

Definition 2.1 Let (M, g) be a complete manifold. Two Riemannian submersions $\pi, \overline{\pi} : (M, g) \to (B, g')$ with connected and totally geodesic fibres are *equivalent* if there exists a bundle isometry $(f, \overline{f})$ between π and $\overline{\pi}$, *i.e.* a pair of isometries $f : (M, g) \to (M, g)$, $\overline{f} : (B, g') \to (B, g')$ such that $\overline{f} \circ \pi = \overline{\pi} \circ f$. A *bundle automorphism* of a Riemannian submersion π is a bundle isometry $(f, \overline{f})$ between π and π; π is *homogeneous* if for any p, q in M there exists a bundle automorphism $(f, \overline{f})$ of π such that $f(p) = q$.

Now, we are able to state the classification theorem ([86; 246]).

Theorem 2.6 *Let $\pi : (\mathbf{S^m}, g) \to (B, g')$ be a Riemannian submersion with connected, totally geodesic fibres whose dimension ranges in $\{1, \ldots, m-1\}$. Then, as a fibre bundle, π is one of the following:*

a) $\pi : \mathbf{S^{2n+1}} \to \mathbf{P_n(C)}$, $n \geq 1$, *with fibres isometric to* $\mathbf{S^1}$*;*
b) $\pi : \mathbf{S^{4n+3}} \to \mathbf{P_n(Q)}$, $n \geq 1$, *with fibres isometric to* $\mathbf{S^3}$*;*
c) $\pi : \mathbf{S^{15}} \to \mathbf{S^8}(\frac{1}{2})$, *with fibres isometric to* $\mathbf{S^7}$.

Moreover, in the case a) $\mathbf{P_n(C)}$ *has the canonical metric of constant holomorphic sectional curvature* 4*, and in the case* b) $\mathbf{P_n(Q)}$ *has the canonical metric of constant quaternionic sectional curvature* 4*. Finally in* c), $\mathbf{S^8}(\frac{1}{2})$ *is the sphere of sectional curvature* 4.

The proof is divided into two steps:
Step A. Determine the manifolds which can occur as the base space of π.
Step B. Determine the dimension of the fibres and the sectional curvatures of the base manifold.

To this aim, we shall apply the following result ([246]).

Proposition 2.6 *Let $\pi : (\mathbf{S^m}, g) \to (B, g')$ be a Riemannian submersion with connected, totally geodesic fibres, and $m \geq 3$. Then B is a compact, simply connected, locally symmetric space of rank* 1.

Proof. In fact, Theorem 2.4 implies that (B, g') is locally symmetric; moreover, B is compact, since π is onto. Let F be a fibre of π; obviously it is complete. This, together with the total geodesicity, implies that F is a q-dimensional sphere, $1 \leq q < m$. Then, considering the long exact sequence of homotopy groups induced by the fibration π:

$$\cdots \to \pi_2(B) \to \pi_1(F) \to \pi_1(\mathbf{S^m}) \to \pi_1(B) \to 0$$

since $\pi_1(\mathbf{S^m}) = 0$, one has: $\pi_1(B) = 0$, *i.e.* B is simply connected. Finally, B has rank one; in fact, we prove that any geodesic $c : \mathbf{R} \to B$ is periodic. Let x_0 be a point on the geodesic $c : \mathbf{R} \to B$. Since the horizontal distribution of π is Ehresmann-complete, ($\mathbf{S^m}$ being complete), given p_0 in $\pi^{-1}(x_0)$, the horizontal lift $\tilde{c}_{p_0}$ is a periodic curve, and so is $c = \pi \circ \tilde{c}_{p_0}$.

The previous proposition helps to carry out **Step A**. In fact, Theorem 2.4 and Proposition 2.6 imply that (B, g') is a compact, simply connected and locally symmetric Riemannian space. Since B has rank 1, it is a two-point homogeneous space (Chap. X in [139]). Via Proposition 2.4, one has: $\dim B > \frac{1}{2}m$, and, in particular, $\dim B \geq 2$. Then, being a compact, simply connected, two-point homogeneous space, up to isometries, (B, g') has to be one of the following spaces: the n-sphere of radius r, $\mathbf{S^n}(r)$, the complex projective space $\mathbf{P_n}(\mathbf{C})$, the quaternionic projective space $\mathbf{P_n}(\mathbf{Q})$, or, possibly, the projective plane $\mathbf{P_2}(\mathbf{Cay})$ over the Cayley numbers. Each of the projective spaces $\mathbf{P_2}(\mathbf{Cay}), \mathbf{P_n}(\mathbf{C}), \mathbf{P_n}(\mathbf{Q})$, $n \geq 2$, is canonically equipped with a metric whose sectional curvatures range between k and $4k$, $k > 0$. Moreover, in this case, one has $k = 1$, as follows from (2.8). Finally, if $n = 1$, we obtain the projective spaces $\mathbf{P_1}(\mathbf{C})$, $\mathbf{P_1}(\mathbf{Q})$, respectively diffeomorphic to $\mathbf{S^2}$ and $\mathbf{S^4}$.

Step B. If (B, g') is isometric to $\mathbf{S^n}(r)$, for the fibration $\pi : \mathbf{S^m} \to \mathbf{S^n}(r)$ with fibres isometric to the q-sphere $\mathbf{S^q}$, $1 \leq q \leq m-1$, we consider the exact sequence of homotopy groups, starting at $\pi_n(\mathbf{S^m})$, *i.e.*

$$\pi_n(\mathbf{S^m}) \to \pi_n(\mathbf{S^n}) \to \pi_{n-1}(\mathbf{S^q}) \to \pi_{n-1}(\mathbf{S^m}) \to \cdots$$

Since for any $h \in \{1, \ldots, n-1\}$ we have $\pi_h(\mathbf{S^m}) = \pi_h(\mathbf{S^n}) = 0$, the exactness of the sequence implies $\pi_h(\mathbf{S^q}) = 0$ for any $h \leq n-2$ and $\pi_{n-1}(\mathbf{S^q}) = \pi_n(\mathbf{S^n}) = \mathbf{Z}$. We conclude that $q = n-1, m = n+q = 2n-1$, *i.e.* $\pi : \mathbf{S^{2n-1}} \to \mathbf{S^n}(r)$ is a $\mathbf{S^{n-1}}$-fibration. Now, we prove that $\mathbf{S^{n-1}}$ is a parallelizable sphere. In fact, given an orthonormal basis $\{X'_1, \ldots, X'_n\}$ of the tangent space $T_b\mathbf{S^n}$ at a point b, let $\{X_1, \ldots, X_n\}$ be basic orthonormal vector fields defined on $F = \pi^{-1}(b)$ and projecting on the X'_i's, $i = 1, \ldots, n$. We prove that $\{A_{X_n}X_1, \ldots, A_{X_n}X_{n-1}\}$ is a global orthonormal frame on F. First of all, we consider a horizontal unit vector X at a point p in F.

Since $\mathbf{S^{2n-1}}$ has sectional curvature 1, the first formula in (2.7) implies:

$$A_X A_X W = -v(R_p(W, X, X)) = -W ,$$
$$g(A_X W, A_X W) = -g(A_X^2 W, W) = g(W, W) ,$$

for any $W \in \mathcal{V}_p$. Since $g(A_X W, X) = -g(A_X X, W) = 0$, A_X acts as an isometry between the vertical space $\mathcal{V}_p$ and the $\mathcal{H}_p$-orthogonal complement $< X >^{\perp}$ of the linear space spanned by X. Since $\dim \mathcal{V}_p = n - 1 = \dim < X >^{\perp}$, A_X is also an isomorphism and $A_X \circ A_X = -I$ on $\mathcal{V}_p$ implies that $A_X :< X >^{\perp} \to \mathcal{V}_p$ is an isometry. In particular, applying this property to the unit vector $X_n(p)$, we have that $\{A_{X_n(p)} X_i(p)\}_{1 \le i \le n-1}$ is an orthonormal basis of $\mathcal{V}_p$. This means that $\{A_{X_n} X_i\}_{1 \le i \le n-1}$ is a global orthonormal frame on the fibre F, which is isometric with $\mathbf{S^{n-1}}$. Thus, $n - 1 \in \{1, 3, 7\}$, *i.e.* $n \in \{2, 4, 8\}$. Moreover, the previous discussion also implies that the base manifold $\mathbf{S^n}(r)$ has radius $r = \frac{1}{2}$. Indeed, given an orthonormal basis $\{X, Y\}$ of a horizontal 2-plane α at a point p in $\mathbf{S^{2n-1}}$, since $A_X :< X >^{\perp} \to \mathcal{V}_p$ is an isometry, one has $||A_X Y|| = 1$, hence the plane α' spanned by $\{\pi_* X, \pi_* Y\}$ has sectional curvature $K'(\alpha') = K(\alpha) + 3||A_X Y||^2 = 4$. Therefore, we obtain three submersions:

$$\pi : \mathbf{S^3} \to \mathbf{S^2}\left(\frac{1}{2}\right), \quad \pi : \mathbf{S^7} \to \mathbf{S^4}\left(\frac{1}{2}\right), \quad \pi : \mathbf{S^{15}} \to \mathbf{S^8}\left(\frac{1}{2}\right),$$

with fibres isometric to $\mathbf{S^1}$, $\mathbf{S^3}$, $\mathbf{S^7}$, respectively. Moreover, the first submersion falls in the type a), for $n = 1$, since $\mathbf{S^2}(\frac{1}{2})$ is isometric to the projective space $\mathbf{P_1(C)}$ equipped with the Fubini–Study metric of constant holomorphic sectional curvature 4, while the submersion $\pi : \mathbf{S^7} \to \mathbf{S^4}(\frac{1}{2})$ falls in the type b), since $\mathbf{S^4}(\frac{1}{2})$ and $\mathbf{P_1(Q)}$ are isometric. Now, we assume that (B, g') is isometric to $\mathbf{P_n(C)}$, $n \ge 2$ and consider the exact sequence:

$$\to \pi_2(\mathbf{S^m}) \to \pi_2(\mathbf{P_n(C)}) \to \pi_1(\mathbf{S^q}) \to \pi_1(\mathbf{S^m})$$

induced by π. Since $m > 2n \ge 4$, we have $\pi_2(\mathbf{S^m}) = \pi_1(\mathbf{S^m}) = 0$, $\pi_2(\mathbf{P_n(C)}) = \mathbf{Z}$, and thus $\pi_1(\mathbf{S^q}) = \mathbf{Z}$, *i.e.* $q = 1, m = 2n + 1$. Therefore, π can be considered as an $\mathbf{S^1}$-fibration from $\mathbf{S^{2n+1}}$ onto $\mathbf{P_n(C)}$, *i.e.* π falls in the type a). When (B, g') is isometric to $\mathbf{P_n(Q)}$, $n \ge 2$, one has $m > 4n > 4$ and considering the exact sequence

$$\pi_4(\mathbf{S^m}) \to \pi_4(\mathbf{P_n(Q)}) \to \pi_3(\mathbf{S^q}) \to \pi_3(\mathbf{S^m}) \to \cdots$$

one has: $\pi_3(\mathbf{S^q}) = \pi_4(\mathbf{P_n(Q)})$, $\pi_2(\mathbf{S^q}) = \pi_3(\mathbf{P_n(Q)})$, $\pi_1(\mathbf{S^q}) = \pi_2(\mathbf{P_n(Q)})$. Now, since the homotopy exact sequence of the generalized Hopf fibration

(Example 1.4) gives $\pi_4(\mathbf{P_n(Q)}) = \mathbf{Z}$ and $\pi_3(\mathbf{P_n(Q)}) = \pi_2(\mathbf{P_n(Q)}) = 0$, one gets $q = 3$, $m = 4n + 3$. Thus, π falls in the class b), since it can be considered as an $\mathbf{S^3}$-fibration from $\mathbf{S^{4n+3}}$ onto $\mathbf{P_n(Q)}$. Finally, with analogous procedure, one obtains $q = 7$ for the dimension of the fibre of π, when its base space is isometric to $\mathbf{P_2(Cay)}$; in this case $m = 23$, but Riemannian submersions $\pi : \mathbf{S^{23}} \to \mathbf{P_2(Cay)}$ with totally geodesic fibres do not exist (see Proposition 5.1 [246] or Theorem 3.5 [86]).

Remark 2.2 The Hopf submersion considered in Sec. 1.2 and the generalized Hopf fibration (Example 1.4) provide examples in the classes a), b) of Theorem 2.6, respectively. Also the class c) is non-empty. Namely, let $C = \mathbf{Cay} = \mathbf{Q} \times \mathbf{Q}$ be the real linear space of the Cayley numbers, equipped with the multiplication defined by:

$$(q_1, q_2)(q_1', q_2') = (q_1 q_1' - \overline{q_2'} q_2, q_2' q_1 + q_2 \overline{q_1'}) \ ,$$

which makes C a non-associative algebra. Let $X = C \cup \{\infty\}$ be the compactification of C and $p : (C^2)^* \to X$ the map such that $p(c, 0) = \infty$, $c \in C^*$, $p(c, d) = cd^{-1}, d \in C^*$; p acts as the projection of a principal bundle with fibre C^* and structure group $GL(8, \mathbf{R})$ (see 20.6 [276]). Since this bundle is linear, it is equivalent in $GL(8, \mathbf{R})$ to a bundle P with structure group $O(8)$. Then, looking at the sphere $\mathbf{S^7} = \mathbf{S^7}(1)$ as a submanifold of C^* which is invariant under the action of $O(8)$, one obtains the existence of a subbundle P' of P with fibres diffeomorphic to $\mathbf{S^7}$ and total space diffeomorphic to $\mathbf{S^{15}}$. The base space is also a sphere; this sphere can be considered of radius $\frac{1}{2}$, since it is equipped with the metric which makes the projection of P' a Riemannian submersion. An elementary and detailed study of this submersion, as well as of the Hopf submersions, is made in [119].

Let $\pi : (\mathbf{S^m}, g) \to (B, g')$ be a Riemannian submersion and $c : \mathbf{R} \to B$ a geodesic with unit speed. In [227], O' Neill defines the *norm* $\alpha(c)$ *of the invariant* A *of* π *along* c as the maximum of the norms $||A_X Y||$, for X, Y horizontal lifts of unit vectors tangent at a point of c, such that $\pi_* X$ is tangent to c. This notion is useful in determining the following condition for the total geodesicity of the fibres of π ([247]).

Proposition 2.7 *Let* $\pi : (\mathbf{S^m}, g) \to (B, g')$ *be a Riemannian submersion such that* $\alpha(c) \leq 1$, *for any geodesic* c *on* B *with unit speed. Then the fibres are totally geodesic.*

Proof. Let $\gamma : \mathbf{R} \to \mathbf{S^m}$ be a horizontal geodesic with unit tangent vector field X and let $\{V_i\}_{1\leq i\leq r}$ be an orthonormal frame of vertical vector fields, defined along γ and parallel with respect to the Schouten connection. Then, via (1.29), one has, along γ:

$$\begin{aligned} r = \textstyle\sum_{i=1}^r R(X,V_i,X,V_i) &= \textstyle\sum_{i=1}^r \{g((\nabla_X T)(V_i,V_i),X) \\ &\quad -g(T_{V_i}X, T_{V_i}X) + g(A_X V_i, A_X V_i)\} \\ &= rg(\nabla_X H, X) + \textstyle\sum_{i=1}^r (\|A_X V_i\|^2 - \| T_{V_i} X \|^2) \ , \end{aligned}$$

H denoting the mean curvature vector field of the fibres. Since X is parallel along γ, we have $g(\nabla_X H, X) = X(g(H,X))$. Moreover, since $\pi \circ \gamma$ is a geodesic on B, the skew-symmetry of A_X, combined with the hypothesis, implies:

$$\|A_X V_i\|^2 = -g(A_X A_X V_i, V_i) \leq \|A_X A_X V_i\| \leq \alpha(\pi\circ\gamma) \leq 1 \ ,$$

so from the above formula we obtain:

$$\sum_{1\leq i\leq r} \|T_{V_i}X\|^2 \leq rX(g(H,X)) \ .$$

Now, the curve γ is periodic with period 2π, as a geodesic on $\mathbf{S^m}$, thus, integrating, we have:

$$\int_0^{2\pi} \sum_i \|T_{V_i}X\|^2 dt \leq r\int_0^{2\pi} X(g(H,X))dt = 0 \ .$$

This implies the vanishing of $T_V X$ for any vertical vector V at any point of γ, and then the vanishing of T, since $\mathbf{S^m}$ is complete.

2.4 The Uniqueness Theorem

We begin with some basic definitions on Clifford algebras and their representations ([184]).

Let (V,q) be a real vector space equipped with a quadratic form q, let $\mathcal{T}(V)$ be the tensor algebra of V and define $\mathcal{I}_q V$ as the ideal of $\mathcal{T}(V)$ generated by the elements of the form $v\otimes v + q(v)1$, for any vector v in V. The Clifford algebra associated with (V,q) is the quotient algebra $Cl(V,q) = \mathcal{T}(V)/\mathcal{I}_q(V)$. Let $\pi_q : \mathcal{T}(V) \to Cl(V,q)$ be the canonical projection. Since its restriction to V is injective, it provides a natural embedding of V into $Cl(V,q)$ and $Cl(V,q)$ can be regarded as the algebra generated by the unit 1 and the vector space V subject to the relations:

$$v \cdot v = -q(v) \cdot 1, \quad v \in V . \tag{2.13}$$

Denoting again with q the bilinear form given by $2q(v,w) = q(v+w) - q(v) - q(w)$, one has:

$$v \cdot w + w \cdot v = -q(v,w) \cdot 1, \quad v, w \in V . \tag{2.14}$$

Proposition 2.8 **(Extension Property)** *Let $f : V \to \mathcal{A}$ be a linear map from (V,q) into a unit, associative $\mathbf{R}$-algebra such that*

$$f(v) \cdot f(v) = -q(v) \cdot 1, \quad v \in V . \tag{2.15}$$

Then f uniquely extends to an algebra homomorphism $\overline{f} : Cl(V,q) \to \mathcal{A}$.

A (real) *representation* of the Clifford algebra $Cl(V,q)$ is an algebra homomorphism $\rho : Cl(V,q) \to End(W)$, W being a real vector space. The linear space W is called a *$Cl(V,q)$-module.* The representation ρ is *reducible* if there exist non-trivial subspaces W_1, W_2 of W such that $W = W_1 \oplus W_2$ and $\rho(\tau)(W_j) \subseteq W_j$, $\tau \in Cl(V,q)$, $j = 1,2$. In this case, one writes $\rho = \rho_1 \oplus \rho_2$, where ρ_1, ρ_2 are the representations defined by $\rho_j(\tau) = \rho(\tau)_{|W_j}$, $\tau \in Cl(V,q)$, $j = 1,2$. Any representation ρ can be expressed as $\rho = \rho_1 \oplus \rho_2 \oplus \cdots \oplus \rho_k$, where each ρ_j is an irreducible representation. Two representations $\rho_1 : Cl(V,q) \to W_1$, $\rho_2 : Cl(V,q) \to W_2$ are *equivalent* if there exists an isomorphism $F : W_1 \to W_2$ such that:

$$F \circ \rho_1(\tau) \circ F^{-1} = \rho_2(\tau) \ , \quad \tau \in Cl(V,q) .$$

Now, let $\pi : (\mathbf{S^m}, g) \to (B, g')$ be a Riemannian submersion satisfying the hypothesis of Theorem 2.6. For a p in $\mathbf{S^m}$, we denote again by g_p the quadratic form on the vertical space $\mathcal{V}_p$ associated with the inner product g_p. We are going to define a representation $\overline{\rho} : Cl(\mathcal{V}_p, g_p) \to End(\mathcal{H}_p)$. Here, a detailed study of $\overline{\rho}$ is made in the case $\pi : \mathbf{S^{4n+3}} \to \mathbf{P_n(Q)}$, (*i.e.* when π falls in the class b) of Theorem 2.6), but the same results are still valid in the other cases. Thus, fixed $p \in \mathbf{S^{4n+3}}$, let $\rho : \mathcal{V}_p \to End(\mathcal{H}_p)$ be the linear map such that:

$$\rho(v)(X) = A_X v, \quad v \in \mathcal{V}_p, \quad X \in \mathcal{H}_p . \tag{2.16}$$

Putting, for any $v \in \mathcal{V}_p, A^v = \rho(v)$, A^v is a skew-symmetric operator (see Lemma 1.1 and Proposition 1.5).

Proposition 2.9 *The linear map ρ uniquely extends to a representation $\overline{\rho} : Cl(\mathcal{V}_p, g_p) \to End(\mathcal{H}_p)$.*

Proof. It suffices to prove the formula

$$A^v \circ A^w + A^w \circ A^v = -2g_p(v,w)I_{\mathcal{H}_p} \ , \quad v, w \in \mathcal{V}_p \ . \tag{2.17}$$

Indeed, since $\mathbf{S^{4n+3}}$ has constant curvature 1, via Lemmas 1.1 and 2.1, we obtain, for any $v, w \in \mathcal{V}_p, X, Y \in \mathcal{H}_p$,

$$\begin{aligned} g_p((A^v \circ A^w + A^w \circ A^v)(X), Y) &= -g_p(A_Y v, A_X w) - g_p(A_X v, A_Y w) \\ &= g_p((A_X \circ A_Y + A_Y \circ A_X)(v), w) \\ &= -2g_p(v,w)g_p(X,Y) \ , \end{aligned}$$

and (2.17) follows.

Lemma 2.3 *Let $\{v_1, v_2, v_3\}$ be an orthonormal basis of $\mathcal{V}_p$. For any unit horizontal vector X at p, $A_X(A_{A_X v_3} v_2)$ is a unit vector.*

Proof. First of all, we prove that for any unit horizontal vector X at p, $\{X, A_X v_1, A_X v_2, A_X v_3\}$ is an orthonormal set. Namely, for $i \in \{1,2,3\}$, $g(A_X v_i, X) = -g(A_X X, v_i) = 0$. Moreover, from (2.17) we have $A_X^2 v_i = -v_i$. So, for any j, $g_p(A_X v_i, A_X v_j) = -g_p(A_X^2 v_i, v_j) = \delta_{ij}$. Furthermore, the 2-plane α_i' spanned by $\{\pi_* X, \pi_* A_X v_i\}$ has sectional curvature $K'(\alpha_i') = 1 + 3||A_X(A_X v_i)||^2 = 4$. On the other hand, there exists a unique projective line $\mathbf{P_1(Q)}$ through $\pi(p)$, embedded in $\mathbf{P_n(Q)}$ as a totally geodesic submanifold, such that:

i) $\pi_* X \in T_{\pi(p)}\mathbf{P_1(Q)}$;
ii) $K'(\alpha') = 4$, for any 2-plane α' in $T_{\pi(p)}\mathbf{P_1(Q)}$;
iii) $K'(\alpha') = 1$, for any 2-plane α' spanned by unit vectors $\{Y, Z\}$, with $Y \in T_{\pi(p)}\mathbf{P_1(Q)}$, $Z \in T_{\pi(p)}\mathbf{P_1(Q)}^\perp$.

Namely, $\mathbf{P_1(Q)}$ is realized as the quaternionic submanifold of $\mathbf{P_n(Q)}$ through $\pi(p)$ whose tangent space at $\pi(p)$ is the quaternionic section $Q(\pi_* X)$ determined by $\pi_* X$ ([153]). The properties ii) and iii) hold since $\mathbf{P_n(Q)}$ has constant quaternionic sectional curvature 4. Moreover, since for any $i \in \{1,2,3\}$ the sectional curvature of the 2-plane α_i' attains the maximum value 4, α_i' is a quaternionic plane, *i.e.* $\pi_*(A_X v_i) \in Q(\pi_* X)$. This also implies $T_{\pi(p)}\mathbf{P_1(Q)} = \pi_* S$, S denoting the linear space spanned by $\{X, A_X v_1, A_X v_2, A_X v_3\}$. Thus, considering the 2-plane α' spanned by $A_X v_2, A_X v_3$, via ii), we have $4 = K'(\alpha') = 1 + 3||A_{A_X v_2} A_X v_3||^2$, and so

$A_{A_X v_2} A_X v_3$ is a unit vector. Finally, via (2.17) we obtain:

$$A_X(A_{A_X v_3} v_2) + A_{A_X v_3} A_X v_2 = -2g(X, A_X v_3) v_2 = 0 ,$$

and then $A_X(A_{A_X v_3} v_2)$ is a unit vector, too.

Lemma 2.4 *There exists an orthonormal basis $\{v_1, v_2, v_3\}$ of the vertical space $\mathcal{V}_p$, such that $A^{v_1} \circ A^{v_2} \circ A^{v_3} = I_{\mathcal{H}_p}$.*

Proof. Let $\{v_1, v_2, v_3\}$ be an orthonormal basis of $\mathcal{V}_p$ and X a unit horizontal vector at p. We prove that the vertical vector $A_X(A_{A_X v_3} v_2)$ is orthogonal to v_2, v_3. Using Lemma 1.1 and (2.17), one has:

$$\begin{aligned} g_p(A_X(A_{A_X v_3} v_2), v_2) &= -g_p(A_{A_X v_3} v_2, A_X v_2) = -g_p(A^{v_2} A^{v_3} X, A^{v_2} X) \\ &= -g_p(A^{v_3} X, X) = 0 , \end{aligned}$$

$$g_p(A_X(A_{A_X v_3} v_2), v_3) = -g_p(A_{A_X v_3} v_2, A_X v_3) = g_p(A_{A_X v_3} A_X v_3, v_2) = 0 .$$

Thus $A_X(A_{A_X v_3} v_2)$ belongs to the linear space spanned by v_1 and is a unit vector (Lemma 2.3), and so $A_X(A_{A_X v_3} v_2) = v_1$ or $A_X(A_{A_X v_3} v_2) = -v_1$. The last case is reducible to the first one, starting from the orthonormal basis $\{-v_1, v_2, v_3\}$. So, we assume that $A_X(A_{A_X v_3} v_2) = v_1$ and applying again Lemma 1.1, we have:

$$\begin{aligned} g_p(A^{v_1}(A^{v_2}(A^{v_3} X)), X) &= -g_p(A_{A_X v_3} v_2, A_X v_1) \\ &= g_p(A_X(A_{A_X v_3} v_2), v_1) = 1 . \end{aligned}$$

Then, the continuity of the map $X \to g_p(A^{v_1} \circ A^{v_2} \circ A^{v_3}(X), X)$ implies that $A^{v_1} \circ A^{v_2} \circ A^{v_3}$ is an isometry of $\mathcal{H}_p$. Moreover, via the Cauchy–Schwarz inequality, one has $A^{v_1} \circ A^{v_2} \circ A^{v_3}(X) = X$, for any unit vector X, and the statement follows.

Lemma 2.4 means that the representation $\overline{\rho}$ considered in Proposition 2.9 is *isotypical*, that is $\overline{\rho} = \overline{\rho}_1 \oplus \overline{\rho}_2 \oplus \cdots \oplus \overline{\rho}_k$, where $\overline{\rho}_1, \overline{\rho}_2, \ldots, \overline{\rho}_k$ are mutually equivalent irreducible representations. In this case each $\overline{\rho}_j$ takes values in $End(W_j)$, where W_j is a 4-dimensional subspace of $\mathcal{H}_p$. Thus, $k = n$, *i.e.* $\overline{\rho} = \overline{\rho}_1 \oplus \cdots \oplus \overline{\rho}_n$. Finally, we remark that there are exactly two nonequivalent irreducible representations of $Cl(\mathcal{V}_p, g_p)$, say θ, θ'. They are determined by the action on the volume element $v_1 v_2 v_3$ of $Cl(\mathcal{V}_p, g_p)$ according to:

$$\theta(v_1 v_2 v_3) = I, \quad \theta'(v_1 v_2 v_3) = -I ,$$

where $\{v_1, v_2, v_3\}$ is an orthonormal basis of $\mathcal{V}_p$ (Prop. 5.9 Chap. 1 [184]). These properties help in proving the main theorem of this section.

Theorem 2.7 *Any two Riemannian submersions in one of the classes* a), b), c) *of Theorem* 2.6 *are equivalent.*

Proof. The proof is carried out by considering two submersions of the type b), $\pi, \pi' : \mathbf{S^{4n+3}} \to \mathbf{P_n(Q)}$. An analogous technique works well in the other cases. We fix two points $p, q \in \mathbf{S^{4n+3}}$ with $\pi(p) = \pi'(q)$ and we prove the existence of a bundle isometry (φ, φ') between π' and π such that $\varphi(q) = p$. First of all, Lemma 2.4 allows to consider orthonormal bases $\{v_1, v_2, v_3\}$ of the vertical space $\mathcal{V}_p$, $\{v'_1, v'_2, v'_3\}$ of $\mathcal{V}'_q$ such that:

$$A^{v_1} \circ A^{v_2} \circ A^{v_3} = I_{\mathcal{H}_p} \ , \quad A'^{v'_1} \circ A'^{v'_2} \circ A'^{v'_3} = I_{\mathcal{H}'_q} \ , \tag{2.18}$$

A' denoting the integrability tensor of π'. Moreover, let $L_1 : \mathcal{V}'_q \to \mathcal{V}_p$ be the linear map such that

$$L_1(v'_i) = v_i \ , \quad i = 1, 2, 3 \ . \tag{2.19}$$

Since L_1 is an isometry, one has:

$$L_1(v) \cdot L_1(v) = -\|L_1(v)\|^2 1 = -\|v\|^2 1 \ ,$$

for any $v \in \mathcal{V}'_q$, and L_1 uniquely extends to an isomorphism of algebras $\overline{L}_1 : Cl(\mathcal{V}'_q, g_q) \to Cl(\mathcal{V}_p, g_p)$. We denote by $\overline{\rho} : Cl(\mathcal{V}_p, g_p) \to End(\mathcal{H}_p)$, and $\overline{\rho}' : Cl(\mathcal{V}'_q, g_q) \to End(\mathcal{H}'_q)$ the representations respectively determined by π, π' as in Proposition 2.9. Then, via (2.18), (2.19) one has:

$$\overline{\rho}'(v'_1 v'_2 v'_3) = I_{\mathcal{H}'_q} \ , \quad (\overline{\rho} \circ \overline{L}_1)(v'_1 v'_2 v'_3) = I_{\mathcal{H}_p} \ . \tag{2.20}$$

Moreover, we put $\overline{\rho}' = \overline{\rho}'_1 \oplus \cdots \oplus \overline{\rho}'_n$, and $\overline{\rho} \circ \overline{L}_1 = \overline{\rho}_1 \oplus \cdots \oplus \overline{\rho}_n$, where, for any $i = 1, 2, ..., n$, $\overline{\rho}'_i$ (resp. $\overline{\rho}_i$) is an irreducible representation into the endomorphism algebra of a 4-dimensional invariant subspace H'_i (H_i) of $\mathcal{H}'_q$ ($\mathcal{H}_p$). Condition (2.20) implies the equivalence between $\overline{\rho}'_i, \overline{\rho}_i$ and thus the existence of an isomorphism $F_i : H'_i \to H_i$ such that

$$F_i \circ A'^{v'_k} = A^{v_k} \circ F_i, \quad k = 1, 2, 3 \ . \tag{2.21}$$

Moreover F_i is an isometry between (H'_i, g_q) and (H_i, g_p). In fact, one can pick a unit vector Y in H'_i such that $\|F_i Y\| = 1$ (otherwise, one considers the isomorphism $\|F_i Y\|^{-1} F_i$). Now, as in the proof of Lemma 2.3, one proves that $\{Y, A'^{v'_1} Y, A'^{v'_2} Y, A'^{v'_3} Y\}$ is an orthonormal basis of H'_i which is mapped, via F_i, in the orthonormal basis $\{F_i Y, A^{v_1} F_i Y, A^{v_2} F_i Y, A^{v_3} F_i Y\}$ of H_i. Therefore, since any F_i is an isometry, the map $L_2 = F_1 \oplus \cdots \oplus F_n$

is an isometry from $\mathcal{H}'_q$ onto $\mathcal{H}_p$ and (2.19), (2.21) allow to prove that the isometry $L = L_1 \oplus L_2 : T_q\mathbf{S^{4n+3}} \to T_p\mathbf{S^{4n+3}}$ satisfies:

$$L(\mathcal{H}'_q) = \mathcal{H}_p\,, \quad A_{L(E)}L(F) = L(A'_E F), \quad E, F \in T_q\mathbf{S^{4n+3}}\,. \qquad (2.22)$$

Since $\mathbf{S^{4n+3}}$ is a symmetric space, there exists a uniquely determined isometry φ of $\mathbf{S^{4n+3}}$ such that $\varphi(q) = p$ and $\varphi_{*q} = L$ (see Theorem 2.3.12 [340]). Moreover, (2.22) is equivalent to:

$$\varphi_{*q}(\mathcal{H}'_q) = \mathcal{H}_p, \quad A_{\varphi_{*q}E}\varphi_{*q}F = \varphi_{*q}(A'_E F), \quad E, F \in T_q\mathbf{S^{4n+3}}\,.$$

Thus, Theorem 2.2 in [86] implies the existence of an isometry φ' of $\mathbf{P_n(Q)}$ such that (φ, φ') is a bundle isometry between π' and π.

Remark 2.3 The statements of Theorems 2.6 and 2.7 are strictly connected with a more general problem concerning the investigation of sets of a round sphere filled by pieces of parallel great spheres. Relevant results avoiding the O'Neill's submersion calculus are due to Wong ([341]), Wolf ([339]), Gluck, Warner and Ziller ([120]). Stronger results concerning smooth fibrations of spheres by parallel, but not necessarily totally geodesic spheres, are obtained by Gromoll and Grove ([133]) and, recently, by Wilking ([338]).

2.5 Submersions from the Complex Projective Spaces

Let P be a closed k-dimensional submanifold of the base space (B, g') of a Riemannian submersion $\pi : (M, g) \to (B, g')$. We are going to state some properties of $\pi^{-1}(P)$, which is a closed $(k + r)$-dimensional submanifold of M ($r = \dim M - \dim B$). Since the vertical space at any point p of $\pi^{-1}(P)$ is a subspace of the tangent space $T_p\pi^{-1}(P)$, any vector ξ normal to $\pi^{-1}(P)$ is horizontal. Fixed a vector ξ in T_pM normal to $\pi^{-1}(P)$ we define the linear map $C_\xi : T_p\pi^{-1}(P) \to \mathcal{H}_p$ such that

$$C_\xi E = h(a_\xi hE), \quad E \in T_p\pi^{-1}(P)\,, \qquad (2.23)$$

a_ξ denoting the *Weingarten operator* on $\pi^{-1}(P)$ with respect to ξ.

Proposition 2.10 *Let P be a submanifold of the base space of a Riemannian submersion $\pi : (M, g) \to (B, g')$ and ξ a vector field normal to $\pi^{-1}(P)$. The following properties hold:*

i) *For any horizontal vector field X tangent to $\pi^{-1}(P)$, $C_\xi X, A_\xi X$ are respectively the horizontal and the vertical component of $a_\xi X$.*

ii) *If ξ is basic and π-related to ξ' and X is a vector field tangent to $\pi^{-1}(P)$, basic and π-related to X', then $C_\xi X$ is the basic vector field π-related to $a'_{\xi'}X'$, $a'_{\xi'}$ denoting the Weingarten operator on P with respect to ξ'.*

iii) *For any vertical vector field V on $\pi^{-1}(P)$ one has:*

$$a_\xi V = \nabla^\perp_V \xi - h(\nabla_V \xi) - T_V \xi \,, \tag{2.24}$$

$\nabla^\perp$ denoting the normal connection of $\pi^{-1}(P)$.

In particular, when ξ is basic, (2.24) *reduces to:*

$$a_\xi V = \nabla^\perp_V \xi - A_\xi V - T_V \xi \,. \tag{2.25}$$

Proof. Let X be a horizontal vector field tangent to $\pi^{-1}(P)$. Via (1.25) and the Weingarten formula we obtain: $a_\xi X = -A_X \xi - h(\nabla_X \xi) + \nabla^\perp_X \xi$, and then $C_\xi X = h(a_\xi X) = -h(\nabla_X \xi) + \nabla^\perp_X \xi$ is the horizontal component of $a_\xi X$. Moreover the vertical vector field $A_\xi X = -A_X \xi$ is the vertical part of $a_\xi X$. When ξ, X are basic, respectively π-related to ξ', X' we know that $h(\nabla_X \xi)$ is π-related to $\nabla'_{X'}\xi'$, and thus $-a'_{\xi'}X'$ is the component of $\nabla'_{X'}\xi'$ tangent to P. On the other hand, applying i), pointwise we get:

$$\pi_*(h(\nabla_X \xi)) = -\pi_*(C_\xi X) + \pi_*(\nabla^\perp_X \xi) \,,$$

where $\pi_*(C_\xi X), \pi_*(\nabla^\perp_X \xi)$ are respectively tangent and normal to P. Hence $C_\xi X$ is basic π-related to $a'_{\xi'}X'$ and this proves ii). Finally, given a vertical vector field V tangent to $\pi^{-1}(P)$, one has:

$$a_\xi V = -\nabla_V \xi + \nabla^\perp_V \xi = -h(\nabla_V \xi) - T_V \xi + \nabla^\perp_V \xi \,.$$

In particular, if ξ is basic, $[\xi, V]$ is a vertical vector field, hence we obtain $A_\xi V = h(\nabla_\xi V) = h(\nabla_V \xi)$.

Theorem 2.8 *Let $\pi : (M, g) \to (B, g')$ be a Riemannian submersion with minimal fibres and P a submanifold of B. Then, P is minimal if and only if $\pi^{-1}(P)$ is minimal.*

Proof. Let q be a point in $\pi^{-1}(P)$, ξ a vector at q normal to $\pi^{-1}(P)$. Proposition 2.10 allows to consider the Weingarten operator a_ξ as the symmetric endomorphism of $T_q\pi^{-1}(P)$ acting on any $E \in T_q\pi^{-1}(P)$ by:

$$a_\xi E = C_\xi(hE) + \nabla^\perp_{vE}\overline{\xi} - h(\nabla_{vE}\overline{\xi}) + A_\xi(hE) - T_{vE}\xi \,,$$

$\overline{\xi}$ denoting a local vector field extending ξ. Then, with respect to an orthonormal basis $\{X_1, ..., X_k, V_1, ..., V_r\}$ of $T_q\pi^{-1}(P)$, where any X_i is a horizontal and any V_j is a vertical vector, one has:

$$\begin{aligned} \text{tr } a_\xi &= \sum_{i=1}^k g_q(a_\xi X_i, X_i) + \sum_{j=1}^r g_q(a_\xi V_j, V_j) \\ &= \sum_{i=1}^k g_q(C_\xi X_i, X_i) + \sum_{j=1}^r g_q(T_{V_j} V_j, \xi) . \end{aligned}$$

Moreover, since π has minimal fibres and T represents the second fundamental form of the fibres of π, the last term in the previous formula vanishes. Finally, considering local basic vector fields $\overline{X}_1, \ldots, \overline{X}_k, \overline{\xi}$ extending $X_1, \ldots, X_k, \xi$ and π-related to $X'_1, \ldots, X'_k, \xi'$, via Proposition 2.10 we get:

$$\text{tr } a_\xi = \sum_{i=1}^k g(C_{\overline{\xi}} \overline{X}_i, \overline{X}_i)(q) = \sum_{i=1}^k g'(a'_{\xi'} X'_i, X'_i)(\pi(q)) = \text{tr } a'_{\xi'}(\pi(q)) ,$$

and the statement follows.

Theorem 2.9 *Let $\pi : (M, g) \to (B, g')$ be a Riemannian submersion with totally geodesic fibres and P a totally geodesic submanifold of B. Then $\pi^{-1}(P)$ is totally geodesic if and only if $A_\xi X$ vanishes for any horizontal vector field X tangent to $\pi^{-1}(P)$ and for any vector field ξ normal to $\pi^{-1}(P)$.*

Proof. In fact, via Proposition 2.10, given the vector fields X, ξ with X horizontal and tangent to $\pi^{-1}(P)$ and ξ normal to $\pi^{-1}(P)$, we have $A_\xi X = v(a_\xi X) = 0$, provided that $\pi^{-1}(P)$ is totally geodesic. Vice versa, we assume that $A_\xi X = 0$, for any vector field ξ normal to $\pi^{-1}(P)$ and for any vector field X horizontal and tangent to $\pi^{-1}(P)$. Since the horizontal distribution is locally spanned by basic vector fields, we can argue using basic vector fields. Thus, fixed ξ normal to $\pi^{-1}(P)$ and π-related to ξ', we prove that the Weingarten operator a_ξ vanishes. To this aim we consider X tangent to $\pi^{-1}(P)$, basic and π-related to X'. Via Proposition 2.10 and the assumption we have $a_\xi X = C_\xi X = 0$, since it is π-related to $a'_{\xi'} X'$, which vanishes. Moreover, given a vertical vector field V on $\pi^{-1}(P)$, since $T = 0$, via (2.25) we obtain that $a_\xi V = \nabla^\perp_V \xi - A_\xi V$ is horizontal. Now, for any horizontal X tangent to $\pi^{-1}(P)$, we get: $g(a_\xi V, X) = -g(A_\xi V, X) = g(A_\xi X, V) = 0$, and $a_\xi V = 0$. This proves that $a_\xi = 0$.

Corollary 2.3 *Let $\pi : \mathbf{S^{2n+1}} \to \mathbf{P_n(C)}$ be the Hopf submersion and consider $\mathbf{P_m(C)}$, $m < n$, as a complex submanifold of $\mathbf{P_n(C)}$. Then, $\pi^{-1}(\mathbf{P_m(C)})$ is totally geodesic in $\mathbf{S^{2n+1}}$. In fact $\pi^{-1}(\mathbf{P_m(C)}) = \mathbf{S^{2m+1}}$.*

Proof. Since the Hopf fibration has totally geodesic fibres (isometric to $\mathbf{S^1}$) and $\mathbf{P_m(C)}$ is totally geodesic in $\mathbf{P_n(C)}$, it is enough to verify that $A_\zeta X = 0$, with ζ vector field normal to $\pi^{-1}(\mathbf{P_m(C)})$, and X horizontal vector field tangent to $\pi^{-1}(\mathbf{P_m(C)})$. Let J denote the complex structure on $\mathbf{P_n(C)}$ and φ the $(1,1)$-tensor field of the contact structure on $\mathbf{S^{2n+1}}$ defined by (1.5). If X is horizontal and tangent to $\pi^{-1}(\mathbf{P_m(C)})$, since $\pi_*(\varphi X) = J(\pi_* X)$ and $\mathbf{P_m(C)}$ is a complex submanifold of $\mathbf{P_n(C)}$, φX is tangent to $\pi^{-1}(\mathbf{P_m(C)})$. Hence, for any ζ normal to $\pi^{-1}(\mathbf{P_m(C)})$, one gets $A_\zeta X = g(\varphi X, \zeta) = 0$. Now, $\pi^{-1}(\mathbf{P_m(C)})$ is totally geodesic in $\mathbf{S^{2n+1}}$; it is also complete, as a compact submanifold of $\mathbf{S^{2n+1}}$. Moreover the restriction of π to $\pi^{-1}(\mathbf{P_m(C)})$ acts as the projection of a fibre bundle on $\mathbf{P_m(C)}$ with connected fibres (homeomorphic to $\mathbf{S^1}$); in particular $\pi^{-1}(\mathbf{P_m(C)})$ is connected. Since the only complete, connected and totally geodesic submanifolds of spheres are spheres, $\pi^{-1}(\mathbf{P_m(C)})$ is the $(2m+1)$-dimensional sphere.

This Corollary, combined with Theorem 2.6, helps in proving the first result, due to R. Escobales ([87]), concerning the classification of the submersions from the projective space $\mathbf{P_m(C)}$.

Proposition 2.11 *Let $\rho : (\mathbf{P_m(C)}, g_4) \to (B, g')$ be a Riemannian submersion with connected, complete and totally geodesic $2h$-dimensional fibres, $1 \le h \le m-1$, g_4 denoting the Fubini–Study metric with constant holomorphic sectional curvature 4. If the fibres of ρ are complex submanifolds of $\mathbf{P_m(C)}$ then, as a fibre bundle, ρ falls in one of the classes:*

i) $\rho : \mathbf{P_{2n+1}(C)} \to \mathbf{P_n(Q)}$, *with fibres isometric to* $\mathbf{S^2}$;
ii) $\rho : \mathbf{P_7(C)} \to \mathbf{S^8}(\frac{1}{2})$ *with fibres isometric to* $\mathbf{P_3(C)}$.

Moreover, $\mathbf{P_n(Q)}$ is equipped with the metric of constant quaternionic sectional curvature 4.

Proof. Let $\pi : (\mathbf{S^{2m+1}}, g) \to (\mathbf{P_m(C)}, g_4)$ be the Hopf submersion considered in Theorem 1.1, and $\rho : (\mathbf{P_m(C)}, g_4) \to (B, g')$ a submersion satisfying the hypothesis. Then $\rho \circ \pi : (\mathbf{S^{2m+1}}, g) \to (B, g')$ is a Riemannian submersion with connected, totally geodesic $(2h+1)$-dimensional fibres. Namely, for any $b \in B$, the fibre $\rho^{-1}(b)$ is isometric to $\mathbf{P_h(C)}$, since it is a complex, connected and totally geodesic submanifold of $\mathbf{P_m(C)}$, and thus $\pi^{-1}(\rho^{-1}(b))$ is totally geodesic in $\mathbf{S^{2m+1}}$ (see Corollary 2.3) and connected. Thus $\rho \circ \pi$ satisfies the hypothesis of Theorem 2.6 and then, as a fibre bundle, it falls in one of the classes a), b), c) considered there. In the case a), $\rho \circ \pi : \mathbf{S^{2m+1}} \to \mathbf{P_m(C)}$ is an $\mathbf{S^1}$-bundle, *i.e.* B is diffeomorphic to $\mathbf{P_m(C)}$ and so $2h = \dim fibre = 2m - 2m = 0$, contradicting the hypothesis. In the case b), $\rho \circ \pi : \mathbf{S^{4n+3}} \to \mathbf{P_n(Q)}$ has fibres isometric to $\mathbf{S^3}$, where $2n = m - 1$, then B is diffeomorphic to $\mathbf{P_n(Q)}$ and the fibres of ρ are identified with $\mathbf{P_1(C)}$ *i.e.* with $\mathbf{S^2}$. Hence, ρ falls in the class i). Finally, when $\rho \circ \pi : \mathbf{S^{15}} \to \mathbf{S^8}(\frac{1}{2})$ has fibres isometric to $\mathbf{S^7}$, one has $m = 7, h = 3$, and then $\rho : \mathbf{P_7(C)} \to \mathbf{S^8}(\frac{1}{2})$ has fibres isometric to $\mathbf{P_3(C)}$.

Example 2.1 As in Example 1.4, we consider the generalized Hopf submersion $\pi' : (\mathbf{S^{4n+3}}, g) \to (\mathbf{P_n(Q)}, g'_4)$, g'_4 denoting the metric of constant quaternionic sectional curvature 4. We recall that π' is the projection of a fibre bundle with structure group $\mathbf{S^3}$ and fibres isometric to $\mathbf{S^3}$. Indeed, the vertical distribution is generated by the vector fields IN, JN, KN, N denoting the outward unit normal to $\mathbf{S^{4n+3}}$ (embedded in $\mathbf{R}^{4n+4}$), I, J, K being the complex structures on $\mathbf{R}^{4n+4}$ induced by the multiplication by $\mathbf{i}, \mathbf{j}, \mathbf{k}$ in $\mathbf{Q}^{n+1}$. The vector field IN spans the vertical distribution of the Hopf fibration $\pi : (\mathbf{S^{4n+3}}, g) \to (\mathbf{P_{2n+1}(C)}, g_4)$ considered in Sec. 1.2. Since $\mathbf{P_{2n+1}(C)}$ is the quotient space of $\mathbf{S^{4n+3}}$ by the $\mathbf{S^1}$-action of the group generated by IN and this action is the restriction of the $\mathbf{S^3}$-action on $\mathbf{S^{4n+3}}$, a Riemannian submersion $\rho : (\mathbf{P_{2n+1}(C)}, g_4) \to (\mathbf{P_n(Q)}, g'_4)$ with $\rho \circ \pi = \pi'$ is well-defined. Obviously, ρ has connected fibres, which turn out to be complex submanifolds of $\mathbf{P_{2n+1}(C)}$. In fact, the vectors $\{\pi_* JN, \pi_* KN\}$ pointwise span the vertical distribution of ρ. Let I' denote the complex structure on $\mathbf{P_{2n+1}(C)}$; by means of (1.15) and (1.5) we get $I'(\pi_* JN) = \pi_*(\varphi(JN)) = \pi_*(I(JN)) = \pi_*(KN)$, *i.e.* the vertical space at any point of $\mathbf{P_{2n+1}(C)}$ is a holomorphic 2-plane. Moreover, since $\rho \circ \pi$ has totally geodesic fibres, via Lemma 2.2, ρ has totally geodesic fibres. Hence, any fibre is isometric to $\mathbf{P_1(C)}$, as a complex, totally geodesic submanifold of the Kähler manifold $\mathbf{P_{2n+1}(C)}$. Thus, ρ provides an example of submersion in the class i) of Proposition 2.11. Indeed, up to equivalence, ρ is the only example in such class. In fact, R. Escobales proved that any two submersions in the class i) are equivalent (Proposition 4.4 in [87]). As far as class ii) is concerned, it is empty. In fact, in [302] J. Ucci proves the non-existence of fibrations $\pi : \mathbf{P_7(C)} \to \mathbf{S^8}(\frac{1}{2})$, with $\mathbf{P_3(C)}$ as typical fibre.

Summing up, one derives the following classification theorem.

Theorem 2.10 *The projection* $\rho : (\mathbf{P_{2n+1}(C)}, g_4) \to (\mathbf{P_n(Q)}, g'_4)$ *considered in Example* 2.1 *is, up to equivalence, the only Riemannian submersion from the complex projective space whose fibres are connected, complete, complex and totally geodesic submanifolds of* $\mathbf{P_{2n+1}(C)}$.

Chapter 3

Almost Hermitian Submersions

The aim of this chapter is the investigation of the Riemannian submersions in the context of Almost Hermitian Geometry.

In Sec. 3.1, we present the basic elements of the theory of almost Hermitian manifolds and their holomorphic submanifolds, which will be useful in the remaining sections. We include the lattice of some fundamental classes in the classification due to A. Gray and L. M. Hervella ([130]). Also the minimality of submanifolds is discussed.

Section 3.2 deals with the almost Hermitian submersions, firstly introduced by B. Watson ([326]). Almost Hermitian submersions are Riemannian submersions between almost Hermitian manifolds, interchanging the almost complex structures. We state the fundamental properties of the invariants T and A and discuss the transference of geometric properties from the total space to the fibres and to the base space.

In Sec. 3.3 we prove that the horizontal distribution of a Kähler submersion is holomorphic. D. L. Johnson ([157]) obtains this result as a consequence of a more general theorem in the context of foliations on Kähler manifolds. We also describe some results, due to D. L. Johnson, in order to illustrate the relationship between the existence of Kähler submersions and of holomorphic connections on principal bundles over Kähler manifolds.

Then, in Sec. 3.4 some relations involving the holomorphic sectional and bisectional curvatures are stated. In particular we examine the case of almost Hermitian submersions having integrable horizontal distribution as it happens for almost Kähler and nearly Kähler submersions. We also give some relations between the Ricci and *-Ricci tensors and the scalar curvatures of the total space, the base space and the fibres.

Following the I. Vaisman's papers ([305; 309; 310; 311]), in Sec. 3.5 we present the basic tools on a locally or globally conformal Kähler manifold

M and on its holomorphic submanifolds. We discuss the minimality of such submanifolds which turns out to be related to the behavior of the Lee vector field on M.

In Secs. 3.6 and 3.7, which are devoted to the locally conformal Kähler submersions, we present the results of J. C. Marrero and J. Rocha ([195; 196]) and a large class of examples on the generalized Hopf manifolds, also known as Vaisman manifolds.

Finally, in Sec. 3.8 we discuss almost complex conformal submersions. They are submersions $\pi : (M, J, g) \to (B, J', g')$ such that $\pi_* \circ J = J' \circ \pi_*$ and $\pi^* g' = e^\sigma g$, for some differentiable function σ on M. Under the hypothesis that M is a locally conformal Kähler manifold, we state some properties; in particular the integrability of the horizontal distribution, its total umbilicity or total geodesicity are investigated. When referred to the Kähler or almost Kähler cases, these results specialize.

3.1 Almost Hermitian Manifolds

An almost complex manifold is a manifold M with an almost complex structure, *i.e.* a tensor field J of type $(1,1)$ such that $J^2 = -I$. Such a manifold is orientable and has even dimension, say $2m$.

Denoting by N_J, or simply by N, the Nijenhuis tensor of J, defined by:

$$N(X,Y) = [JX, JY] - J[JX, Y] - J[X, JY] - [X, Y] \,,$$

a well-known theorem of Newlander and Nirenberg states that the almost complex manifold (M, J) is a complex manifold if and only if $N = 0$, that is J is integrable.

An almost Hermitian manifold (M, J, g) is an almost complex manifold (M, J) with a J-invariant Riemannian metric g. The J-invariance of g means that $g(X, Y) = g(JX, JY)$, for any $X, Y \in \mathcal{X}(M)$.

The Kähler form of the almost Hermitian manifold (M, J, g) is the differential form of bidegree $(1,1)$ defined by: $\Omega(X, Y) = g(X, JY)$, for any $X, Y \in \mathcal{X}(M)$.

Denoting by ∇ the Levi-Civita connection on M, we have the following formulas for the covariant derivative, the exterior differentiation and the codifferential of Ω:

$$(\nabla_X \Omega)(Y, Z) = g(Y, (\nabla_X J)Z), \quad 3d\Omega(X, Y, Z) = \sigma(\nabla_X \Omega)(Y, Z) \,,$$

where σ denotes the cyclic sum over X, Y, Z,

$$(\delta\Omega)(X) = \sum_{i=1}^{m}((\nabla_{E_i}\Omega)(E_i, X) + (\nabla_{JE_i}\Omega)(JE_i, X)) ,$$

where $\{E_1, \ldots, E_m, JE_1, \ldots, JE_m\}$ is a local orthonormal J-frame.

In [130], A. Gray and L. M. Hervella classify the almost Hermitian manifolds (AH-manifolds) in 16 classes. In Table 3.1 we recall some of these classes, with the defining condition. Note that the QK-condition is equivalent to $d\Omega^{(2,1)} = d\Omega^{(1,2)} = 0$ and $K = AK \cap NK = QK \cap H$.

Table 3.1 Classification of almost Hermitian manifolds

Symbol	name	condition
K	Kähler	$\nabla J = 0$
AK	Almost Kähler	$d\Omega = 0$
NK	Nearly Kähler	$(\nabla_X J)X = 0$
QK	Quasi Kähler	$(\nabla_X J)Y + (\nabla_{JX} J)JY = 0$
ASK	Almost semi-Kähler	$\delta\Omega = 0$
SK	Semi-Kähler	$\delta\Omega = 0,\ N = 0$
H	Hermitian	$N = 0$

For the relations between the given classes and examples, see [130]. Furthermore, the above classes are related by the following lattice, where all the inclusions are strict.

Table 3.2 Inclusion relations

Almost Complex	additional condition	Complex
AH	$N = 0$	H
$\cup$		$\cup$
ASK	$N = 0$	SK
$\cup$		$\cup$
QK	$N = 0$	K
$\cup$		$\Vert$
AK NK	$N = 0$	K
$\cup$		
K		

Observe that the integrability condition $N = 0$ forces a QK (or an AK or NK)-manifold to be a Kähler manifold, whereas any ASK-manifold transforms in an SK-manifold.

Let (M, J, g) be an AH-manifold. A submanifold M' of M is called a *holomorphic* (or *invariant* or *almost-complex*) submanifold of M if for any $p \in M'$ the tangent space T_pM' is J-invariant, *i.e.* $J(T_pM') = T_pM'$. The almost complex structure induced on M' is still denoted by J. The first interesting results on holomorphic submanifolds of an AH-manifold are obtained by A. Gray ([125]).

Proposition 3.1 *If M belongs to one of the classes QK, AK, NK, H, K, then any holomorphic submanifold M' of M belongs to the same class. Moreover, M' is necessarily minimal, except in the Hermitian case.*

Proof. The J-invariance of M' implies that N and $d\Omega$ restrict to M' and so the algebraic characterizations of the H and AK structures are preserved. Now, for the remaining cases, from the Gauss equation, we easily obtain:

$$\nabla_X J = \nabla'_X J + \alpha_X.J, \qquad X \in \mathcal{X}(M') , \tag{3.1}$$

where $(\alpha_X.J)Y = \alpha(X, JY) - J\alpha(X, Y)$, for $Y \in \mathcal{X}(M')$, α being the second fundamental form and ∇' the Levi-Civita connection on M'. Then (3.1) implies that $\nabla'_X J$ and $\alpha_X.J$ take values on $\mathcal{X}(M')$, $\mathcal{X}(M')^\perp$, respectively. It follows that M' belongs to the same class of M and the characterization of the considered class holds for α_X, also. Finally, if M is a QK-manifold, we have $\alpha_X.J + (\alpha_{JX}.J) \circ J = 0$, and then

$$\alpha(X, X) + \alpha(JX, JX) = 0 , \tag{3.2}$$

so that M' is minimal and the inclusions in Table 3.2 allow to complete the proof.

Corollary 3.1 *The second fundamental form α of a holomorphic submanifold M' of (M, J, g) satisfies:*

1) $\alpha_{JX} = \alpha_X \circ J$, *if M is QK* ,
2) $\alpha_{JX} = \alpha_X \circ J = J \circ \alpha_X$, *if M is NK* .

Proof. Let M be QK. Then, from (3.2), since α is symmetric, we have $\alpha(X, JY) - \alpha(JX, Y) = 0$, *i.e.* $\alpha_X \circ J = \alpha_{JX}$. If M is NK, then, for any $X \in \mathcal{X}(M')$, one has:

$$0 = (\alpha_X.J)X = (\alpha \circ (K \times J) - J \circ \alpha)(X, X) ,$$

K denoting the Kronecker tensor field. Since M is QK also, $\alpha \circ (K \times J) - J \circ \alpha$ is symmetric, and $\alpha_X \circ J - J \circ \alpha_X$ vanishes.

Remark 3.1 One can easily prove that $\alpha_X \circ J = J \circ \alpha_X$ implies

$$\alpha(X, JY) = \alpha(JX, Y) = J(\alpha(X, Y)) ,$$

for any $X, Y \in \mathcal{X}(M')$.

Now, we discuss the cases of ASK and SK-manifolds, which are excluded from the above proposition, since the definition of the codifferential $\delta\Omega$ requires the choice of local orthonormal frames. Furthermore, in general, it is not known if holomorphic submanifolds of almost semi-Kähler manifolds are minimal. However, if M' is a holomorphic submanifold of an ASK-manifold with codimension 2, then M' is minimal ([326]).

Definition 3.1 Let M' be a holomorphic submanifold of an AH-manifold (M, J, g). The *partial (or tangent) coderivative* of Ω with respect to M' is the linear operator $\overline{\delta}\Omega$, such that for any $X \in \overline{\mathcal{X}}(M)$:

$$(\overline{\delta}\Omega)(X) = \sum_{i=1}^{r} \{(\nabla_{E_i}\Omega)(E_i, X) + (\nabla_{JE_i}\Omega)(JE_i, X)\} ,$$

where $\{E_1, \dots, E_r, JE_1, \dots, JE_r\}$ is a local orthonormal J-frame on TM', $\dim M' = 2r$, $\overline{\mathcal{X}}(M)$ denoting the Lie algebra of the vector fields tangent to M along M'.

By direct computation, using (3.1), we get:

Proposition 3.2 *For any $Z \in \mathcal{X}(M')^{\perp}$ one has:*

$$(\overline{\delta}\Omega)(Z) = 2rg(JH, Z) ,$$

where H denotes the mean curvature vector field of the submanifold M'.

Corollary 3.2 *A holomorphic submanifold M' of an AH-manifold is minimal if and only if $\overline{\delta}\Omega = 0$ on $\mathcal{X}(M')^{\perp}$.*

In particular, the corollary applies in the ASK or SK cases.

Definition 3.2 Let M' be a holomorphic $2r$-dimensional submanifold of an almost Hermitian manifold (M, J, g). The *normal-coderivative* of Ω with respect to M', is the linear operator $\overline{\overline{\delta}}\Omega$ defined by:

$$(\overline{\overline{\delta}}\Omega)(X) = \sum_{i=1}^{m-r} \{(\nabla_{F_i}\Omega)(F_i, X) + (\nabla_{JF_i}\Omega)(JF_i, X)\} ,$$

for any $X \in \overline{\mathcal{X}}(M)$. Here $\{F_1, \dots, F_{m-r}, JF_1, \dots, JF_{m-r}\}$ is a local orthonormal J-frame on $(TM')^{\perp}$.

Proposition 3.3 *Let M' be a holomorphic submanifold of an almost Hermitian manifold M. Then, for any $X \in \mathcal{X}(M')$, one has:*

$$(\delta\Omega)(X) = (\delta'\Omega)(X) + (\bar{\bar{\delta}}\Omega)(X) ,$$

δ' denoting the codifferential operator on M'.

Proof. Let us consider a local orthonormal J-frame on TM, $\{E_1, \dots, E_r, JE_1, \dots, JE_r, F_1, \dots, F_{m-r}, JF_1, \dots, JF_{m-r}\}$, such that $\{E_1, \dots, E_r, JE_1, \dots, JE_r\}$ is a local frame of TM'. According to Definitions 3.1 and 3.2, we have:

$$(\delta\Omega)(X) = (\bar{\delta}\Omega)(X) + (\bar{\bar{\delta}}\Omega)(X) .$$

On the other hand, using (3.1), we obtain $(\nabla_Y \Omega)(Z, X) = (\nabla'_Y \Omega)(Z, X)$, for any $X, Y, Z \in \mathcal{X}(M')$, and so the partial coderivative $\bar{\delta}\Omega$ coincides with the codifferential $\delta'\Omega$ on $\mathcal{X}(M')$ and the statement follows.

Corollary 3.3 *A holomorphic submanifold M' of an ASK or SK-manifold M belongs to the same class of M if and only if $\bar{\bar{\delta}}\Omega$ vanishes on $\mathcal{X}(M')$.*

Definition 3.3 Let (M, J, g) be an almost Hermitian manifold and M' a holomorphic submanifold. M' is called *superminimal* if for any $X \in \mathcal{X}(M')$ one has $\nabla_X J = 0$.

Clearly, using (3.1), we see that the superminimality of M' implies that M' is a minimal Kähler submanifold.

3.2 Almost Hermitian Submersions

Definition 3.4 Let (M^{2m}, J, g) and (B^{2n}, J', g') be almost Hermitian manifolds. A surjective map $\pi : M \to B$ is called an *almost Hermitian submersion* if π is a Riemannian submersion and an almost complex (or a (J, J')-holomorphic) map, *i.e.*

$$\pi_* \circ J = J' \circ \pi_* .$$

Proposition 3.4 *Let $\pi : (M^{2m}, J, g) \to (B^{2n}, J', g')$ be an almost Hermitian submersion. Then, the horizontal and vertical distributions are J-invariant and J commutes with the horizontal and vertical projectors.*

Proof. For any vertical vector field U, we have: $\pi_*(JU) = J'(\pi_* U) = 0$, thus JU is vertical. Obviously, for any horizontal vector field X and any vertical vector field U, we get $g(JX, U) = -g(X, JU) = 0$ and JX is horizontal. Thus, we have $J(\mathcal{V}) \subset \mathcal{V}$, $J(\mathcal{H}) \subset \mathcal{H}$ and so $J(\mathcal{V}) = \mathcal{V}$, $J(\mathcal{H}) = \mathcal{H}$. The last part of the statement follows immediately.

As an obvious consequence of Definition 3.4 we obtain:

Proposition 3.5 *Let $\pi : (M, J, g) \to (B, J', g')$ be an almost Hermitism submersion and let X, Y be basic vector fields on M, π-related to X', Y' on B. Then, we have:*

1) *JX is the basic vector field π-related to $J'X'$;*
2) *$h(N(X, Y))$ is the basic vector field π-related to $N'(X', Y')$;*
3) *$h((\nabla_X J)Y)$ is the basic vector field π-related to $(\nabla'_{X'} J')Y'$.*

We remark that, since $\mathcal{V}$ is J-invariant, any fibre $\pi^{-1}(x)$, $x \in B$, inherits from M an almost Hermitian structure and the inclusion $i : \pi^{-1}(x) \to M$ is an almost complex map. So $\pi^{-1}(x)$ is a closed holomorphic submanifold of M.

Now, we denote by P any of the classes in Table 3.2 and discuss the influence of a given P-structure on the total space, on the fibres and on the base manifold of an almost Hermitian submersion.

Proposition 3.6 **(Watson).** *Let $\pi : (M, J, g) \to (B, J', g')$ be an almost Hermitian submersion and M a P-manifold, with the only exceptions $P = ASK$, $P = SK$. Then, the fibres and B are P-manifolds. Moreover, the fibres are necessarily minimal, except in the Hermitian case.*

Proof. The statement on the fibres follows from Proposition 3.1. Now, if X and Y are basic vector fields on M, π-related to X', Y' on B, then

$$\Omega(X, Y) = g(X, JY) = g'(X', J'Y') \circ \pi = (\pi^* \Omega')(X, Y) \,.$$

Thus, on the horizontal distribution, Ω and $d\Omega$ coincide with $\pi^*\Omega'$ and $\pi^*(d\Omega')$, respectively. Since π^* is a linear isometry and preserves the bidegree of differential forms, the cases $P = QK$ and $P = AK$ are obtained. Furthermore, if $P = NK$ or $P = H$, we apply 3) and 2) of Proposition 3.5. Finally, if M is Kähler, the statement immediately follows, since $K = AK \cap H$.

Definition 3.5 A *P-submersion* $\pi : (M, J, g) \to (B, J', g')$ is an almost Hermitian submersion such that M belongs to the class P.

Obviously, the existence of a P-structure on M gives rise to some restrictions on the fundamental tensors A and T.

Proposition 3.7 *Let $\pi : (M, J, g) \to (B, J', g')$ be an almost Hermitian submersion and V, X vertical and horizontal vector fields, respectively.*
a) *If π is a QK-submersion, then:*

1) $T_{JV} = T_V \circ J$ *on* $\mathcal{X}^v(M)$ *;*
2) $T_{JV} = -J \circ T_V$ *on* $\mathcal{X}^h(M)$ *;*
3) $A_{JX} = A_X \circ J$ *on* $\mathcal{X}^h(M)$ *;*
4) $A_{JX} = -J \circ A_X = A_X \circ J$ *on* $\mathcal{X}^v(M)$ *.*

b) *If π is an NK-submersion, then:*

1) $T_V \circ J = J \circ T_V = T_{JV}$ *on* $\mathcal{X}^v(M)$ *;*
2) $T_V \circ J = J \circ T_V = -T_{JV}$ *on* $\mathcal{X}^h(M)$ *.*

Proof. Since the fibres are holomorphic submanifolds and T acts on them as the second fundamental form, a1) and b1) follow from Proposition 3.4 and Corollary 3.1. Now, for any $U \in \mathcal{X}^v(M)$ and $Y \in \mathcal{X}^h(M)$ one has:

$$g(T_{JV}Y, U) = -g(Y, T_{JV}U) = -g(Y, T_V(JU)) = -g(J(T_V Y), U) ,$$

and a2) follows, since T_V interchanges vertical and horizontal vector fields. Similarly, one achieves b2). To prove a3), taking the vertical part of the QK-condition, one obtains:

$$A_X JX - A_{JX} X - JA_X X - JA_{JX} JX = 0 ,$$

i.e. $A_X JX = 0$, for any $X \in \mathcal{X}^h(M)$, and a3) follows by polarization. The first equality in a4) can be proved as a2). Finally, choosing X basic, taking the horizontal part of the QK-condition $(\nabla_V J)X + (\nabla_{JV} J)JX = 0$ and using Remark 1.5, we have:

$$A_{JX} V - JA_X V - A_X JV - JA_{JX} JV = 0 .$$

Now, the first equality in a4) applied to X and JX gives: $-JA_X V = A_X JV$, which obviously holds for any horizontal vector field X. Finally, note that nearly Kähler and Kähler submersions obviously satisfy $T_E \circ J = J \circ T_E$, for any vector field E.

Before studying ASK and SK-submersions, we remark that, in the context of almost Hermitian submersions, the tangent coderivation and the normal coderivation with respect to the fibres give rise to a vertical coderivative $\overline{\delta}\Omega$ and a horizontal coderivative $\overline{\overline{\delta}}\Omega$ which turn out to be 1-forms on the total space M. Furthermore, one has:

$$\delta\Omega = \overline{\delta}\Omega + \overline{\overline{\delta}}\Omega \, ,$$

where $\overline{\overline{\delta}}\Omega$ can be computed fixing a local orthonormal J-frame $\{F_1, \dots, F_n, JF_1, \dots, JF_n\}$ of basic vector fields.

Proposition 3.8 *Let $\pi : (M, J, g) \to (B, J', g')$ be an ASK-submersion. Then B is an ASK-manifold if and only if the fibres are minimal.*

Proof. For any vector field X' on B, let X be the basic vector field on M, π-related to X'. Then, we have:

$$0 = (\delta\Omega)(X) = (\overline{\delta}\Omega)(X) + (\overline{\overline{\delta}}\Omega)(X) \, ,$$

and the statement follows from the above remark, Corollary 3.2 and the equality $(\overline{\overline{\delta}}\Omega)(X) = (\delta'\Omega')(X') \circ \pi$. Namely, for any basic vector fields X, Y, Z on M, π-related to X', Y', Z' on B, one has: $g(Z, (\nabla_Y J)X) = g'(Z', (\nabla'_{Y'} J')X') \circ \pi$, and then

$$(\nabla_Y \Omega)(Z, X) = (\nabla'_{Y'} \Omega')(Z', X') \circ \pi \, .$$

Corollary 3.4 *Let $\pi : M \to B$ be an SK-submersion. Then, B is an SK-manifold if and only if the fibres are minimal submanifolds of M.*

Proposition 3.9 *Let $\pi : (M, J, g) \to (B, J', g')$ be an ASK-submersion. Then the fibres of π are ASK-manifolds if and only if $\operatorname{tr}\beta = 0$, where, for any $E, F \in \mathcal{X}(M)$, $\beta(E, F) = A_E JF - A_{JE} F$.*

Proof. Via Corollary 3.3, we have only to prove that $\overline{\overline{\delta}}\Omega$ vanishes on $\mathcal{X}^v(M)$ if and only if $\operatorname{tr}\beta = 0$. Let U be a vertical vector field and $\{F_1, \dots, F_n, JF_1, \dots, JF_n\}$ orthonormal basic vector fields. One has:

$$\begin{aligned}
(\overline{\overline{\delta}}\Omega)(U) &= \textstyle\sum_{i=1}^n \{g(F_i, (\nabla_{F_i} J)U) + g(JF_i, (\nabla_{JF_i} J)U)\} \\
&= \textstyle\sum_i \{g(F_i, A_{F_i} JU - JA_{F_i} U) + g(JF_i, A_{JF_i} JU - JA_{JF_i} U)\} \\
&= \textstyle\sum_i \{-g(F_i, JA_{Fi} U) - g(JF_i, JA_{JF_i} U)\} \\
&= \textstyle\sum_i \{-g(A_{F_i} JF_i, U) + g(A_{JF_i} F_i, U)\} \\
&= -2 \textstyle\sum_i g(A_{F_i} JF_i, U) = -\frac{1}{2} g(\operatorname{tr}\beta, U) \, .
\end{aligned}$$

The last equality holds since $A_E = A_{hE}$ and

$$\operatorname{tr}\beta = \sum_{i=1}^n (A_{F_i} JF_i - A_{JF_i} F_i - A_{JF_i} F_i + A_{F_i} JF_i) = 4 \sum_{i=1}^n A_{F_i} JF_i \, .$$

Finally, since β takes values in $\mathcal{V}$, the proof is complete.

Proposition 3.10 *Let $\pi : (M, J, g) \to (B, J', g')$ be an almost Hermitian submersion such that $T_{JU} = T_U \circ J$ on $\mathcal{X}^v(M)$ and $\operatorname{tr}\beta = 0$. Then, if the fibres and B are ASK-manifolds, M is an ASK-manifold, too.*

Proof. We know that $\delta\Omega = \bar{\delta}\Omega + \bar{\bar{\delta}}\Omega$. Now, since the assumption on T implies the minimality of the fibres, via Corollary 3.2, we obtain $(\bar{\delta}\Omega)(X) = 0$, for any horizontal vector field X. On the other hand, since B is an ASK-manifold, we have $(\bar{\bar{\delta}}\Omega)(X) = (\delta'\Omega')(\pi_* X) \circ \pi = 0$. Finally, since the fibres are ASK, using Proposition 3.3 and the computation in Proposition 3.9, we obtain:

$$(\delta\Omega)(U) = (\hat{\delta}\Omega)U + (\bar{\bar{\delta}}\Omega)U = (\bar{\bar{\delta}}\Omega)U = \frac{1}{2}g(\operatorname{tr}\beta, U) = 0 ,$$

for any vertical vector field U, $\hat{\delta}$ being the codifferential operator on the fibres.

Remark 3.2 In [333] B. Watson and L. Vanhecke investigate the influence of the J-symmetries of T and A, *i.e.* of the conditions $T_{JU} = T_U \circ J$ on $\mathcal{X}^v(M)$ and $A_{JX} = A_X \circ J$, on $\mathcal{X}^h(M)$. They prove the above result requiring on A the stronger condition $B = 0$, instead of $\operatorname{tr}\beta = 0$, where $B(X,Y) = A_X JY - A_{JX}Y$, $X, Y \in \mathcal{X}^h(M)$. Note that $\operatorname{tr} B = \operatorname{tr}\beta$, since $A_E = A_{hE}$ for any vector field E.

Example 3.1 On the tangent bundle TM of an almost Hermitian manifold (M, J, g) we consider the Sasakian metric g^D defined in Example 1.3 and the almost complex structure $\tilde{J}$, introduced by Tanno, such that:

$$\tilde{J}(X^V) = (JX)^V , \quad \tilde{J}(X^H) = (JX)^H , \quad X \in \mathcal{X}(M) ,$$

X^V, X^H respectively denoting the vertical and the horizontal lift of X ([284]). It is easy to prove that $(TM, g^D, \tilde{J})$ is an almost Hermitian manifold and that $\pi : (TM, \tilde{J}, g^D) \to (M, J, g)$ is an almost Hermitian submersion, but in general π is neither an ASK nor an SK-submersion, even if (M, J, g) is a Kähler manifold. Indeed, the codifferential of the fundamental 2-form $\tilde{\Omega}$ of $(TM, \tilde{J}, g^D)$ satisfies:

$$\delta\tilde{\Omega}(X^H) = \delta\Omega(X) \circ \pi ,$$

$$\delta\tilde{\Omega}(X^V)_\xi = \tfrac{1}{2} \sum_{i=1}^{2n} g_p(R(e_i, Je_i, \xi), X) = -\rho^*_p(X, J\xi) ,$$

where $\xi \in TM$, $\pi(\xi) = p$, ρ^* denoting the *-Ricci tensor on (M, J, g) and $\{e_i\}_{1\leq i\leq 2n}$ being a local orthonormal frame on M. Thus TM is almost semi-Kähler if and only if M is almost semi-Kähler and *-Ricci flat. Further examples of AH-submersions are described in details in [334].

Now, we prove a result stated in [94] and in [331], independently.

Theorem 3.1 *Let $\pi : (M, J, g) \to (B', J', g')$ be an AK-submersion. Then:*

a) *the horizontal distribution is integrable and totally geodesic;*
b) $(L_X J)V = 2J(T_V X)$, $V \in \mathcal{X}^v(M), X \in \mathcal{X}^b(M)$;
c) $(\nabla_X J)V = 0$, $X \in \mathcal{X}^h(M), V \in \mathcal{X}^v(M)$.

Proof. To achieve a), it suffices to prove that $v([X,Y]) = 0$, for basic vector fields X, Y on M. The AK-condition implies $d\Omega(X, Y, V) = 0$, for any vertical vector field V. Then, one obtains:

$$\begin{aligned} &X(\Omega(Y,V)) - Y(\Omega(X,V)) + V(\Omega(X,Y)) \\ &-\Omega([X,Y],V) + \Omega([X,V],Y) - \Omega([Y,V],X) = 0 \,. \end{aligned}$$

Since $[X,V], [Y,V]$ are vertical and the two distributions are J-invariant, the last two and the first two terms vanish. Thus, one gets:

$$g([X,Y], JV) = V(g(X, JY)) = V(g'(X', J'Y') \circ \pi) = 0 \,,$$

X, Y being π-related to X', Y'. Then, a) follows. Now, consider X basic and V, W vertical vector fields. By direct computation, one has:

$$0 = 3d\Omega(W, JV, X) = g(JW, (L_X J)V - 2J(T_V X)) \,.$$

Then the vertical vector field $(L_X J)V - 2J(T_V X)$ turns out to be horizontal and so $(L_X J)V = 2J(T_V X)$, thus proving b). To obtain c) observe that, since $A = 0$, $(\nabla_X J)V$ is vertical when X is basic, and then

$$\begin{aligned} (\nabla_X J)V &= \nabla_X JV - J\nabla_X V = \nabla_{JV} X + [X, JV] - J\nabla_V X - J[X,V] \\ &= T_{JV} X - J(T_V X) + (L_X J)V = -2J(T_V X) + (L_X J)V = 0 \,. \end{aligned}$$

Obviously, $(\nabla_X J)V = 0$, for any horizontal vector field X.

Corollary 3.5 *In the previous hypothesis, if $T = 0$, then the horizontal lift of a holomorphic vector field $X' \in \mathcal{X}(B)$ is holomorphic too.*

Proposition 3.11 *Let $\pi : (M, J, g) \to (B, J', g')$ be an AK-submersion, with superminimal fibres and B a Kähler manifold. Then, M is a Kähler manifold and π is a Kähler submersion.*

Proof. The superminimality of the fibres implies that $\nabla_V J = 0$ for any $V \in \mathcal{X}^v(M)$, and c) in Theorem 3.1 entails $(\nabla_X J)V = 0$ for any X in $\mathcal{X}^h(M)$ and V in $\mathcal{X}^v(M)$. Finally, if X and Y are basic vector fields, π-related to X' and Y' on B, we have $v((\nabla_X J)Y) = 0$, since $A = 0$. Furthermore, $\pi_*(h((\nabla_X J)Y)) = (\nabla'_{X'} J')Y' = 0$ and the statement follows.

Recently, in [332], B. Watson determined conditions, involving the superminimality of the fibres, which are sufficient to state that the almost Hermitian structure on the total space of an almost Hermitian submersion belongs to the same class of the base and of the fibres. In particular, an alternative proof of Proposition 3.11 is given (see Theorem 11.2 in [332]).

Theorem 3.2 *Let $\pi : (M, J, g) \to (B, J', g')$ be an NK-submersion. Then, the horizontal distribution is integrable and totally geodesic.*

Proof. Let X be a basic vector field and V a vertical vector field on M. Then, since $h(\nabla_V X) = h(\nabla_X V) = A_X V$, we have the decomposition $\nabla_V X = T_V X + A_X V$. Moreover, since $T_V JX = JT_V X$, we obtain $\nabla_V JX - J\nabla_V X = A_X JV - JA_X V$, and using a4) of Proposition 3.7, we have $(\nabla_V J)X = -2JA_X V$. On the other hand, the NK-condition implies $(\nabla_X J)V = -(\nabla_V J)X$. Taking the horizontal part and using a4) of Proposition 3.7, we get $-2JA_X V = 2JA_X V$, so $A = 0$.

Theorem 3.3 *Let $\pi : (M, J, g) \to (B, J', g')$ be an NK-submersion. If B and the fibres are Kähler manifolds, then (M,J,g) is Kähler, too.*

Proof. The vanishing of A and the condition $T_V \circ J = J \circ T_V$ imply $(\nabla_X J)V = (\nabla_V J)X = 0$, for any $X \in \mathcal{X}^h(M)$, $V \in \mathcal{X}^v(M)$. Furthermore, for vertical vector fields we have $(\nabla_U J)V = (\hat{\nabla}_U J)V$, $\hat{\nabla}$ denoting the Levi-Civita connection on the fibres, whereas for basic vector fields X, Y, π-related to X', Y' on B, $(\nabla_X J)Y$ is basic and π-related to $(\nabla'_{X'} J')Y'$. Thus, $\hat{\nabla} J = 0$ and $\nabla' J' = 0$ imply the vanishing of ∇J also on the pairs of vector fields which are both vertical or horizontal.

Remark 3.3 It is well known that nearly Kähler, non-Kähler, manifolds exist only in dimension greater than or equal to 6 ([130]). Thus the above proposition implies that nearly Kähler (but non-Kähler) submersions with total space of dimension 6 do not exist. Namely, in this case, the fibres and the base manifold should be 2 or 4-dimensional Kähler manifolds.

Corollary 3.6 **(Watson).** *The horizontal distribution of a Kähler submersion is integrable and totally geodesic.*

Corollary 3.7 *Let $\pi : M \to B$ be an AK (or NK, or K)-submersion with totally geodesic fibres. Then M is locally a Riemannian product of almost Kähler (or nearly Kähler or Kähler) manifolds. Furthermore, if M is simply connected, then M is a product manifold and π is a holomorphic projection onto one of the factors.*

We recall that a Riemannian submersion $\pi : M \to B$ with minimal fibres is said to be harmonic (in the sense of Eells and Sampson), since π is a harmonic map. Thus QK, AK, NK, K-submersions, and, under suitable hypotheses, also ASK, SK-submersions, are harmonic. Applying results stated in the context of the theory of manifold mappings commuting with the Laplacian or with the codifferential operator, ([122; 324; 325]), we have the following statement involving the Betti numbers.

Proposition 3.12 *Let $\pi : (M, J, g) \to (B, J', g')$ be a QK, NK, AK, K-submersion, or an ASK, SK-submersion with minimal fibres. Then:*

1) $\pi^*\Delta_B f = \Delta_M \pi^* f$, *for any smooth function f on B ;*
2) $\pi^*\delta_B\omega = \delta_M\pi^*\omega$, $\omega \in \Lambda^1(B)$;
3) *if π is an NK, AK, or a K-submersion, then* $\pi^*\Delta_B\omega = \Delta_M\pi^*\omega$, $\omega \in \Lambda^p(B)$, $1 \leq p \leq \dim B$;
4) *if M is compact and QK (ASK, SK), then $b_1(B) \leq b_1(M)$;*
5) *if M is compact, NK, AK or K, then one obtains $b_p(B) \leq b_p(M)$,* $0 \leq p \leq \dim B$.

3.3 Holomorphic Distributions in Kähler Submersions

Proposition 3.13 *Let $\pi : (M, J, g) \to (B, J', g')$ be a Kähler submersion. Then the horizontal distribution is holomorphic if and only if the fibres are totally geodesic.*

Proof. Since B is a Kähler manifold, $\mathcal{X}(B)$ is locally spanned by holomorphic vector fields, *i.e.* by vector fields X' such that $L_{X'}J' = 0$. Thus, we consider a basic vector field X on M, π-related to a holomorphic vector field X' on B, and we prove that X is holomorphic. The integrability of $\mathcal{H}$ implies that, for any basic vector field Y, π-related to Y', $(L_X J)Y$ is horizontal and then it vanishes, since it is π-related to $(L_{X'}J')Y'$. Finally, the condition b) in Theorem 3.1 implies that X is holomorphic if and only if $T = 0$. Then, the statement follows since the basic vector fields locally span $\mathcal{H}$.

Remark. Proposition 3.13 was obtained by Johnson ([157]) as a consequence of the following theorem in the context of foliations in Kähler manifolds.

Theorem 3.4 *Let M be a Kähler manifold and $\mathcal{V}$ a holomorphic distribution on M. The orthogonal distribution $\mathcal{H} = \mathcal{V}^\perp$ is also a holomorphic distribution if and only if both $\mathcal{V}$ and $\mathcal{H}$ are not only integrable, but totally geodesic.*

To illustrate the relationship between the existence of Kähler submersions and of holomorphic connections on principal bundles over Kähler manifolds, conditions which are both extremely restrictive, we describe some results due to Johnson ([157]).

Let (M, J', g') be a Kähler manifold with Levi-Civita connection ∇'. We consider a complex Lie group G with a fixed left-invariant Hermitian metric $<,>_G$ determined by a Hermitian inner product $<,>_{\mathfrak{g}}$ on the Lie algebra $\mathfrak{g}$ of G. We write $<,>$ instead of $<,>_G$ or $<,>_{\mathfrak{g}}$. From now on, we fix a complex analytic principal bundle $\pi : P \to M$ with G as fibre and as structural group. Denoting by J the complex structure on P, we have $\pi_* \circ J = J' \circ \pi_*$. Finally, we put $\mathcal{V} = \ker \pi_*$, so obtaining a holomorphic subbundle consisting of the vectors tangent to the fibres. For any smooth connection $\mathcal{H}$ on P ([172] Vol. I), we can define a Riemannian metric g on P in such a way that $\pi : (P, g) \to (M, g')$ becomes a Riemannian submersion with $\mathcal{V}$ and $\mathcal{H}$ as vertical and horizontal distributions, respectively. Namely, we put, for any $p \in P$

$$\begin{aligned} g_p(U,X) &= 0\,, & U \in \mathcal{V}_p,\ X \in \mathcal{H}_p\,; \\ g_p(U,V) &= <\omega_p(U), \omega_p(V)>\,, & U, V \in \mathcal{V}_p\,; \\ g_p(X,Y) &= g'_{\pi(p)}(\pi_* X, \pi_* Y)\,, & X, Y \in \mathcal{H}_p\,, \end{aligned} \tag{3.3}$$

where ω is the connection form related to the connection $\mathcal{H}$. We denote by ∇ the Levi-Civita connection on (P, g). We also recall that the Lie algebra $\mathfrak{g}$ gives rise to the subalgebra of $\mathcal{X}(P)$ consisting of the so-called fundamental vector fields, which locally span the vertical distribution $\mathcal{V}$ of π; any $W \in \mathfrak{g}$ determines a fundamental vector field W^* such that $\omega(W^*) = W$.

Lemma 3.1 *For any basic vector field X and any fundamental vector field W, one has $[W, X] = 0$.*

Proof. From the theory of connections on a principal fibre bundle, we know that $[W, X]$ is horizontal. On the other hand, $[W, X]$ is also vertical, since π is a Riemannian submersion. Hence, $[W, X] = 0$.

Proposition 3.14 *The fibres of π are totally geodesic.*

Proof. It suffices to get $T_V W = 0$, or equivalently $\nabla_V W \in \mathcal{X}^v(P)$, for any fundamental vector fields V, W. Indeed, fixed a horizontal vector field X, using (1.3) and Lemma 3.1, we obtain $2g(\nabla_V W, X) = -X(g(V,W)) = 0$, since $g(V,W)$ is constant on the fibres.

Proposition 3.15

a) *The metric g is Hermitian if and only if at any point $p \in P$, $\mathcal{H}_p$ is a complex subspace of $T_p(P)$;*

b) *$\mathcal{H}$ is holomorphic if and only if g is Hermitian and $\nabla_y J = 0$ for any horizontal vector y .*

Proof. The statement a) is easily verified. To prove b), firstly we assume that $\mathcal{H}$ is holomorphic. Then, obviously g is Hermitian and $\pi : P \to M$ becomes an almost Hermitian submersion. Let y be a horizontal vector and Y a basic vector field extending y and π-related to Y' on M. For any basic vector field Z, π-related to Z', we have:

$$h(\nabla_Y JZ) = h(J\nabla_Y Z) . \tag{3.4}$$

In fact $h((\nabla_Y J)Z) = 0$, since it is π-related to $(\nabla'_{Y'} J')Z'$ which vanishes. On the other hand, choosing Y as a holomorphic extension, we get

$$v(\nabla_Y JZ) = \frac{1}{2} v([Y, JZ]) = \frac{1}{2} v(J[Y,Z]) = v(J\nabla_Y Z) . \tag{3.5}$$

Thus, (3.4) and (3.5) imply the vanishing of $\nabla_y J$ on the horizontal vectors. Now, we prove that $\nabla_y JU = J\nabla_y U$, for any vertical vector field U. Indeed, if $y \in \mathcal{H}_p$, for any horizontal vector field Z, one has:

$$\begin{aligned} g_p(\nabla_y JU, Z_p) &= -g_p((JU)_p, \nabla_y Z) = g_p(U_p, J_p(\nabla_y Z)) \\ &= g_p(U_p, \nabla_y JZ) = -g_p(\nabla_y U, (JZ)_p) \\ &= g_p(J_p(\nabla_y U), Z_p). \end{aligned} \tag{3.6}$$

On the other hand, taking a holomorphic horizontal extension Y of y, for any vertical vector field W, one gets:

$$\begin{aligned} g(\nabla_Y JU, W) &= g(\nabla_{JU} Y, W) + g([Y, JU], W) \\ &= -g(Y, \nabla_{JU} W) + g(J[Y,U], W) \\ &= g(J[Y,U], W) , \end{aligned}$$

since the fibres are totally geodesic. Analogously,

$$\begin{aligned} g(J\nabla_Y U, W) &= g(J\nabla_U Y, W) + g(J[Y,U], W) \\ &= g(Y, \nabla_U JW) + g(J[Y,U], W) . \end{aligned}$$

So, we have:

$$g_p(\nabla_y JU, W_p) = g_p(J_p \nabla_y U, W_p) , \tag{3.7}$$

and (3.6), (3.7) imply $\nabla_y J = 0$ on the vertical distribution. Vice versa, we suppose that g is Hermitian and $\nabla_y J = 0$ for any horizontal vector y. Since M is Kähler, $\mathcal{X}(M)$ is locally spanned by holomorphic vector fields and their horizontal lifts locally span the distribution $\mathcal{H}$. Then, it is enough to prove that the horizontal lift X of a holomorphic vector field X' on M is holomorphic. Now, for any basic vector field Y, π-related to Y', $h([X, JY])$ is basic π-related to $[X', J'Y']$, $h(J[X,Y])$ is basic π-related to $J'[X', Y']$ and then $h([X, JY]) = h(J[X,Y])$. Furthermore, $v([X, JY]) = 2v(\nabla_X JY) = 2v(J\nabla_X Y) = v(J[X,Y])$, since $\nabla_X J = 0$. It follows:

$$[X, JY] = J[X, Y] , \tag{3.8}$$

for any horizontal vector field Y. Finally, for any fundamental vector field U, we have $[X, JU] = 0 = J[X,U]$, hence $[X, JV] = J[X,V]$ for any vertical vector field V. Combining this with (3.8), we have $L_X J = 0$, *i.e.* X is holomorphic.

Proposition 3.16 *Suppose that the metric* $<,>$ *on* G *is Kähler. Then* $\pi : (P, J, g) \to (M, J', g')$ *is a Kähler submersion if and only if* $\mathcal{H}$ *is a holomorphic and flat connection.*

Proof. Firstly we prove that if $\mathcal{H}$ is holomorphic, then (J, g) is a Kähler structure if and only if the curvature form Ω of $\mathcal{H}$ vanishes. Therefore, from this equivalence and Proposition 3.15, the statement easily follows. Now, let $\mathcal{H}$ be holomorphic. By Proposition 3.15 b), g is Hermitian and $\nabla_X J = 0$ for any horizontal vector field X. Furthermore, for any V, W vertical vector fields one has $\nabla_V JW = J\nabla_V W$, since the fibres are Kähler and totally geodesic. Hence, as regards the evaluation of $\nabla_V JX - J\nabla_V X$ for X horizontal and V vertical vector fields, for any vertical vector field W, we obtain:

$$g(\nabla_V JX - J\nabla_V X, W) = -g(\nabla_V JW - J\nabla_V W, X) = 0 .$$

Finally, the statement follows from the equality

$$g(\nabla_V JX - J\nabla_V X, Y) = -2g(\Omega(X,Y)^*, JV)\ ,$$

for any X, Y basic holomorphic vector fields and V holomorphic vertical vector field. Indeed, one has:

$$\begin{aligned} g(\nabla_V JX - J\nabla_V X, Y) &= g(\nabla_{JX} V + [V, JX] - J(\nabla_X V + [V, X]), Y) \\ &= g(\nabla_{JX} V - J\nabla_X V, Y) \\ &= -g(V, \nabla_{JX} Y) + g(\nabla_X V, JY) \\ &= -g(V, J(\nabla_Y X + \nabla_X Y) + J[X,Y]) \\ &= -2g(V, J\nabla_X Y) = g(JV, v[X,Y]) \\ &= < \omega(JV), \omega(v[X,Y]) > \\ &= -2g(JV, \Omega(X,Y)^*)\ . \end{aligned}$$

Note that, if $\mathcal{H}$ is a holomorphic and flat connection, then P is locally a product of a Kähler manifold and the group G.

3.4 Curvature Properties

We begin this section relating the holomorphic bisectional and the holomorphic sectional curvatures of the total space, the base and the fibres of an almost Hermitian submersion.

Let $\pi : (M, J, g) \to (M', J', g')$ be an almost Hermitian submersion and denote by B the holomorphic bisectional curvature of M, which is the function pointwise defined, for any pair of nonzero tangent unit vectors E, F by $B(E,F) = R(E, JE, F, JF)$. Thus the holomorphic sectional curvature, defined as the sectional curvature of the J-invariant 2-planes, *i.e.* $H(E) = R(E, JE, E, JE)$ for any unit vector E, can be expressed as $H(E) = B(E,E)$. We shall write B', H' and $\hat{B}, \hat{H}$ when we refer to the base and to the fibres of π, respectively. As usual we denote by U, V, W vertical vectors and by X, Y, Z horizontal vectors.

Proposition 3.17 *Let $\pi : (M, g, J) \to (M', g', J')$ be an almost Hermitian submersion. Then, the holomorphic bisectional curvatures and the holomorphic sectional curvatures verify:*

$$\begin{aligned} 1)\ & B(V,U) = \hat{B}(V,U) + g(T_V JU, T_{JV} U) - g(T_V U, T_{JV} JU)\ ; \\ 2)\ & B(X,V) = g((\nabla_V A)_X JX, JV) - g(A_X JV, A_{JX} V) \\ & \qquad + g(A_X V, A_{JX} JV) - g((\nabla_{JV} A)_X JX, V) \\ & \qquad + g(T_{JV} X, T_V JX) - g(T_V X, T_{JV} JX)\ ; \end{aligned}$$

3) $B(X,Y) = B'(\pi_*X, \pi_*Y) - 2g(A_X JX, A_Y JY) + g(A_{JX}Y, A_X JY)$
$-g(A_X Y, A_{JX} JY)$;
4) $H(V) = \hat{H}(V) + \|T_V JV\|^2 - g(T_V V, T_{JV} JV)$;
5) $H(X) = H'(\pi_* X) - 3\|A_X JX\|^2$.

Proof. The relations 1), 2), 3) are just an application of (1.27), (1.28) and (1.29); 4) and 5) follow from 1) and 2).

Remark 3.4 As well as for the sectional curvatures, an almost Hermitian submersion is holomorphic sectional curvature increasing on the horizontal holomorphic 2-planes. Moreover, the above relations simplify if the horizontal distribution is integrable *i.e.* $A = 0$ or if the fibres are totally geodesic *i.e.* $T = 0$, respectively reducing to:

$$B(X,Y) = B'(\pi_*X, \pi_*Y) , \quad H(X) = H'(\pi_*X) ,$$
$$B(U,V) = \hat{B}(V,U) , \quad H(V) = \hat{H}(V) .$$

Obviously, these equations also hold in some special cases. For example, it is easy to prove that $A_{JX} = A_X \circ J$ on $\mathcal{X}^h(M)$ if and only if π preserves the horizontal holomorphic sectional curvatures. Consequently, combining with Proposition 3.9, we can state:

Proposition 3.18 *Let $\pi : (M, J, g) \to (M', J', g')$ be an almost Hermitian submersion with M an ASK-manifold. If π preserves the horizontal holomorphic sectional curvatures then the fibres are ASK.*

Applying Propositions 3.17 and 3.7, Theorems 3.1 and 3.2, we have the following results.

Proposition 3.19 *Let $\pi : (M, J, g) \to (M', J', g')$ be a QK-submersion. Then:*

1) *$B(V,U) \geq \hat{B}(V,U)$, $H(V) \geq \hat{H}(V)$ and the equalities hold if and only if the fibres are totally geodesic;*
2) *$B(X,Y) \geq B'(\pi_*X, \pi_*Y)$ and the equality holds if and only if the horizontal distribution is integrable;*
3) *$H(X) = H'(\pi_*X)$.*

Proposition 3.20 *Let $\pi : (M, J, g) \to (M', J', g')$ be an almost Kähler submersion. Then:*

1) *π preserves the horizontal holomorphic bisectional curvatures and the horizontal holomorphic sectional curvatures;*
2) *$B(X,U) = -2g(T_U X, T_{JU} JX)$;*

3) $B(U,V) \geq \hat{B}(U,V)$, $H(V) \geq \hat{H}(V)$ *and the equalities hold if and only if* $T = 0$, *or, equivalently, if and only if* M *is locally a product of almost Kähler manifolds and* π *acts as the projection onto a factor.*

Proposition 3.21 *Let* $\pi : (M, J, g) \to (B', J', g')$ *be a nearly Kähler submersion. Then:*

1) π *preserves the horizontal holomorphic bisectional curvatures and the horizontal holomorphic sectional curvatures;*
2) $B(X,U) = -2||T_U X||^2 \leq 0$;
3) $B(U,V) \geq \hat{B}(U,V)$, $H(U) \geq \hat{H}(U)$ *and the equalities hold if and only if* $T = 0$ *or, equivalently, if and only if* $B(X,U) = 0$. *Each of the last two conditions is equivalent to the local decomposition of* M *as a product of nearly Kähler manifolds.*

The above results hold *a fortiori* for Kähler submersions, so we have:

Proposition 3.22 *Any Kähler submersion preserves the horizontal holomorphic bisectional and sectional curvatures, and satisfies:*

$$B(X,U) \leq 0, \quad \hat{B}(U,V) \leq B(U,V) .$$

Moreover, $B(X,U) = 0$ *if and only if* $T = 0$ *or, equivalently, if and only if* $\hat{B}(U,V) = B(U,V)$.

Proposition 3.23 *Any Kähler manifold* (M, J, g) *with (pointwise) constant positive holomorphic sectional curvature cannot be the total space of a Kähler submersion with non-discrete fibres.*

Proof. Let us suppose that M is a Kähler manifold with (pointwise) constant holomorphic sectional curvature μ. A well-known result (Proposition 7.3 Chap. IX in [172], Vol. II), implies that $B(X,U) = \frac{1}{2}\mu$, for any unit vectors X, U horizontal and vertical, respectively. Thus, when $\mu > 0$, one has $B(X,U) > 0$ for any nonzero X and U. This contradicts Proposition 3.22 unless $\dim M = \dim B$; but this would imply that the fibres are zero-dimensional.

Corollary 3.8 *If a Kähler manifold* M *with non-negative holomorphic bisectional curvatures is the total space of a Kähler submersion, then* M *is locally a product of Kähler manifolds. Moreover, if* M *is simply connected, then it is a product of Kähler manifolds.*

Proof. The hypothesis $B(E,F) \geq 0$ implies $B(X,U) = 0$ and then $T = 0$.

Remark 3.5 Let $\pi : (M,J,g) \to (M',J',g')$ be a Kähler submersion and suppose that M has constant holomorphic sectional curvature c. Then, M' has constant holomorphic sectional curvature c and $c = 2B(X,U) = -4\|T_U X\|^2$. Now, if $c = 0$, then $T = 0$ and M, M' and the fibres are locally biholomorphic to $\mathbf{C^n}$ (for suitable n). If $c < 0$, then T never vanishes and locally M and B are complex hyperbolic spaces.

In the context of almost Hermitian geometry, several authors ([129; 294; 252]) studied special classes of almost Hermitian manifolds with curvature tensor satisfying the so-called Kähler identities:

$$\begin{aligned}
&K_1\colon\ R(X,Y,Z,W) = R(X,Y,JZ,JW), \text{ or, equivalently, } R_{XY}.J = 0\ ,\\
&K_2\colon\ R(X,Y,Z,W) = R(JX,JY,Z,W) + R(JX,Y,JZ,W)\\
&\qquad\qquad\qquad\qquad\quad + R(JX,Y,Z,JW)\ ,\\
&K_3\colon\ R(X,Y,Z,W) = R(JX,JY,JZ,JW)\ ,
\end{aligned}$$

for any vector fields X, Y, Z, W. Hence, given an almost Hermitian submersion $\pi : (M,J,g) \to (M',J',g')$, it is natural to ask if and under which conditions the K_i-curvature identities transfer from the total space to the base or to the fibres of π. To this aim, the J-symmetries or J-antisymmetries of the tensors T and A play an important role. For the details we refer to [333; 334].

Now, we pass to establish some relations between the Ricci and *-Ricci tensors, in the case of an almost Hermitian submersion with integrable horizontal distribution. We put $\dim M = 2m, \dim B = 2n,\ r = m - n$. Formulas (1.27), (1.28), (1.29), when $A = 0$, reduce to:

$$\begin{aligned}
&R(X,Y,Z,H) = R^*(X,Y,Z,H)\ ;\\
&R(X,Y,U,V) = g(T_V X, T_U Y) - g(T_U X, T_V Y)\ ;\\
&R(X,U,Y,V) = g((\nabla_X T)(U,V), Y) - g(T_U X, T_V Y)\ ;\\
&R(U,V,F,W) = \hat{R}(U,V,F,W) + g(T_U W, T_V F) - g(T_V W, T_U F)\ ,
\end{aligned} \tag{3.9}$$

for any X, Y, Z, H horizontal, U, V, W, F vertical vector fields. Moreover (1.36) becomes:

$$\begin{aligned}
&\rho(X,Y) = \rho'(X',Y') \circ \pi - \textstyle\sum_{j=1}^{2r} g(T_{U_j} X, T_{U_j} Y)\ ;\\
&\rho(U,V) = \hat{\rho}(U,V) + \textstyle\sum_{i=1}^{2n} g((\nabla_{X_i} T)(U,V), X_i)\ ;\\
&\rho(X,U) = -\textstyle\sum_{j=1}^{2r} g((\nabla_{U_j} T)(U_j, U), X)\ .
\end{aligned} \tag{3.10}$$

Finally, the scalar curvatures τ, τ' of M, M' are related by:

$$\tau = \tau' \circ \pi + \hat{\tau} - ||T||^2 , \tag{3.11}$$

$\hat{\tau}$ being the scalar curvature of the fibres.

We recall that the Ricci tensor of a nearly Kähler manifold is J-invariant, *i.e.* $\rho(E,F) = \rho(JE,JF)$ ([128]). The same property does not hold for almost Kähler manifolds. In fact, in this case, the J-anti-invariant component of ρ, which explicitly depends on the covariant derivatives $\nabla J, \nabla^2 J$, determines the W_8-component of R according to the splitting of the algebraic curvature tensor fields defined by F. Tricerri and L. Vanhecke ([294]). Thus, an investigation of the Ricci tensor properties is also interesting in the context of submersions ([94]).

Proposition 3.24 *Given an almost Kähler submersion, one has:*

$$\rho(U,V) + \rho(JU,JV) = \hat{\rho}(U,V) + \hat{\rho}(JU,JV), \quad U,V \in \mathcal{X}^v(M) .$$

Proof. Fix a vertical vector field U. Since $A = 0$ and $T_U V + T_{JU} JV = 0$ for any vertical V, using c) in Proposition 3.9, we obtain:

$$(\nabla_X T)(U,U) + (\nabla_X T)(JU,JU) = -2T_{JU}((\nabla_X J)U) = 0 ,$$

for any horizontal X. Since $A = 0$, we also have that $\nabla_X U$ is vertical and the symmetry of T on the vertical distribution implies $(\nabla_X T)(U,V) + (\nabla_X T)(JU,JV) = 0$. Then, via (3.10) we get the statement.

The previous result can be applied to Kähler submersions. Since moreover the Ricci tensor of a Kähler manifold is J-invariant, we obtain:

Corollary 3.9 *Given a Kähler submersion, for any U,V vertical vector fields, one has:*

$$\rho(U,V) = \hat{\rho}(U,V) .$$

We recall that the **-Ricci tensor* of an almost Hermitian manifold (M^{2m}, J, g) is defined by:

$$\rho^*(X,Y) = \sum_{i=1}^{2m} R(X,e_i,JY,Je_i) ,$$

$\{e_i\}_{1 \le i \le 2m}$ being a local orthonormal frame, and, even if ρ^* is J-invariant, in general ρ^* is not symmetric. The skew-symmetric part of ρ^* determines the W_9-component of R according to the splitting considered in [294]. Moreover, linear combinations of ρ and ρ^* and the corresponding traces also play a relevant role in the study of algebraic curvature tensor fields.

Proposition 3.25 *Let $\pi : (M, J, g) \to (M', J', g')$ be a quasi Kähler submersion with integrable horizontal distribution. The *-Ricci tensors $\hat{\rho}^*$ of the fibres, ρ'^* of B and ρ^* of M are related by:*

$$\rho^*(X,Y) = \rho'^*(X',Y') \circ \pi - \textstyle\sum_{j=1}^{2r} g(T_{U_j}X, T_{JU_j}JY) \ ;$$

$$\rho^*(U,V) = \hat{\rho}^*(U,V) + \textstyle\sum_{j=1}^{2r} g(T_{U_j}U, T_{U_j}V)$$

$$+ \textstyle\sum_{i=1}^{2n} \{g((\nabla_{X_i}T)(U,JV), JX_i) - g(T_U X_i, T_{JV}JX_i)\},$$

for any $X, Y \in \mathcal{X}^b(M)$, π-related to X', Y', and $U, V \in \mathcal{X}^v(M)$, $\{X_i, U_j\}$ being a local orthonormal frame with any X_i horizontal and any U_j vertical.

Proof. Consider X, Y basic vector fields, π-related to X', Y'. Since $\operatorname{tr} T = 0$, we also have $\operatorname{tr}(\nabla_X T) = 0$ and $\sum_{j=1}^{2r} g((\nabla_X T)(U_j, JU_j), JY) = 0$. Then a direct application of (3.9) gives the first relation. Using again (3.9), we get the last formula, since $g(T_U JU_j, T_{U_j} JV) = g(T_{U_j}U, T_{U_j}V)$ for any $j \in \{1, \dots, 2r\}$.

Corollary 3.10 *If π is a nearly Kähler submersion, we have:*

$$(\rho - \rho^*)(X,Y) = (\rho' - \rho'^*)(X',Y') \circ \pi \ ,$$
$$\tau - \tau^* = (\tau' - \tau'^*) \circ \pi + \hat{\tau} - \hat{\tau}^* \ ,$$

X, Y being basic vector fields π-related to X', Y'.

Proof. Since $T_{JU}JY = T_U Y$, comparing (3.10) with the first relation in Proposition 3.25, we obtain the first statement. Moreover, for the *-scalar curvatures, we have:

$$\tau^* = \tau'^* \circ \pi + \hat{\tau}^* - \|T\|^2 \ ,$$

and, combining with (3.11), we get the last formula.

Remark. Since $\tau - \tau^* = \|\nabla J\|^2$ in the nearly Kähler case ([129]), from the second relation in the previous corollary we obtain an alternative proof of Theorem 3.3.

Proposition 3.26 *Given an almost Kähler submersion, we have:*

$$\rho^*(U,V) - \rho^*(V,U) = \hat{\rho}^*(U,V) - \hat{\rho}^*(V,U), \quad U, V \in \mathcal{X}^v(M) \ ,$$
$$\tau - \tau^* = (\tau' - \tau'^*) \circ \pi - 2(\|T\|^2 - \textstyle\sum_{i,j} g(T_{U_j}X_i, T_{JU_j}JX_i)) \ .$$

Proof. We apply directly the second formula in Proposition 3.25 and observe that $(\nabla_X T)(U, JV) = (\nabla_X T)(V, JU)$, for any X horizontal (see the proof of Proposition 3.24). Moreover, considering the local horizontal orthonormal frame $\{X_h, JX_h\}_{1\leq h\leq n}$, since $T_{JU} = -J \circ T_U$ on the horizontal distribution, the last addendum in the same formula vanishes and this implies the first relation. Furthermore, via Proposition 3.25, since

$$\sum_{i,h} g(T_{U_j}U_h, T_{U_j}U_h) = \sum_{i,j} g(T_{U_j}X_i, T_{U_j}X_i) ,$$

we have

$$\tau^* = \tau'^* \circ \pi + \hat{\tau}^* - 2\sum_{i,j} g(T_{U_j}X_i, T_{JU_j}JX_i) + \|T\|^2 ,$$

and combining with (3.11) we obtain the last formula.

A recent result on AK-submersions is due to B. Watson ([332]).

Theorem 3.5 *Let $\pi : (M, J, g) \to (B, J', g')$ be an AK-submersion with $\dim M = 4$, $\dim B = 2$. If g is Einstein, then the fibres of π are superminimal, totally geodesic submanifolds of M. Thus, M is Kähler and π acts as a locally product projection map.*

3.5 Locally Conformal Kähler Manifolds

Among the subclasses of almost Hermitian manifolds the one of locally (globally) conformal Kähler manifolds (l.c.K., g.c.K.) is particularly interesting and largely studied. It appears as the class $\mathcal{W}_4$ in the classification of $2n$-dimensional almost Hermitian manifolds, $n \geq 3$, due to A. Gray and L. M. Hervella, while is a proper subclass of $\mathcal{W}_4$ in the 4-dimensional case ([130; 310; 311]) and represents the most natural link between conformal geometry and Hermitian geometry.

Now, we recall some basic tools in view of the analysis of locally conformal Kähler submersions, referring to [78] for more details. Furthermore, as long as possible, we consider locally (globally) conformal almost Kähler manifolds (l.c.AK., g.c.AK.).

Definition 3.6 A $2m$-dimensional almost Hermitian manifold (M, J, g) is called a *locally conformal almost Kähler* (*locally conformal Kähler*) manifold if M admits an open covering $(U_\alpha)_{\alpha\in\Lambda}$ and a family $(\sigma_\alpha : U_\alpha \to \mathbf{R})_{\alpha\in\Lambda}$ of differentiable functions such that, for any $\alpha \in \Lambda$, $(U_\alpha, e^{\sigma_\alpha} g_{|U_\alpha}, J_{|U_\alpha})$ is an almost Kähler (Kähler) manifold. M is called *globally conformal almost*

Kähler (*globally conformal Kähler*) if there exists a differentiable function $\sigma : M \to \mathbf{R}$ such that $(M, J, e^\sigma g)$ is an almost Kähler (Kähler) manifold.

To simplify the notation, we shall say that the metric g is a locally (globally) conformal AK (K) metric to mean that g is conformal to an almost Kähler (Kähler) metric $\tilde{g}$. Such a metric $\tilde{g}$ is determined up to homotheties. Namely, we have the following result, that we prove in the global case, without loss of generality.

Proposition 3.27 *Let (M, J, g) and $(\tilde{M}, \tilde{J}, \tilde{g})$ be almost Kähler manifolds. If $f : (M, J, g) \to (\tilde{M}, \tilde{J}, \tilde{g})$ is a conformal, almost complex diffeomorphism, then it is a homothety with positive factor.*

Proof. Since $f^*(\tilde{g}) = e^\sigma g$, for some smooth function σ on M, and $f_* \circ J = \tilde{J} \circ f_*$, the Kähler forms are related by $f^*(\tilde{\Omega}) = e^\sigma \Omega$. Then, $0 = f^*(d\tilde{\Omega}) = d(e^\sigma \Omega) = de^\sigma \wedge \Omega$ implies that e^σ is a positive constant.

Now, we recall some useful characterizations ([305]).

Proposition 3.28 *Let (M, J, g) be an almost Hermitian manifold. Then:*

1) *M is l.c.AK. if and only if there exists a closed 1-form ω such that $d\Omega = \omega \wedge \Omega$;*
2) *M is l.c.K. if and only if M is Hermitian and there exists a closed 1-form ω such that $d\Omega = \omega \wedge \Omega$.*

In the theory of l.c.K. manifolds the 1-form

$$\omega = -\frac{1}{m-1}\delta\Omega \circ J \, , \tag{3.12}$$

which is defined for any almost Hermitian manifold of dimension $2m \geq 4$, plays a fundamental role. It is called the *Lee form* of (M, J, g) and its dual (with respect to g) vector field B is called the *Lee vector field* of M. Namely, if M^{2m}, $m \geq 2$, is an l.c.AK. manifold, then the 1-form ω in Proposition 3.28 turns out to be uniquely determined since it coincides with the Lee form. Moreover, locally, we have:

$$B = -\frac{1}{m-1} J\Big(\sum_{i=1}^{m}((\nabla_{E_i} J)E_i + (\nabla_{JE_i} J)JE_i)\Big) \, ,$$

where $\{E_i, JE_i\}_{i\in\{1,\dots,m\}}$ is a local orthonormal J-frame. Thus, B can be viewed as the trace of the linear operator

$$(X, Y) \in \mathcal{X}(M)^2 \mapsto -\frac{1}{m-1} J(\nabla_X J)Y \, .$$

Consider, now, a fixed l.c.AK. manifold (M, J, g) with Levi-Civita connection ∇, Kähler form Ω, Lee form ω and Lee vector field B. Since homothetic metrics give rise to the same Riemannian connection, and connections are localizable, we obtain a unique connection $\tilde{\nabla}$, globally defined on M, called the *Weyl connection* of (M, J, g).

Proposition 3.29 *Let (M, J, g) be an l.c.AK. manifold. Then, for any $X, Y \in \mathcal{X}(M)$, one has:*

$$\tilde{\nabla}_X Y = \nabla_X Y - \frac{1}{2}\{\omega(X)Y + \omega(Y)X - g(X,Y)B\} .$$

Moreover, $\tilde{\nabla}$ is torsion-free and metric with respect to the local metrics conformal to g.

Remark 3.6 Weyl connections exist apart from the almost complex structure. They depend on the conformal class of (M, g) and are related to the Levi-Civita connection ∇ by:

$$D = \nabla - \frac{1}{2}\{\omega \otimes K + K \otimes \omega - g \otimes B\} ,$$

where ω is an arbitrary 1-form with dual vector field B and K denotes the Kronecker tensor field. Furthermore, D is torsion-free and $Dg = \omega \otimes g$. Therefore, for an l.c.AK. manifold (M, J, g) the Weyl connection of M is the Weyl connection determined by the Lee form ω. Weyl connections allow to characterize l.c.K. manifolds ([305]).

Proposition 3.30 *Let (M, J, g) be an almost Hermitian manifold. Then M is l.c.K. if and only if there exists a 1-form ω such that $d\omega = 0$ and the Weyl connection D determined by ω is almost complex, i.e. $DJ = 0$. Such a form ω is the Lee form.*

Corollary 3.11 *Let (M, J, g) be an almost Hermitian manifold. Then M is l.c.K. if and only if for any $X, Y \in \mathcal{X}(M)$,*

$$(\nabla_X J)Y = \frac{1}{2}(\omega(JY)X - \omega(Y)JX + g(X,Y)JB - \Omega(X,Y)B) ,$$

where ω is the Lee form with dual vector field B.

Definition 3.7 An l.c.K. manifold (M, J, g) with Lee form ω is a *Vaisman manifold* (or a *generalized Hopf manifold*, briefly g.H.m.) if ω never vanishes and $\nabla\omega = 0$.

Obviously, we have that M is a Vaisman manifold if and only if B never vanishes and $\nabla B = 0$, and this implies $||\omega|| = ||B|| =$ constant.

Remark 3.7 Let (M, J, g) be an l.c.K. manifold such that the Lee form ω is parallel, *i.e.* $\nabla\omega = 0$. Then, we have only one of the following possibilities:

1) $\omega = 0$, *i.e.* M is Kähler;
2) $\omega \neq 0$, *i.e.* M is Vaisman.

Now, we present some results about submanifolds of l.c.AK. manifolds which extend the analogous ones for the l.c.K. case.

Proposition 3.31 *Let (M^{2m}, J, g) be an l.c.AK. manifold and S a holomorphic submanifold of M^{2m},* $\dim S = 2r$. *Then, we have:*

1) *S is l.c.AK. and if $r \geq 2$, then $\hat{\omega} = \omega_{|S}, \hat{B} = B^t$, where $\hat{\omega}$ is the Lee form of S, $\hat{B}$ is the Lee vector field of S and B^t denotes the component of $B_{|S}$ tangent to S;*
2) *the mean curvature vector field H of S is given by $H = -\frac{1}{2}B^{\perp}$, where $B^{\perp}$ denotes the component of $B_{|S}$ normal to S;*
3) *S is a minimal submanifold if and only if B is tangent to S along its points.*

Proof. Without loss of generality, we can argue assuming that M is a globally conformal almost Kähler manifold. Let $\tilde{g}$ be an AK-metric conformal to g, $\tilde{g} = e^{\sigma}g$. Then, we consider the two immersions:

$$\text{a) } (S, J, g) \to (M, J, g)\,, \qquad \text{b) } (S, J, \tilde{g}) \to (M, J, \tilde{g})\,.$$

Since $(S, J, \tilde{g})$ is almost Kähler, we get that (S, J, g) is g.c.AK. Now, assuming $r \geq 2$, we observe that for any $X, Y, Z \in \mathcal{X}(S)$,

$$d\hat{\Omega}(X, Y, Z) = d\Omega(X, Y, Z) = (\omega \wedge \Omega)(X, Y, Z) = (\omega' \wedge \hat{\Omega})(X, Y, Z)\,,$$

where ω' denotes the restriction of ω to S. Since $d\omega = 0$, ω' is also closed, and we conclude that $\omega' = \hat{\omega}$. Finally, for any $X \in \mathcal{X}(S)$, we get: $g(X, \hat{B}) = \hat{\omega}(X) = g(X, B^t)$, thus proving that $\hat{B} = B^t$. Note that, if $r = 1$, then S is Kähler, so it can be considered l.c.AK. To prove 2), denoting with α, α', H, H' the second fundamental forms and the mean curvature vector fields of the immersions a), b), respectively, we apply the Gauss formula and the link between the Weyl and the Levi-Civita connections given in Proposition 3.29. We obtain, for any $X, Y \in \mathcal{X}(S)$,

$$\alpha(X, Y) = \alpha'(X, Y) - \frac{1}{2}g(X, Y)B^{\perp}\,.$$

Hence,

$$H = \frac{1}{2r}\mathrm{tr}\alpha = \frac{1}{2r}e^{\sigma}\mathrm{tr}\alpha' - \frac{1}{4r}2rB^{\perp} = e^{\sigma}H' - \frac{1}{2}B^{\perp} .$$

Since the immersion b) is minimal, *i.e.* $H' = 0$, we get $H = -\frac{1}{2}B^{\perp}$. The statement 3) is a trivial consequence of 2).

The behavior of submanifolds of a Vaisman manifold is quite different, as the following result shows.

Proposition 3.32 *Let S be a holomorphic submanifold of a generalized Hopf manifold (M, J, g), and put $\dim S = 2r$, $r \neq 1$. Then, the following conditions are equivalent:*

a) *S is a g.H.m.;*
b) *S is minimal;*
c) *the Lee vector field B of M is tangent to S in its points.*

Proof. For any $X \in \mathcal{X}(S)$, $\nabla_X B = 0$ implies $\nabla_X B^t = -\nabla_X B^{\perp}$. Hence, applying the Gauss formula, we get:

$$\hat{\nabla}_X B^t + \alpha(X, B^t) = a_{B^{\perp}}X - \nabla^{\perp}_X B^{\perp} ,$$

a denoting the Weingarten operator. Separating the tangent and the normal parts with respect to S, we obtain $\hat{\nabla}_X B^t = a_{B^{\perp}}X$, $\alpha(X, B^{\perp}) = -\nabla^{\perp}_X B^{\perp}$. Furthermore, S is Hermitian and l.c.AK., because of Proposition 3.31, so it is l.c.K. with Lee vector field B^t. Now, using 2) in Proposition 3.31, for any $X, Y \in \mathcal{X}(S)$ we have:

$$g(\hat{\nabla}_X B^t, Y) = g(a_{B^{\perp}}X, Y) = g(B^{\perp}, \alpha(X, Y)) = -2g(H, \alpha(X, Y)) ,$$

which implies that a) and b) are equivalent. The equivalence between b) and c) is the statement 3) in Proposition 3.31.

Remark 3.8 The above proposition is meaningful for submanifolds S with real dimension at least 4. On the other hand, any Vaisman manifold M does admit 2-dimensional holomorphic totally geodesic submanifolds which turn out to be Kähler and does not admit Kähler submanifolds whose (real) dimension is at least 4.

We end this section describing a special class of generalized Hopf manifolds, the so-called k-g.H.m., $k \in \mathbf{R}$, introduced by Marrero and Rocha ([195]), which turn out to be related to Hermitian–Weyl structures (see next Sec. 5.6) and to Sasakian manifolds.

Let (M, J, g) be a g.H.m. with Lee form ω and Lee vector field B. Then

$||B|| = ||\omega||$ and we put $c = \frac{1}{2}||\omega||$, $\omega_0 = \frac{1}{||\omega||}\omega$, $B_0 = \frac{1}{||B||}B$, $\overline{B} = JB_0$ and $\overline{\omega} = -\omega_0 \circ J$. Obviously, ω_0, B_0 and $\overline{\omega}, \overline{B}$ are dual with respect to g. Denote by $\mathcal{F}$ the canonical foliation defined by $\omega = 0$; $\mathcal{F}$ has clearly codimension one, its normal bundle is spanned by B and, since $\nabla B = 0$, it is totally geodesic. Namely, for any $X, Y \in \mathcal{X}(S)$, S being a leaf of $\mathcal{F}$, we have

$$\omega(\nabla_X Y) = g(\nabla_X Y, B) = -g(Y, \nabla_X B) = 0 \ ,$$

and the second fundamental form of S vanishes.

Proposition 3.33 *Via the canonical immersion in M, each leaf S of $\mathcal{F}$ inherits a c-Sasakian structure (φ, ξ, η, g), where*

$$\varphi = J_{|S} + \overline{\omega}_{|S} \otimes B_{0|S}, \quad \xi = \overline{B}_{|S}, \quad \eta = \overline{\omega}_{|S} \ ,$$

g denotes the restriction to S of the metric g and $c = \frac{1}{2}||\omega||$.

Proof. Since $\overline{\omega}(B_0) = 0$, for any $X \in \mathcal{X}(S)$ we have:

$$\begin{aligned}\varphi^2(X) &= \varphi(JX + \overline{\omega}(X)B_0) = J^2X + \overline{\omega}(JX)B_0 + \overline{\omega}(X)JB_0 \\ &= -X + \omega_0(X)B_0 + \eta(X)\overline{B}_{|S} = (-I + \eta \otimes \xi)(X) \ .\end{aligned}$$

Furthermore, for any $X, Y \in \mathcal{X}(S)$:

$$\begin{aligned}g(\varphi X, \varphi Y) &= g(JX, JY) + g(JX, \overline{\omega}(Y)B_0) + g(\overline{\omega}(X)B_0, JY) \\ &\quad + \overline{\omega}(X)\overline{\omega}(Y)g(B_0, B_0) = g(X,Y) - \overline{\omega}(Y)\overline{\omega}(X) \\ &= g(X,Y) - \eta(X)\eta(Y) \ .\end{aligned}$$

This means that (φ, ξ, η, g) is an almost contact metric structure. To prove that it is c-Sasakian *i.e.* normal with $d\eta = c\,\Phi$ (cf. Sec. 4.1), we use the following characterization in terms of the Levi-Civita connection:

$$(\nabla_X \varphi)Y = c\{g(X,Y)\xi - \eta(Y)X\}, \quad c \neq 0, \quad X, Y \in \mathcal{X}(S) \ .$$

Now, $\overline{\omega} = -\omega_0 \circ J$, implies $\nabla_X \overline{\omega} = -\omega_0 \circ \nabla_X J$ and using Corollary 3.11, we have:

$$\begin{aligned}(\nabla_X \varphi)(Y) &= (\nabla_X (J + \overline{\omega} \otimes B_0))(Y) \\ &= (\nabla_X J)Y - (\omega_0 \circ \nabla_X J)(Y) \otimes B_0 \\ &= \tfrac{1}{2}\{\omega(JY)X + g(X,Y)JB - \Omega(X,Y)B + \Omega(X,Y)\omega_0(B)B_0\} \\ &= \tfrac{1}{2}\{||\omega||\omega_0(JY)X + ||\omega||g(X,Y)JB_0\} \\ &= c\{g(X,Y)\xi - \eta(Y)X\} \ .\end{aligned}$$

Definition 3.8 A g.H.m. (M, J, g) is called a *k-g.H.m.*, $k \in \mathbf{R}$, if each leaf of the canonical foliation is a c-Sasakian manifold of constant φ-sectional curvature k.

We recall that the φ-sectional curvature is the sectional curvature of the 2-planes generated by $\{X, \varphi X\}$.

Remark 3.9 The previous properties suggest how to construct examples of k-g.H.m. using c-Sasakian structures ([306]). Let N be a c-Sasakian manifold with structure (φ, ξ, η, h) and define an almost Hermitian structure (J, g) on $M = N \times \mathbf{R}$, putting:

$$\begin{aligned} J(X, a\tfrac{\partial}{\partial t}) &= (\varphi X + a\xi, -\eta(X)\tfrac{\partial}{\partial t}) \,, \\ g((X, a\tfrac{\partial}{\partial t}), (Y, b\tfrac{\partial}{\partial t})) &= h(X, Y) + ab \,, \end{aligned} \tag{3.13}$$

for any $X, Y \in \mathcal{X}(M), a, b \in \mathcal{F}(N \times \mathbf{R})$. Then, (M, J, g) turns out to be a g.H.m. with Lee form $\omega = 2cdt$ and Lee vector field $B = 2c\frac{\partial}{\partial t}$. Moreover, if N has constant φ-sectional curvature $k \in \mathbf{R}$, then M is a k-g.H.m. and, since ω is exact, we obtain a globally conformal Kähler manifold. Consequently, M admits Kähler metrics.

Now, we describe three fundamental examples of simply connected globally c.K. manifolds with parallel non-vanishing Lee form.

Example 3.2 Consider the sphere $\mathbf{S^{2m-1}}$ in $\mathbf{R^{2m}} \cong \mathbf{C^m}$, $m \geq 2$ and let α, c be real numbers with $\alpha > 0$, $c \neq 0$. Denoting by (J_0, g_0) the flat Kähler structure on $\mathbf{C^m}$ and by N the unit normal vector field on $\mathbf{S^{2m-1}}$ in $\mathbf{R^{2m}}$, we put:

$$\xi = -\frac{c}{\alpha} J_0 N, \quad \eta = -\frac{\alpha}{c} \eta_{0|\mathbf{S^{2m-1}}}, \quad \varphi = J_{0|\mathbf{S^{2m-1}}} - \eta_{0|\mathbf{S^{2m-1}}} \otimes N \,,$$

where $\eta_0 = \sum_{i=1}^m (y^i dx^i - x^i dy^i)$, (x^i, y^i) being coordinates in $\mathbf{R^{2m}}$. Finally, considering the metric g induced by g_0 on $\mathbf{S^{2m-1}}$, we obtain on $\mathbf{S^{2m-1}}$ the metric $h = \frac{\alpha}{c^2} g + \frac{\alpha^2 - \alpha}{c^2} \eta \otimes \eta$, such that (φ, ξ, η, h) is a c-Sasakian structure of constant φ-sectional curvature $k = (\frac{4}{\alpha} - 3)c^2 > -3c^2$. The resulting c-Sasakian manifold is denoted by $\mathbf{S^{2m-1}}(c, k)$. It follows that $\mathbf{S^{2m-1}}(c, k) \times \mathbf{R}$, $k > -3c^2$, with the complex structure defined by (3.13) is a simply connected k-g.H.m. Passing to the quotient under the action of $\{I\} \times \mathbf{Z}$, we obtain that $\mathbf{S^{2m-1}}(c, k) \times \mathbf{S^1}$ is a compact k-g.H.m., which is an l.c.K., but not g.c.K., manifold and then does not admit Kähler metrics.

The following two examples come from the same general construction described in [195; 224]. Let (M, J, g) be a Kähler manifold of constant holomorphic sectional curvature H and Kähler form $\Omega = \frac{1}{c} d\omega$, for some $c \in \mathbf{R}^*$ and $\omega \in \Lambda^1(M)$. Denoting by $\pi : \mathbf{R} \times M \to M$ the canonical projection, we obtain a c-Sasakian structure on $\mathbf{R} \times M$ putting:

$$\begin{aligned} &\varphi = J \circ \pi_* - \pi^*(\omega \circ J) \otimes \xi, \;\; \xi = \tfrac{\partial}{\partial t}\,, \\ &\eta = dt + \pi^*\omega, \;\; h = g + \eta \otimes \eta\,. \end{aligned} \tag{3.14}$$

The c-Sasakian manifold so obtained, denoted by $(\mathbf{R}\times M)(c,k)$ has constant φ-sectional curvature $k = H - 3c^2$.

Example 3.3 Choosing $M = \mathbf{C^{m-1}}$, the flat Kähler space, since the Kähler 2-form is exact, we obtain that $(\mathbf{R}\times\mathbf{C^{m-1}})(c,-3c^2) \cong \mathbf{R^{2m-1}}(c)$ is a c-Sasakian manifold with φ-sectional curvature $k = -3c^2$. Consequently, $\mathbf{R^{2m-1}}(c) \times \mathbf{R}$ with the complex structure defined as in (3.13) is a simply connected k-g.H.m.

Example 3.4 Choose $M = \mathbf{C}D^{m-1}(H)$ as the unit ball in $\mathbf{C^{m-1}}$ with Kähler structure of constant holomorphic sectional curvature $H < 0$ ([172]). Then $(\mathbf{R} \times \mathbf{C}D^{m-1})(c,k)$ is a c-Sasakian manifold of constant φ-sectional curvature $k = H - 3c^2 < -3c^2$. Hence $(\mathbf{R} \times \mathbf{C}D^{m-1})(c,k) \times \mathbf{R}$ is a simply connected k-g.H.m.

The relevance of these three examples is stated by the following theorem due to I. Vaisman ([306]) and J. Marrero, J. Rocha ([195]).

Theorem 3.6 *Let (M,J,g) be a $2m$-dimensional complete k-g.H.m. with Lee form ω and $\overline{M}$ its universal covering space with the induced almost Hermitian structures $(\overline{J},\overline{g})$. Then, putting $c = \frac{\|\omega\|}{2}$, we have:*

1) *if $k > -3c^2$, then $(\overline{M},\overline{J},\overline{g})$ and $\mathbf{S^{2m-1}}(c,k) \times \mathbf{R}$ are holomorphically isometric;*
2) *if $k = -3c^2$, then $(\overline{M},\overline{J},\overline{g})$ and $\mathbf{R^{2m-1}}(c) \times \mathbf{R}$ are holomorphically isometric;*
3) *if $k < -3c^2$, then $(\overline{M},\overline{J},\overline{g})$ and $(\mathbf{R} \times \mathbf{C}D^{m-1})(c,k) \times \mathbf{R}$ are holomorphically isometric.*

3.6 Locally Conformal Kähler Submersions

Definition 3.9 An almost Hermitian submersion is called a *locally conformal* almost Kähler (locally conformal Kähler) submersion if the total space is an l.c.AK. (l.c.K.) manifold.

As in the previous sections, we use the symbols $'$ and $\hat{}$ for the geometric objects on the base manifold and on the fibres, respectively. For example,

ω', B' and $\hat{\omega}, \hat{B}$ denote the Lee form and the Lee vector field on the base and on the fibres.

Proposition 3.34 *Let $\pi : (M^{2m}, J, g) \to (M'^{2n}, J', g')$ be an l.c.AK. submersion and suppose $n \neq 1$. Then we have:*

1) *The mean curvature vector field of the fibres is given by* $H = -\frac{1}{2}h(B)$;
2) $\omega(X) = \omega'(X') \circ \pi$, *for any basic vector field X on M, π-related to X' on M' ;*
3) $h(B)$ *is the basic vector field π-related to B' .*

Proof. Since the fibres are holomorphic submanifolds of M^{2m}, Proposition 3.31 implies $B^{\perp} = h(B)$ and $H = -\frac{1}{2}h(B)$. To prove 2), we consider a basic vector field X, π-related to X'. Then, using the tangent and the normal coderivatives with respect to the fibres, we can write:

$$\omega(X) = -\frac{1}{m-1}(\delta\Omega)JX = -\frac{1}{m-1}(\bar{\delta}\Omega)JX - \frac{1}{m-1}(\bar{\bar{\delta}}\Omega)JX ,$$

where $(\bar{\bar{\delta}}\Omega)(JX) = (\delta'\Omega')(J'X') \circ \pi$. Now, 1) and Proposition 3.2 imply:

$$(\bar{\delta}\Omega)JX = 2rg(JH, JX) = rg(2H, X) = -r\omega(X) ,$$

where $2r = 2m - 2n$ is the dimension of the fibres. Then, we have:

$$\omega(X) = \frac{r}{m-1}\omega(X) + \frac{n-1}{m-1}\omega'(X') \circ \pi ,$$

and $\omega(X) = \omega'(X') \circ \pi$ follows immediately. Finally, let Y be the basic vector field π-related to the Lee vector field B' on M'. For any basic vector field X, π-related to X', we get:

$$g(hB, X) = g(B, X) = \omega(X) = \omega'(X') \circ \pi = g'(B', X') \circ \pi = g(Y, X) ,$$

and we conclude $hB = Y$.

Since we are interested in investigating the almost Hermitian structure of the fibres of an l.c.AK. submersion, we consider $\pi : M^{2m} \to M'^{2n}$ with $m > n$, remarking that the case $m = n$ means that π acts as a local biholomorphism.

Proposition 3.35 *Let $\pi : (M^{2m}, J, g) \to (M'^{2n}, J', g')$ be an l.c.AK. (l.c.K.) submersion. Then the fibres and M' are l.c.AK. (l.c.K.) manifolds.*

Proof. Since the fibres are almost Hermitian manifolds, Proposition 3.31 allows to conclude that any fibre F is an l.c.AK. manifold. Analogously, M' is an almost Hermitian manifold, and using the relations $\omega(X) = \omega'(X') \circ \pi$, $\Omega(X, Y) = \Omega'(X', Y') \circ \pi$, where X, Y are basic vector fields, π-related to X', Y', we get $d\omega' = 0$, $d\Omega' = \omega' \wedge \Omega'$, and M' is l.c.AK., too. Finally, if M is l.c.K., then it is Hermitian and l.c.AK; consequently the fibres and M' are Hermitian and l.c.AK, that is they are l.c.K.

Proposition 3.36 *Let $\pi : (M^{2m}, J, g) \to (M'^{2n}, J', g')$ be an l.c.AK. submersion. Then the integrability tensor A of π satisfies:*

$$A_X Y = -\frac{1}{2}\Omega(X, Y)v(JB) ,$$

for any horizontal vector fields X, Y.

Moreover, the horizontal distribution $\mathcal{H}$ is integrable if and only if B is a horizontal vector field.

Proof. Let X, Y be basic vector fields π-related to X', Y'. Then, for any vertical vector field U, we have:

$$\begin{aligned} 3d\Omega(X, Y, U) &= X(\Omega(Y, U)) - Y(\Omega(X, U)) + U(\Omega(X, Y)) \\ &\quad -\Omega([X, Y], U) + \Omega([X, U], Y) - \Omega([Y, U], X) \\ &= -\Omega([X, Y], U) , \end{aligned}$$

since $[X, U], [Y, U]$ are vertical and $\Omega(X, Y) = \Omega'(X', Y') \circ \pi$ implies that $\Omega(X, Y)$ is constant on the fibres. On the other hand, since $d\Omega = \omega \wedge \Omega$, we have:

$$\begin{aligned} 3d\Omega(X, Y, U) &= \omega(X)\Omega(Y, U) + \omega(Y)\Omega(U, X) + \omega(U)\Omega(X, Y) \\ &= \omega(U)\Omega(X, Y) . \end{aligned}$$

Hence, we get $\Omega([X, Y], U) = -\omega(U)\Omega(X, Y)$, or, equivalently,

$$g([X, Y], JU) = -g(B, U)\Omega(X, Y) = -g(\Omega(X, Y)JB, JU) .$$

It follows that

$$v([X, Y]) = -\Omega(X, Y)v(JB), \quad \text{and} \quad A_X Y = -\frac{1}{2}\Omega(X, Y)v(JB) .$$

Finally, $\mathcal{H}$ is integrable if and only if $A = 0$, *i.e.* if and only if B is a horizontal vector field.

Proposition 3.37 *Let $\pi : (M^{2m}, J, g) \to (M'^{2n}, J', g')$, $n \neq 1$, be an l.c.AK. (l.c.K.) submersion. Then, the following conditions are equivalent:*

a) *the fibres are minimal submanifolds;*
b) M'^{2n} *is an AK (K)-manifold;*
c) *the Lee vector field* B *is vertical.*

Proof. Relation 1) in Proposition 3.34 implies that a) and c) are equivalent. Now, we have: B is vertical if and only if $B' = 0$ if and only if $\omega' = 0$ if and only if $d\Omega' = 0$, so b) and c) are equivalent.

Corollary 3.12 *Let* $\pi : (M^{2m}, J, g) \to (M'^{2n}, J', g')$, $n \neq 1$, *be an l.c.AK. (l.c.K.) submersion, with minimal fibres and integrable horizontal distribution. Then* π *is an AK (K) submersion.*

Proof. Propositions 3.37 and 3.36 imply $v(B) = B = h(B)$, *i.e.* $B = 0$.

3.7 Submersions from Generalized Hopf Manifolds

We begin this section considering l.c.K. submersions with total space a g.H.m., that we call *Vaisman submersions*.

Proposition 3.38 *Let* $\pi : (M, J, g) \to (M', J', g')$ *be a Vaisman submersion. Then* M' *is an l.c.K. manifold with parallel Lee form* ω' *and only one of the following cases can occur:*

1) $\omega' = 0$, *M' is Kähler, B is vertical and the fibres are minimal Vaisman submanifolds;*
2) *ω' never vanishes, M' is Vaisman, $h(B)$ never vanishes, the fibres are neither minimal nor Vaisman submanifolds.*

Proof. Since π is an l.c.K. submersion, we only have to prove that $\nabla'\omega' = 0$ or, equivalently, $\nabla' B' = 0$. Namely, this implies $||\omega'|| = ||B'|| =$ *const.*, and we have $\omega' = 0$ *aut* ω' never vanishes. In the first case M' is Kähler and in the latter one M' is Vaisman. The other statements in 1) and in 2) follow immediately from Propositions 3.34 and 3.37. Now, let X' be a vector field on M' and X the basic vector field on M π-related to X'. Since $h(B)$ is basic, π-related to B' and $\nabla_X B = 0$, we have:

$$\nabla'_{X'} B' = \pi_*(h(\nabla_X hB)) = -\pi_*(h(\nabla_X vB)) = -\pi_*(A_X(vB)) \ .$$

On the other hand, for any horizontal vector field Y, we get:

$$g(A_X(vB), Y) = -g(vB, A_X Y) = \frac{1}{2}\Omega(X, Y) g(vB, J(vB)) = 0 \ ,$$

and, since $A_X(vB)$ is a horizontal vector field, we obtain $A_X(vB) = 0$, which implies $\nabla'_{X'}B' = 0$.

Proposition 3.39 *Let $\pi : (M, J, g) \to (M', J', g')$ be a Vaisman submersion and suppose $\dim M > \dim M'$. Then the horizontal distribution $\mathcal{H}$ is not integrable.*

Proof. Arguing by contradiction, let us suppose that $\mathcal{H}$ is integrable. Then $h(B) = B$ and consequently $\hat{B} = 0$, so that the fibres are Kähler, non-minimal, submanifolds. Thus, for any $V \in \mathcal{X}^v(M)$, we have $\omega(V) = \hat{\omega}(V) = 0$ and

$$(d\Omega)(B, V, JV) = (\omega \wedge \Omega)(B, V, JV) = -\|\omega\|^2 g(V, V) . \qquad (3.15)$$

On the other hand, using $\nabla B = 0$, we get:

$$\begin{aligned} 3(d\Omega)(B, V, JV) &= B(\Omega(V, JV)) + V(\Omega(JV, B)) + JV(\Omega(B, V)) \\ &\quad -\Omega([B, V], JV) - \Omega([V, JV], B) - \Omega([JV, B], V) \\ &= -B(g(V, V)) + g(\nabla_B V, V) + g(\nabla_B JV, JV) \\ &= -B(g(V, V)) + \tfrac{1}{2}B(g(V, V)) + \tfrac{1}{2}B(g(JV, JV)) = 0 . \end{aligned}$$

Comparing with (3.15), we obtain $V = 0$, thus contradicting the hypothesis on the dimensions.

Proposition 3.40 *Let $\pi : (M, J, g) \to (M', J', g')$ be an l.c.K. submersion with minimal fibres from a k-generalized Hopf manifold. Then M' is a Kähler manifold of constant holomorphic sectional curvature $H' = k + \frac{3}{4}\|\omega\|^2$, ω being the Lee form on M.*

Proof. Since π is an almost Hermitian submersion, from 5) in Proposition 3.17 we have $H(X) = H'(\pi_*X) - 3\|A_X JX\|^2$, with $\|X\| = 1$. The minimality of the fibres implies that B is a vertical vector field and, consequently, the canonical foliation defined by $\omega = 0$ contains the horizontal distribution $\mathcal{H}$. Thus, via Proposition 3.36, we get:

$$H'(\pi_*X) = k + 3\|A_X JX\|^2 = k + \frac{3}{4}\|v(JB)\|^2 = k + \frac{3}{4}\|B\|^2 = k + \frac{3}{4}\|\omega\|^2 .$$

Now, we describe some examples of Vaisman submersions closely related to Examples 3.2, 3.3, 3.4.

Example 3.5 For $c \in \mathbf{R}^*$, $k \in \mathbf{R}$, $k > -3c^2$, we consider the Riemannian submersions given by the projections:

$$\pi : \mathbf{S^{2m-1}}(c, k) \times \mathbf{R} \to \mathbf{S^{2m-1}}(c, k) ,$$

$$\tilde{\tau} : \mathbf{S^{2m-1}}(c,k) \to \mathbf{P_{m-1}}(\mathbf{C})(k+3c^2) .$$

Then the map $\tau(c,k,m) = \tilde{\tau} \circ \pi : \mathbf{S^{2m-1}}(c,k) \times \mathbf{R} \to \mathbf{P_{m-1}}(\mathbf{C})(k+3c^2)$ is a Vaisman submersion with totally geodesic fibres from the k-g.H.m. $\mathbf{S^{2m-1}}(c,k) \times \mathbf{R}$ on the Kähler manifold $\mathbf{P_{m-1}}(\mathbf{C})$ of constant holomorphic sectional curvature $k+3c^2$. Namely, $\tau(c,k,m)$ is a Riemannian submersion and for any vector field X on $\mathbf{S^{2m-1}}(c,k) \times \mathbf{R}$ we have

$$\begin{aligned}(\tau(c,k,m)_* \circ J)(X) &= \tilde{\tau}_*(\pi_*(JX)) = \tilde{\tau}_*(\varphi(\pi_* X)) = J'(\tilde{\tau}_*(\pi_*(X)) \\ &= (J' \circ \tau(c,k,m)_*)(X) ,\end{aligned}$$

J' denoting the complex structure on $\mathbf{P_{m-1}}(\mathbf{C})$. Thus $\tau(c,k,m)$ induces an l.c.K. submersion with totally geodesic fibres from $\mathbf{S^{2m-1}}(c,k) \times \mathbf{S^1}$ over $\mathbf{P_{m-1}}(\mathbf{C})(k+3c^2)$. Moreover, $\mathbf{S^{2m-1}}(c,k) \times \mathbf{S^1}$ falls in the class of compact strongly regular g.H.m. and then can be viewed as a differentiable principal bundle over a Kähler manifold, with fibre the complex 1-dimensional torus ([310]).

Example 3.6 Let $c \in \mathbf{R}^*$ and consider the g.H.m. $\mathbf{R^{2m-1}}(c) \times \mathbf{R}$ described in Example 3.3. Then we obtain a basis of $\mathcal{X}(\mathbf{R^{2m-1}}(c))$ putting:

$$X_i = \frac{\partial}{\partial x^i} \; , \; Y_i = \frac{\partial}{\partial y^i} + 2cx^i \frac{\partial}{\partial z} \; , \; 1 \le i \le m-1 \; , \; Z = \frac{\partial}{\partial z} \; ,$$

where $\{x^i, y^i, z\}$ are the usual coordinates in $\mathbf{R^{2m-1}}$. So, in $\mathbf{R^{2m-1}}(c) \times \mathbf{R}$ we have:

$$JX_i = Y_i \; , \; JY_i = -X_i \; , \; JZ = -\frac{\partial}{\partial t} \; , \; J\frac{\partial}{\partial t} = Z \; ,$$

and $\{X_i, Y_i, Z, \frac{\partial}{\partial t}\}$ is an orthonormal frame with respect to the metric g which is expressed by:

$$g = \sum_{i=1}^{m-1} (\alpha_i \otimes \alpha_i + \beta_i \otimes \beta_i) + \gamma \otimes \gamma + dt \otimes dt \; ,$$

where $\alpha_i = dx^i$, $\beta_i = dy^i$, $\gamma = dz - 2c\sum_{j=1}^{m-1} x^j dy^j$ form the dual basis of $\{X_i, Y_i, Z\}$ in $\mathbf{R^{2m-1}}(c)$. For any $m' \le m-1$, the projection

$$\pi(c,m,m') : \mathbf{R^{2m-1}}(c) \times \mathbf{R} \to \mathbf{C^{m'}} \cong \mathbf{R^{2m'}}(0) \; ,$$

such that

$$\pi(c,m,m')(x^1, \ldots, x^{m-1}, y^1, \ldots, y^{m-1}, z, t) = (x^1, \ldots, x^{m'}, y^1, \ldots, y^{m'}) \; ,$$

is an almost Hermitian submersion, with totally geodesic fibres, from the k-g.H.m. $(\mathbf{R^{2m-1}}(c) \times \mathbf{R}, J, g)$, $k = -3c^2$, onto the flat Kähler manifold $\mathbf{C^{m'}}$.

Example 3.7 Given $c \in \mathbf{R}^*$ and $k \in \mathbf{R}$ such that $k < -3c^2$, the natural projection

$$\gamma(c,k,m) : (\mathbf{R} \times \mathbf{C}D^{m-1})(c,k) \times \mathbf{R} \to (\mathbf{C}D^{m-1})(k+3c^2)$$

is an almost Hermitian submersion, with totally geodesic fibres, from the k-g.H.m. $(\mathbf{R} \times \mathbf{C}D^{m-1})(c,k) \times \mathbf{R}$ onto the Kähler manifold $\mathbf{C}D^{m-1}(k+3c^2)$. This is achieved by applying the general constructions given in Sec. 3.5 to introduce Examples 3.3, 3.4 followed by the one described in Remark 3.9. The map γ is obtained by composition via the related natural projections which are both Riemannian submersions; via (3.13), (3.14) one easily proves that γ is almost Hermitian. Finally, the vertical distribution of γ is spanned by the Lee and the anti-Lee vector fields B, JB and since $\nabla B = 0$, using Corollary 3.11, we get that the fibres are totally geodesic.

The above three examples are again relevant since they describe classifying submersions for a certain class of Vaisman submersions, investigated by Marrero and Rocha in [195].

Definition 3.10 Let $\pi_i : (M_i, J_i, g_i) \to (M'_i, J'_i, g'_i)$, $i = 1, 2$, be almost Hermitian submersions. Then, π_1 and π_2 are said to be *equivalent* if there exist two almost complex isometries $\tau : (M_1, J_1, g_1) \to (M_2, J_2, g_2)$ and $\tau' : (M'_1, J'_1, g'_1) \to (M'_2, J'_2, g'_2)$ such that $\tau' \circ \pi_1 = \pi_2 \circ \tau$.

Theorem 3.7 *Let $\pi : (M^{2m}, J, g) \to ({M'}^{2m'}, J', g')$ be an almost Hermitian submersion with connected, totally geodesic fibres from a simply connected, complete, k-g.H.m. with Lee form ω. Putting $c = \frac{1}{2}||\omega||$, one has:*

1) *if $k > -3c^2$, then π and $\tau(c,k,m)$ are equivalent and $m' = m - 1$;*
2) *if $k = -3c^2$, then π and $\pi(c,m,m')$ are equivalent;*
3) *if $k < -3c^2$, then π and $\gamma(c,k,m)$ are equivalent and $m' = m - 1$.*

Proof. Assuming that $k > -3c^2$, according to Theorem 3.6, there exists a holomorphic isometry $\tau : (M, J, g) \to (\mathbf{S^{2m-1}}(c,k) \times \mathbf{R}, \tilde{J}, \tilde{g})$, $(\tilde{J}, \tilde{g})$ denoting the Hermitian structure considered in Example 3.2. Then, the vector fields $\tilde{B}$ and $\tilde{J}\tilde{B}$, which span the vertical distribution of $\tau(c,k,m)$, are vertical fields for the almost Hermitian submersion $\pi \circ \tau^{-1}$, also. Since the fibres of $\tau(c,k,m)$ are connected, one can construct an almost Hermitian submersion $\tau' : \mathbf{P_{m-1}(C)}(k+3c^2) \to (M', J', g')$ such that

$\tau' \circ \tau(c,k,m) = \pi \circ \tau^{-1}$. It is easy to verify that τ' is a Kähler submersion. Furthermore, the fibres of $\tau' \circ \tau(c,k,m)$ are connected and totally geodesic and by Lemma 2.2, the same property also holds for the fibres of τ'. Thus, taking into account Proposition 3.23, τ' must be an isometry. Hence, we obtain 1). Via Theorem 3.6, 3) follows in a similar way. Finally, if $k = -3c^2$, Theorem 3.6 implies the existence of a holomorphic isometry $\tau : \mathbf{R^{2m-1}}(c) \times \mathbf{R} \to (M, J, g)$ and then, as above, we get a Kähler submersion with connected, totally geodesic fibres $\tau'' : \mathbf{C^{m-1}} \to (M', J', g')$ such that $\tau'' \circ \pi(c,m,m-1) = \pi \circ \tau$. By Corollary 3.7, τ'' is equivalent to the projection $\tau_2(m-1, m')$ over a factor $\mathbf{C^{m'}}$, so there exists a holomorphic isometry $\tau' : \mathbf{C^{m'}} \to M'$ such that $\tau' \circ \tau_2(m-1,m') = \tau''$ and $\pi \circ \tau = \tau'' \circ \pi(c,m,m-1) = \tau' \circ \tau_2(m-1,m') \circ \pi(c,m,m-1) = \tau' \circ \pi(c,m,m')$.

Remark 3.10 The above theorem classifies the almost Hermitian submersions from a simply connected, and then globally conformal Kähler, k-g.H.m. Note that in the cases 1) and 3) the fibres are 2-dimensional and the base manifolds are (simply connected) projective or hyperbolic spaces. Case 2) contains more possibilities, but the base space is always a flat Kähler manifold.

We end this section with some results, due to F. Narita, concerning Riemannian submersions from a compact k-g.H.m. ([218]), obtained linking the above theorems to the ones due to R. H. Escobales and A. Ranjan. As proved by Vaisman ([310]), a connected compact homogeneous g.H.m. (M^{2m}, J, g) carries a natural complex analytic foliation $\mathcal{L}$ spanned by the Lee vector field B and the anti-Lee vector field JB. Thus, one can consider the natural projection $\mu : M^{2m} \to M^{2m}/\mathcal{L}$ which turns out to be an almost Hermitian submersion with totally geodesic fibres and base space a compact homogeneous simply connected Kähler manifold.

Proposition 3.41 *Let (M^{2m}, J, g) be a compact homogeneous k-g.H.m. and $c = \|\omega\|/2$. Then the canonical projection μ is a Vaisman submersion with totally geodesic fibres, $k + 3c^2 > 0$, and $M^{2m}/\mathcal{L}$ is holomorphically isometric to the complex projective space $\mathbf{P_{m-1}(C)}(k+3c^2)$.*

Proof. Since the fibres of μ are totally geodesic, Proposition 3.40 implies that $M^{2m}/\mathcal{L}$ is a Kähler manifold of constant holomorphic sectional curvature $k + 3c^2$. Moreover $M^{2m}/\mathcal{L}$ is compact and then complete, which implies $k + 3c^2 > 0$. It follows that $M^{2m}/\mathcal{L}$ is simply connected and holomorphically isometric to $\mathbf{P_{m-1}(C)}(k+3c^2)$ ([172], Vol. II, p. 170).

Theorem 3.8 (Narita). *Let $\pi : (M^{2m}, J, g) \to (N^{2n}, g')$ be a Riemannian submersion from a compact homogeneous k-g.H.m. with connected, totally geodesic and holomorphic fibres. Then π falls in one of the following equivalence classes of Riemannian submersions:*

a) $\pi : (M^{2m}, g) \to (\mathbf{P_{m-1}}(\mathbf{C}), g')$, $m \geq 2$,
b) $\pi : (M^{4l+4}, g) \to (\mathbf{P_l}(\mathbf{Q}), g')$, $l \geq 1$,

with g' of constant holomorphic (quaternionic) sectional curvature $k + 3c^2$, respectively.

Proof. We consider the projection $\mu : M^{2m} \to M^{2m}/\mathcal{L} \cong \mathbf{P_{m-1}}(\mathbf{C})$ and, firstly, we suppose $n = m - 1$. By 3) in Proposition 3.31, the fibres of π are submanifolds of real dimension 2 with the vector fields B and JB tangent to them. Then, the vertical distributions of π and μ coincide since, in both cases, the fibres are connected, compact, totally geodesic with the same tangent spaces. One can define a map $f : \mathbf{P_{m-1}}(\mathbf{C}) \to N^{2(m-1)}$ such that $\pi = f \circ \mu$ and, since π, μ are Riemannian submersions, f is an isometry. Thus π belongs to class a). Now, suppose that $n < m - 1$. For any $y \in M^{2m}$, putting $\pi(y) = x$ and $\mu(y) = z$, we obtain $\mu^{-1}(z) \subset \pi^{-1}(x)$, since the tangent space $T_y(\mu^{-1}(z))$ is contained in $T_y(\pi^{-1}(x))$. Again, we can define a map $\rho : \mathbf{P_{m-1}}(\mathbf{C}) \to N^{2n}$ such that $\pi = \rho \circ \mu$, and ρ turns out to be a Riemannian submersion. We prove that the fibres of ρ are connected totally geodesic holomorphic submanifolds of $\mathbf{P_{m-1}}(\mathbf{C})$, and using the Escobales–Ranjan theorem (Proposition 2.11, Theorem 2.10) we have $\rho : \mathbf{P_{2l+1}}(\mathbf{C}) \to \mathbf{P_l}(\mathbf{Q})$ and, consequently, $\pi : M^{4l+4} \to \mathbf{P_l}(\mathbf{Q})$ belongs to class b). Namely, for any $x \in N^{2n}$, $\pi^{-1}(x) = \mu^{-1}(\rho^{-1}(x))$, so $\rho^{-1}(x)$ is connected, since $\pi^{-1}(x)$ is connected. Let $V', W' \in \mathcal{X}(\mathbf{P_{m-1}}(\mathbf{C}))$ be vertical vector fields with respect to ρ. Then their horizontal lifts V, W in μ are vertical in π, since $\pi_*(V) = \rho_*(\mu_*(V)) = \rho_*(V') = 0$. Thus, since the fibres of π are totally geodesic, using superscripts ρ, π for the geometric objects on the total spaces of ρ and π, we get:

$$0 = T_V W = h^\pi(\nabla^\pi_V W) \ ,$$

i.e. $\nabla^\pi_V W$ is π-vertical. Consequently, $\mu_*(\nabla^\pi_V W)$ is ρ-vertical and, since $\mu_*(\nabla^\pi_V W) = \mu_*(h^\mu(\nabla^\pi_V W)) = \nabla^\rho_{V'} W'$, we get

$$T^\rho_{V'} W' = h^\rho(\nabla^\rho_{V'} W') = 0 \, ,$$

so obtaining the total geodesicity of the fibres of ρ. Finally, since μ is an almost Hermitian submersion and $\mathcal{V}^\pi$ is J^π-invariant, for any ρ-vertical vector field V', we have:

$$\rho_*(J^\rho V') = \rho_*(J^\rho(\mu_* V)) = \rho_*(\mu_*(J^\pi V)) = \pi_*(J^\pi V) = 0 \ ,$$

and this completes the proof, ensuring that the fibres of ρ are holomorphic submanifolds.

For any $\lambda \in \mathbf{C}^*, |\lambda| \neq 1$, let $\mathbf{H}^n_\lambda = (\mathbf{C}^n)^*/\Delta_\lambda$ be the *Hopf manifold*, Δ_λ being the cyclic group generated by the transformation

$$z \in (\mathbf{C^n})^* \mapsto \lambda z \in (\mathbf{C^n})^* \ .$$

Then $\mathbf{H}^n_\lambda$ carries a structure of 1-g.H.m., explicitly described in [306; 310].

Corollary 3.13 *Let $\pi : \mathbf{H}^m_\lambda \to N^{2n}$ be a Riemannian submersion from a Hopf manifold $\mathbf{H}^m_\lambda$ with connected, totally geodesic and holomorphic fibres. Then π falls in one of the following classes:*

a) $\pi : \mathbf{H}^m_\lambda \to \mathbf{P_{m-1}}(\mathbf{C}),\ m \geq 2$;

b) $\pi : \mathbf{H}^{2m+2}_\lambda \to \mathbf{P_m}(\mathbf{Q}),\ m \geq 1$.

Proof. Simply observe that a Hopf manifold is a compact homogeneous 1-g.H.m.

3.8 Almost Complex Conformal Submersions

We begin by explaining some effects determined by a conformal change of the metric of a Riemannian manifold on its submanifolds.

Proposition 3.42 *Let S be a submanifold of a Riemannian manifold (M^n, g) and $\tilde{g} = e^\sigma g$, σ differentiable function on M, a metric conformal to g. Then the second fundamental forms $\tilde{\alpha}$ and α of $(S, \tilde{g})$ and (S, g), as submanifolds of $(M, \tilde{g})$ and (M, g), respectively, are related by:*

$$\tilde{\alpha}(X, Y) = \alpha(X, Y) - \frac{1}{2} g(X, Y)(\mathrm{grad}_g \sigma)^\perp_{|S},\ X, Y \in \mathcal{X}(S) \ .$$

Furthermore, the mean curvature vector fields are related by:

$$\tilde{H} = e^{-\sigma}(H - \frac{1}{2}(\mathrm{grad}_g \sigma)^\perp_{|S}) \ .$$

Proof. Obviously, for any $x \in S$ we have $(T_x S)^{\perp_g} = (T_x S)^{\perp_{\tilde{g}}}$ and $\mathrm{grad}_g(\sigma_{|S}) = (\mathrm{grad}_g \sigma)^t_{|S}$, where the superscript t means that we consider the component of $(\mathrm{grad}_g\sigma)_{|S}$ tangent to S. Thus, the Levi-Civita connections $\tilde{\nabla}$ and ∇ determined on M by $\tilde{g}$ and g, respectively, verify:

$$\tilde{\nabla}_X Y = \nabla_X Y + \frac{1}{2}(X(\sigma)Y + Y(\sigma)X - g(X,Y)\mathrm{grad}_g\sigma) ,$$

and an analogous formula holds for the induced Riemannian connections on $(S,\tilde{g})$ and (S,g). Hence, for any $X,Y \in \mathcal{X}(S)$, we get:

$$\begin{aligned}\tilde{\alpha}(X,Y) &= \alpha(X,Y) - \tfrac{1}{2}g(X,Y)\mathrm{grad}_g\sigma + \tfrac{1}{2}g(X,Y)(\mathrm{grad}_g\sigma)^t_{|S} \\ &= \alpha(X,Y) - \tfrac{1}{2}g(X,Y)(\mathrm{grad}_g\sigma)^{\perp}_{|S} .\end{aligned}$$

Finally, if $(E_1,\dots,E_s)$ is a local $\tilde{g}$-orthonormal frame of $\mathcal{X}(S)$, then a local g-orthonormal frame of $\mathcal{X}(S)$ is given by $(\exp(\frac{\sigma}{2})E_1,\dots,\exp(\frac{\sigma}{2})E_s)$ and we obtain:

$$\begin{aligned}\tilde{H} &= \tfrac{1}{s}\textstyle\sum_{i=1}^{s}\tilde{\alpha}(E_i,E_i) = \tfrac{1}{s}\sum_{i=1}^{s}(\alpha(E_i,E_i) - \tfrac{1}{2}g(E_i,E_i)(\mathrm{grad}_g\sigma)^{\perp}_{|S}) \\ &= \tfrac{1}{s}\textstyle\sum_{i=1}^{s} e^{-\sigma}\alpha(e^{\sigma/2}E_i, e^{\sigma/2}E_i) - \tfrac{1}{2}e^{-\sigma}(\mathrm{grad}_g\sigma)^{\perp}_{|S} \\ &= e^{-\sigma}(H - \tfrac{1}{2}(\mathrm{grad}_g\sigma)^{\perp}_{|S}) .\end{aligned}$$

Corollary 3.14 *In the same hypotheses of the above proposition, if $(S,\tilde{g})$ is totally geodesic in $(M,\tilde{g})$, then (S,g) is totally umbilical in (M,g) with mean curvature vector field $H = \frac{1}{2}(\mathrm{grad}_g\sigma)^{\perp}_{|S}$.*

Proof. Since $\tilde{\alpha} = 0$, we get $\alpha(X,Y) = \frac{1}{2}g(X,Y)(\mathrm{grad}_g\sigma)^{\perp}_{|S}$. For any $\xi \in \mathcal{X}(S)^{\perp}$, the related Weingarten operator satisfies:

$$g(a_\xi X, Y) = g(\alpha(X,Y),\xi) = \frac{1}{2}g(g((\mathrm{grad}_g\sigma)^{\perp}_{|S},\xi)X,Y) ,$$

so $a_\xi X = fX$, $f = \frac{1}{2}g((\mathrm{grad}_g\sigma)^{\perp}_{|S},\xi)$ and the statement follows.

Definition 3.11 An *almost complex σ-conformal submersion* is an almost complex submersion $\pi : (M,J,g) \to (M',J',g')$ such that $\pi^*g' = e^\sigma g$, where σ is a differentiable function on M.

Remark 3.11 In the situation of the above definition, we can define the vertical and the horizontal distributions $\mathcal{V}(M)$ and $\mathcal{H}(M)$ and TM splits as orthogonal sum of $\mathcal{V}(M)$ and $\mathcal{H}(M)$. Moreover, $\pi_{*p} : \mathcal{H}_p \to T_{\pi(p)}M'$ is a homothety, since $e^{\sigma(p)}g_p(X,Y) = g'_{\pi(p)}(\pi_{*p}X, \pi_{*p}Y)$ for any horizontal vectors X,Y at $p \in M$.

Definition 3.12 If the total space of an almost complex σ-conformal submersion π is an l.c.*AK.* (l.c.*K.*) manifold, then π is called an *l.c.AK.* (*l.c.K.*) *σ-conformal submersion.*

Now, we are going to state some results which extend the analogous ones proved in [196] for l.c.K. σ-conformal submersions to the l.c.AK. context.

Proposition 3.43 *Let (M^{2m}, J, g) be an l.c.AK. (l.c.K.) manifold with Lee form ω and Lee vector field B. Then $(M^{2m}, J, \tilde{g})$ with $\tilde{g} = e^{\sigma} g$, σ differentiable function on M, is an l.c.AK. (l.c.K.) manifold with Lee form $\tilde{\omega} = d\sigma + \omega$, Lee vector field $\tilde{B} = e^{-\sigma}(\text{grad}\sigma + B)$, where* grad$\sigma$ *is the gradient of σ with respect to g.*

Proof. Putting $\tilde{g} = e^{\sigma} g$, the two Kähler forms Ω and $\tilde{\Omega}$ are related by $\tilde{\Omega} = e^{\sigma}\Omega$, and we get $d\tilde{\Omega} = e^{\sigma} d\sigma \wedge \Omega + e^{\sigma} d\Omega = (d\sigma + \omega) \wedge \tilde{\Omega}$. So $(M, J, \tilde{g})$ is l.c.AK. with Lee form $\tilde{\omega} = d\sigma + \omega$. Furthermore, for any $X \in \mathcal{X}(M)$,

$$\begin{aligned}\tilde{g}(X, \tilde{B}) &= \tilde{\omega}(X) = X(\sigma) + \omega(X) = g(X, \text{grad}\sigma) + g(X, B) \\ &= e^{-\sigma}\tilde{g}(X, \text{grad}\sigma + B)\ .\end{aligned}$$

Thus $\tilde{B} = e^{-\sigma}(\text{grad}\sigma + B)$.

Remark 3.12 It is obvious that an l.c.AK. (l.c.K.) σ-conformal submersion $\pi : (M, J, g) \to (M', J', g')$ can be viewed as an l.c.AK. (l.c.K.) submersion $\pi : (M, J, e^{\sigma} g) \to (M', J', g')$ with the same vertical and horizontal distributions.

Proposition 3.44 *Let $\pi : (M, J, g) \to (M', J', g')$ be an l.c.AK. (l.c.K.) σ-conformal submersion. Then any fibre $(\pi^{-1}(x), J, \tilde{g})$ and M' are l.c.AK. manifolds.*

Proof. The statement follows from the above remark and Proposition 3.43.

Proposition 3.45 *Let $\pi : (M^{2m}, J, g) \to (M'^{2m'}, J', g')$ be an l.c.AK. σ-conformal submersion and let B the Lee vector field on M^{2m}. Then, the horizontal distribution is integrable if and only if the vector field* grad$\sigma + B$ *is horizontal.*

Moreover, in this case, the horizontal integral submanifolds are totally umbilical with mean curvature vector field $H' = \frac{1}{2}v(\text{grad}\sigma) = -\frac{1}{2}v(B)$.

Proof. Consider the submersion $\sigma : (M^{2m}, J, \tilde{g}) \to (M'^{2m'}, J', g')$, where $\tilde{g} = e^{\sigma} g$ and the Lee vector field is $\tilde{B} = e^{-\sigma}(\text{grad}\sigma + B)$. Then, applying Proposition 3.36, we obtain that the horizontal distribution $\mathcal{H}$ is integrable if and only if grad$\sigma + B$ is a horizontal vector field. Moreover, when $\mathcal{H}$ is integrable, the integral manifolds are totally geodesic in $(M^{2m}, \tilde{g})$ and, by Corollary 3.14, they are totally umbilical in (M^{2m}, g) with mean curvature vector field $H' = \frac{1}{2}v(\text{grad}\sigma) = -\frac{1}{2}v(B)$, since $\text{grad}\sigma + B$ is horizontal.

Corollary 3.15 *Let $\pi : (M^{2m}, J, g) \to (M'^{2m'}, J', g')$ be an l.c.AK. σ-conformal submersion and B the Lee vector field on M^{2m}. Then the horizontal distribution is integrable with totally geodesic integral manifolds if and only if B and* $\text{grad}\sigma$ *are both horizontal.*

Proposition 3.46 *Let $\pi : (M^{2m}, J, g) \to (M'^{2m'}, J', g')$ be an l.c.AK. (l.c.K.) σ-conformal submersion, and let B the Lee vector field on M^{2m}. Then, the following conditions are equivalent:*

a) $\text{grad}\sigma + B$ *is vertical;*
b) $(M'^{2m'}, J', g')$ *is almost Kähler (Kähler);*
c) *the mean curvature vector field of the fibres is given by* $\frac{1}{2}h(\text{grad}\sigma)$.

Proof. Consider π also as an l.c.AK. (l.c.K.) submersion from $(M^{2m}, J, \tilde{g})$, $\tilde{g} = e^{\sigma}g$, onto $(M'^{2m'}, J', g')$. Then, by Proposition 3.42, the mean curvature vector fields of the fibres are related by:

$$\tilde{H} = e^{-\sigma}(H - \frac{1}{2}h(\text{grad}\sigma)) , \tag{3.16}$$

and

$$\tilde{H} = -\frac{1}{2}e^{-\sigma}h(\text{grad}\sigma + B) . \tag{3.17}$$

Now, the statement follows applying Proposition 3.37 and comparing the above relations.

Corollary 3.16 *Let $\pi : (M^{2m}, J, g) \to (M'^{2m'}, J', g')$ be an l.c.AK. (l.c.K.) σ-conformal submersion, and let B the Lee vector field on M^{2m}. Then the fibres of π are minimal if and only if B is a vertical vector field.*

Proof. From (3.16) and (3.17) we get $H = -\frac{1}{2}h(B)$ and the statement follows.

Corollary 3.17 *Let $\pi : (M^{2m}, J, g) \to (M'^{2m'}, J', g')$ be an l.c.AK. (l.c.K.) σ-conformal submersion, and let B the Lee vector field on M^{2m}. Then the following conditions are equivalent:*

a) *the fibres are minimal and $(M'^{2m'}, J', g')$ is almost Kähler (Kähler);*
b) $\text{grad}\sigma$ *and B are both vertical.*

Since Kähler and almost Kähler manifolds are l.c.K. and l.c.AK. with vanishing Lee form, the above results can be improved.

Proposition 3.47 *Let* $\pi : (M^{2m}, J, g) \to (M'^{2m'}, J', g')$ *be an* AK *(*K*)*σ*-conformal submersion. Then, we have:*

1) *the horizontal distribution is integrable if and only if* gradσ *is horizontal and, in this case, the integral manifolds are totally geodesic;*
2) *the following statements are equivalent:*
   a) M' *is almost Kähler (Kähler);*
   b) *the fibres are minimal;*
   c) gradσ *is vertical.*

Examples of l.c.K. σ-conformal submersions are given by J. C. Marrero and J. Rocha in [196]. We shortly present them.

Example 3.8 Let (M_1, J_1, g_1) be a g.c.K. manifold with Lee form ω_1 such that $\omega_1 = d(\ln f)$, f being a positive real differentiable function on M_1. For a given Kähler manifold (M_2, J_2, g_2) we consider the warped product $M = M_1 \times_f M_2$ with metric $g = p_1^* g_1 + p_1^*(f) p_2^* g_2$, where p_i, $i = 1, 2$, are the natural projections on the factors. Then, the almost Hermitian manifold $(M, J, e^{-\sigma} g)$, $J = J_1 \times J_2$, σ differentiable function on M, is g.c.K. with Lee form $p_1^* \omega_1 = d\sigma$ and $p_1 : (M, J, e^{-\sigma} g) \to (M_1, J_1, g_1)$ is a g.c.K. σ-conformal submersion whose horizontal distribution is integrable with totally umbilical integral manifolds. These integral manifolds are totally geodesic if and only if $\sigma = \sigma_1 \circ p_1$ for some differentiable function σ_1 on M_1. Furthermore, $p_2 : (M, J, g) \to (M_2, J_2, g_2)$ is an l.c.K. τ-conformal submersion, with $\tau = -\ln(p_1^* f)$ and totally geodesic fibres.

The following three examples are related to Examples 3.5–3.7.

Example 3.9 Consider the Vaisman submersion with totally geodesic fibres

$$\tau(c, k, m) : \mathbf{S^{2m-1}}(c, k) \times \mathbf{R} \to \mathbf{P_{m-1}(C)}(k + 3c^2)\ ,$$

where $c \neq 0, k > -3c^2$. We put $g = e^{-\sigma}\overline{g}$, σ differentiable function on $\mathbf{S^{2m-1}}(c, k) \times \mathbf{R}$, $\overline{g}$ being the metric on $\mathbf{S^{2m-1}}(c, k) \times \mathbf{R}$. Then we obtain an l.c.$K$. σ-conformal submersion

$$\tau(c, k, m, \sigma) : (\mathbf{S^{2m-1}}(c, k) \times \mathbf{R}, J, g) \to \mathbf{P_{m-1}(C)}(k + 3c^2)$$

with totally umbilical fibres and non-integrable horizontal distribution. Moreover, the fibres are totally geodesic with respect to g if and only if $\sigma = p^*\sigma'$ for some differentiable function σ' on $\mathbf{R}$, where p is the canonical projection from $\mathbf{S^{2m-1}}(c,k) \times \mathbf{R}$ onto $\mathbf{R}$.

Example 3.10 Consider the Vaisman submersion with totally geodesic fibres

$$\pi(c,m,m') : (\mathbf{R^{2m-1}}(c) \times \mathbf{R}, J, \overline{g}) \to \mathbf{C^{m'}} ,$$

where $m' \leq m-1$, $c \neq 0$. Given a differentiable function σ on $\mathbf{R^{2m-1}}(c) \times \mathbf{R}$ and putting $g = e^{-\sigma}\overline{g}$, we get an l.c.$K$. σ-conformal submersion

$$\pi(c,m,m',\sigma) : (\mathbf{R^{2m-1}}(c) \times \mathbf{R}, J, g) \to \mathbf{C^{m'}} ,$$

with totally umbilical fibres and non-integrable horizontal distribution. Moreover, the fibres are totally geodesic with respect to g if and only if

$$\frac{\partial \sigma}{\partial z} = 0 \quad \frac{\partial \sigma}{\partial x^i} = \frac{\partial \sigma}{\partial y^i} = 0, \quad i \in \{1, \ldots, m'\} .$$

Example 3.11 Consider $c \neq 0$, $k < -3c^2$ and the almost Hermitian submersion

$$\gamma(c,k,m) : ((\mathbf{R} \times \mathbf{C}D^{m-1})(c,k) \times \mathbf{R}, J, \overline{g}) \to \mathbf{C}D^{m-1}(k+3c^2) .$$

Let σ be a differentiable function on $(\mathbf{R} \times \mathbf{C}D^{m-1})(c,k) \times \mathbf{R}$ and put $g = e^{-\sigma}\overline{g}$. Then,

$$\gamma(c,k,m,\sigma) : ((\mathbf{R} \times \mathbf{C}D^{m-1})(c,k) \times \mathbf{R}, J, g) \to \mathbf{C}D^{m-1}(k+3c^2) ,$$

is an l.c.K. σ-conformal submersion with totally umbilical fibres and non-integrable horizontal distribution. The fibres are totally geodesic with respect to g if and only if $\sigma = p^*\sigma'$, for some differentiable function σ' on $\mathbf{R}$, p being the canonical projection of $(\mathbf{R} \times \mathbf{C}D^{m-1})(c,k) \times \mathbf{R}$ onto $\mathbf{R}$.

Chapter 4

Riemannian Submersions and Contact Metric Manifolds

This chapter deals with the study of Riemannian submersions in contact geometry. Basically, we treat the same problems as in the previous chapter: actually, the contact structures are the natural odd dimensional analogue of the almost complex structures. However, there are many specific properties, which justify recent research and make the geometry of Riemannian contact manifolds interesting.

The fundamental results in this domain are due to the Japanese school and reference books are the well-known monographs of D. Blair ([33; 34]). The recent remarkable results of Boyer, Galicki, Mann and others on the geometry of Sasakian manifolds stimulate interest in this direction. In analogy with the Gray–Hervella classification of almost Hermitian manifolds, Chinea with Gonzales ([68]) and Oubiña ([233]) gave a classification of almost contact metric manifolds. In Sec. 4.1 we present some classes within this classification, considered more relevant from the viewpoint of the Riemannian submersion theory. Three classes of Riemannian submersions are considered in the context of contact and almost Hermitian manifolds.

In particular, in Sec. 4.2 we study the Riemannian submersions between some remarkable almost contact metric manifolds, showing how the structure of the total space influences the geometry of the fibres and the base.

Section 4.3 is dedicated to Riemannian submersions from an almost contact metric manifold to an almost Hermitian one. The Boothby–Wang fibration is the standard example.

In Sec. 4.4, the concept, due to Blair, Ludden and Yano of semi-invariant submanifold in an almost Hermitian manifold is used to define the complex-contact Riemannian submersions. At the end of this section some examples clarify the consistency of the definition.

Section 4.5 is dedicated to basic relations between the φ-holomorphic sectional and φ-holomorphic bisectional curvatures of the manifolds involved in the submersions previously considered.

Riemannian submersions from a (almost) 3-contact metric manifold onto a (almost) quaternionic one have been recently considered in the literature. Section 4.6 deals with this subject.

More generally, manifolds endowed with an f-structure are relevant in the context of submersions. In particular, in Sec. 4.7, it is proved that, under suitable conditions, a manifold carrying a $pk.f$-structure (hyper f-structure) fibers over a complex (hypercomplex) manifold. Finally, a compact $\mathcal{PS}$-manifold is regarded as the total space of two Riemannian submersions, both having fibres which are tori, onto a Sasakian manifold and a hyperKähler one, respectively.

4.1 Remarkable Classes of Contact Metric Manifolds

Definition 4.1 A $(2n+1)$-dimensional manifold M has an *almost contact structure* (φ, ξ, η) if it admits a vector field ξ (the so-called *characteristic vector field*), a 1-form η and a field φ of endomorphisms of the tangent spaces satisfying:

$$\eta(\xi) = 1 \,, \qquad \varphi^2 = -I + \eta \otimes \xi \,. \tag{4.1}$$

These conditions imply that $\varphi(\xi) = 0$ and $\eta \circ \varphi = 0$; moreover, φ has rank $2n$ at every point of M ([33]). Equivalently, as proved by J. Gray ([131]), a $(2n+1)$-dimensional manifold M carries an almost contact structure if and only if the structural group of its tangent bundle is reducible to $U(n) \times \{I\}$.

A manifold M endowed with an almost contact structure (φ, ξ, η) is called an *almost contact manifold*. It is known that M admits a Riemannian metric g such that:

$$g(\varphi X, \varphi Y) = g(X, Y) - \eta(X)\eta(Y) \,, \tag{4.2}$$

for any vector fields X, Y and, when such a metric is chosen, M is said to have an *almost contact metric structure (ACM-structure)* and g is called a metric *compatible* (or *associated*) with the almost contact structure (φ, ξ, η). The *Sasakian form* of an ACM-manifold $(M, \varphi, \xi, \eta, g)$ is the 2-form Φ defined by $\Phi(X, Y) = g(X, \varphi Y)$, for $X, Y \in \mathcal{X}(M)$. Since such a form satisfies $\eta \wedge \Phi^n \neq 0$, M is orientable and one can construct a volume form on M starting from the $(2n+1)$-form $\eta \wedge \Phi^n$.

On the other hand, a $(2n+1)$-dimensional manifold M is said to be a *contact manifold*, if there exists a 1-form η on M such that $\eta \wedge (d\eta)^n \neq 0$ everywhere. It is well known that a contact manifold can be equipped with an almost contact metric structure (φ, ξ, η, g). When $\Phi = d\eta$, $(M, \varphi, \xi, \eta, g)$ is called a *contact metric manifold* or an *almost Sasakian manifold* (*AS*-manifold).

A contact metric manifold is a K-*contact manifold* if ξ is a Killing vector field. The first basic property of a K-contact manifold is:

$$\nabla_X \xi = -\varphi X, \quad X \in \mathcal{X}(M) , \tag{4.3}$$

where ∇ is the Levi-Civita connection of g ([33] p. 64, [34] pp. 67 and 70).

Now, we recall the concept of *normality* for an almost contact structure, which is the analogue of integrability in almost complex geometry. Let (M, φ, ξ, η) be an almost contact manifold and consider the product manifold $M \times \mathbf{R}$. We denote by $(X, f\frac{d}{dt})$ a vector field on $M \times \mathbf{R}$, where X is tangent to M, and f is a C^∞-function on $M \times \mathbf{R}$. Then, we can define an almost complex structure J on $M \times \mathbf{R}$ by:

$$J(X, f\frac{d}{dt}) = (\varphi X - f\xi, \eta(X)\frac{d}{dt}) . \tag{4.4}$$

The almost contact structure (φ, ξ, η) is said to be *normal* if J is integrable, and the normality condition is expressed by:

$$N_\varphi + 2d\eta \otimes \xi = 0 , \tag{4.5}$$

where N_φ is the Nijenhuis tensor of φ, also denoted by $[\varphi, \varphi]$ ([33] p. 49, [34] p. 65).

The normality tensor field $S = N_\varphi + 2d\eta \otimes \xi$, is also named the *Sasaki–Hatakeyama tensor field* associated with the almost contact structure (φ, ξ, η).

Remark 4.1 In the general theory of G-structures defined by a tensor field φ of type (1,1), the concept of integrability means the vanishing of the Nijenhuis tensor N_φ. In the case of an almost contact structure this implies that M is locally a product of a real curve with an almost complex manifold. For this geometric reason the concept of normality, just introduced, is more interesting.

A *Sasakian* manifold is a normal *AS*-manifold. Equivalently, an almost contact metric manifold $(M, \varphi, \xi, \eta, g)$ is a Sasakian manifold if and only if the following formula holds:

$$(\nabla_X \varphi)Y = g(X,Y)\xi - \eta(Y)X\ , \quad X, Y \in \mathcal{X}(M)\ . \tag{4.6}$$

We refer to ([33; 34]) for the proof.

Furthermore, a *c-Sasakian* manifold, $c \in \mathbf{R}$, $\neq 0$, is an almost contact metric manifold $(M, \varphi, \xi, \eta, g)$ which is normal and satisfies $d\eta = c\,\Phi$. An almost contact metric manifold is c-Sasakian if and only if the following formula holds:

$$(\nabla_X \varphi)Y = c(g(X,Y)\xi - \eta(Y)X)\ ,\ c \neq 0\ , \quad X, Y \in \mathcal{X}(M)\ .$$

Remark 4.2 It is easy to see that any Sasakian manifold is a K-contact manifold. The converse also holds in the 3-dimensional case.

Remark 4.3 Let (M, g) be an odd-dimensional Riemannian manifold and ξ a unit Killing vector field such that $R(X, \xi, \xi) = X$ for X orthogonal to ξ. We denote by η the 1-form on M such that $\eta(X) = g(\xi, X)$, and by φ the tensor field defined by $\varphi X = -\nabla_X \xi$ for any vector field X. Then (φ, ξ, η, g) is a K-contact structure on M. Moreover, if (4.6) is satisfied (or (φ, ξ, η) is normal), then M is a Sasakian manifold.

In contrast to the almost Sasakian structures, for which the 1-form η is of maximal rank and $\Phi = d\eta$, we define an *almost cosymplectic* structure (AC-structure) as an almost contact metric structure, such that Φ and η are closed.

A *cosymplectic structure* (C-structure) is a normal AC-structure. Equivalently, (φ, ξ, η, g) is a C-structure if and only if $\nabla\varphi = 0$.

Another interesting almost contact metric structure has been defined by K. Kenmotsu [166].

Definition 4.2 (φ, ξ, η, g) is a *Kenmotsu structure* (KN-structure) if and only if:

$$(\nabla_X \varphi)Y = g(\varphi X, Y)\xi - \eta(Y)\varphi X\ , \tag{4.7}$$

for any vector fields X, Y on M. A Kenmotsu structure is normal and the 1-form η is closed. Moreover, a Kenmotsu manifold cannot be compact since $\operatorname{div}\xi = 2n$.

D. Chinea and C. Gonzales ([68]) and J. Oubiña ([233]) classified the almost contact metric manifolds into several classes. In Table 4.1, some of these classes, with the defining conditions, are listed.

Table 4.1 Classification of almost contact metric manifolds

Symbol	name	condition
N	normal	$N_\varphi + d\eta \otimes \xi = 0$
S	Sasakian	$\Phi = d\eta$, normal
AS	almost Sasakian	$\Phi = d\eta$
KC	K-contact	$\Phi = d\eta$, ξ Killing
NKS	nearly-K-Sasakian	$(\nabla_X\varphi)X = g(X,X)\xi - \eta(X)X$,$\nabla_X\xi = -\varphi X$
QS	quasi-Sasakian	$d\Phi = 0$, normal
C	cosymplectic	$d\Phi = 0$, $d\eta = 0$, normal
AC	almost cosymplectic	$d\Phi = 0$, $d\eta = 0$
NKC	nearly-K-cosymplectic	$(\nabla_X\varphi)X = 0$, $\nabla_X\xi = 0$
KN	Kenmotsu	$(\nabla_X\varphi)Y = g(\varphi X, Y)\xi - \eta(Y)\varphi X$

Let $(M, \varphi, \xi, \eta, g)$ be an almost contact metric manifold. On the manifold $\tilde{M} = M \times \mathbf{R}$ we consider the almost complex structure J defined by (4.4) and the Riemannian metrics $\tilde{g}$ and $\tilde{g}_c$ such that:

$$\tilde{g}((X, f\frac{d}{dt}), (Y, f'\frac{d}{dt})) = g(X,Y) + ff' , \tag{4.8}$$

where f, f' are differentiable functions on $\tilde{M}$ and

$$\tilde{g}_c = e^{2\sigma}\tilde{g} , \tag{4.9}$$

where $\sigma : \tilde{M} \to \mathbf{R}$, $\sigma(p,t) = t$ for any $(p,t) \in \tilde{M}$. Then, the following conditions are equivalent:

i) g is compatible with the (φ, ξ, η)-structure,
ii) $\tilde{g}$ is Hermitian on $(\tilde{M}, J)$,
iii) $\tilde{g}_c$ is Hermitian on $(\tilde{M}, J)$.

There is a standard procedure to compare different types of almost contact metric manifolds in Table 4.1 with the almost Hermitian ones in Table 3.1. We illustrate this procedure with some examples, where $\tilde{M} = M \times \mathbf{R}$:

a) M is an N-manifold if and only if $(\tilde{M}, J, \tilde{g})$ is an H-manifold or, equivalently, $(\tilde{M}, J, \tilde{g}_c)$ is an H-manifold,
b) M is an (almost) S-manifold if and only if $(\tilde{M}, J, \tilde{g}_c)$ is a (almost) K-manifold,
c) M is a (almost) C-manifold if and only if $(\tilde{M}, J, \tilde{g})$ is a (almost) K-manifold,
d) M is an NKS-manifold if and only if $(\tilde{M}, J, \tilde{g}_c)$ is an NK-manifold,

e) M is an NKC-manifold if and only if $(\tilde{M}, J, \tilde{g})$ an NK-manifold.

J. Oubiña ([233]) defined many other classes of almost contact metric manifolds: *trans-Sasakian, quasi-K-Sasakian, quasi-K-cosymplectic, semi-Sasakian, semi-cosymplectic, G_i-Sasakian, G_i-cosymplectic* (i=1,2) etc., and gave examples of manifolds in each of these classes which do not belong to the others. We remark that $(M, \varphi, \xi, \eta, g)$ is a trans-Sasakian manifold if and only if $(\tilde{M}, J, \tilde{g}_c)$ belongs to the class W_4 in the classification of Gray and Hervella ([130]). D. Chinea and C. Gonzales ([68]) constructed examples of different types of almost metric structures (φ, ξ, η, g) on the product of an almost Hermitian manifold (M, J, h) with $\mathbf{R}$ putting:

$$\varphi(X, f\frac{d}{dt}) = (JX, 0) \ , \quad \xi = (0, \frac{d}{dt}) \ , \quad \eta(X, f\frac{d}{dt}) = f \ , \tag{4.10}$$

and defining the metric g as in (4.8) by means of h. They also gave many interesting examples of almost contact metric structures on the hyperbolic spaces, on the generalized Heisenberg groups $H(p,1)$ and $H(1,r)$ and on the Lie groups of matrices occurring in Kowalski's classification of generalized Riemannian symmetric spaces of dimension $n \leq 5$ ([176]).

Locally conformal cosymplectic manifolds, introduced by Vaisman ([308]) and then investigated by Chinea and Marrero ([69]), set up another interesting class of almost contact metric manifolds also considered in the context of Riemannian submersions ([70]).

Definition 4.3 A *locally conformal cosymplectic (l.c.C.) manifold* is an almost contact metric manifold $(M, \varphi, \xi, \eta, g)$ such that any $p \in M$ admits an open neighborhood U and a C^∞-function $\sigma : U \to \mathbf{R}$ such that the local ACM-structure:

$$\varphi' = \varphi_{|U} \ , \quad \xi' = e^{-\sigma}\xi_{|U} \ , \quad \eta' = e^{\sigma}\eta_{|U} \ , \quad g' = e^{2\sigma}g_{|U} \ , \tag{4.11}$$

is cosymplectic. If the open set U is M, then M is *globally conformal cosymplectic (g.c.C.)*.

Several examples of l.c.C. manifolds are well known ([308; 69; 70]). We only recall that, for any $\lambda \in \mathbf{R}^*$, $\lambda \neq 1$, the $(2n+1)$-dimensional real Hopf manifold $\mathbf{H}^{2n+1} = (\mathbf{R^{2n+1}})^*/\Delta_\lambda$, Δ_λ being the cyclic group generated by the transformation $x \in (\mathbf{R^{2n+1}})^* \mapsto \lambda x \in (\mathbf{R^{2n+1}})^*$, carries an l.c.$C$. structure explicitly described in [69]. If $n \geq 2$, this manifold, which is diffeomorphic to $\mathbf{S^{2n}} \times \mathbf{S^1}$, does not admit cosymplectic structures.

As in the locally conformal Kähler case, the Lee form plays a relevant role in the theory of l.c.C. manifolds.

Definition 4.4 Let $(M, \varphi, \xi, \eta, g)$ be an ACM-manifold having dimension $2n+1$. The *Lee form* of M is the 1-form ω defined by:

$$\omega = \frac{1}{2(n-1)}(\delta\Phi \circ \varphi - \nabla_\xi \eta) - \frac{\delta\eta}{2n}\eta\,, \quad \text{if} \quad n \geq 2,$$

$$\omega = \nabla_\xi \eta - \frac{\delta\eta}{2}\eta\,, \quad \text{if} \quad n = 1\,,$$

δ denoting the codifferential operator. The *Lee vector* field is the vector field B dual to ω with respect to g.

We remark that, if M is an l.c.C. manifold, in any open set U where the conformal change given by (4.11) is defined, one has $d\sigma = \omega_{|U}$, ω being the Lee form of M. A useful characterization of the l.c.C. condition is due to Chinea and Marrero ([69]).

Proposition 4.1 *Given an ACM-manifold $(M, \varphi, \xi, \eta, g)$, the following conditions are equivalent:*

i) *M is an l.c.C. (g.c.C.) manifold;*
ii) *the Lee form ω is closed (exact) and for any $X, Y \in \mathcal{X}(M)$*

$$(\nabla_X \varphi)Y = \omega(Y)\varphi X - \omega(\varphi Y)X + \Phi(X,Y)B - g(X,Y)\varphi B\,;$$

iii) *the Lee form ω is closed (exact) and $d\Phi = -2\Phi \wedge \omega$, $d\eta = \eta \wedge \omega$, $N_\varphi = 0$.*

4.2 Contact Riemannian Submersions

D. Chinea ([63; 65]) and B. Watson ([330]) studied the differential geometric properties of Riemannian submersions between almost contact metric manifolds.

Let M^{2m+1} and B^{2n+1} be manifolds carrying the almost contact metric structures (φ, ξ, η, g) and $(\varphi', \xi', \eta', g')$ respectively.

Definition 4.5 A Riemannian submersion $\pi : M^{2m+1} \to B^{2n+1}$ between the almost contact metric manifolds M^{2m+1} and B^{2n+1} is called a *contact Riemannian submersion* if:

a) $\pi_* \xi = \xi'$,
b) $\pi_* \circ \varphi = \varphi' \circ \pi_*$.

The condition b) is also called the (φ, φ')-*holomorphicity* condition.

Proposition 4.2 *Let $\pi : M \to B$ be a contact Riemannian submersion. Then, $\mathcal{V}$ and $\mathcal{H}$ are φ-invariant. Moreover, $\pi^*\eta' = \eta$ and on $\mathcal{H}$ the Sasakian 2-form Φ acts as the pull-back of Φ' under π.*

Proof. If U is a vertical vector at a point $p \in M$, we have $\pi_*\varphi(U) = \varphi'\pi_*(U) = 0$, so that φU belongs to $\mathcal{V}_p$. Let X be a horizontal vector at p. Then, $g(\varphi X, U) = -g(X, \varphi U) = 0$, for any $U \in \mathcal{V}_p$, so φX is a horizontal vector, *i.e.* $\varphi(\mathcal{H}_p) \subseteq \mathcal{H}_p$. Now, let X and Y be horizontal vectors at a point $p \in M$. Using the definition of a contact Riemannian submersion, we have:

$$(\pi^*\Phi')(X,Y) = \Phi'(\pi_*X, \pi_*Y) = g'(\pi_*X, \varphi'\pi_*Y) = g(X, \varphi Y) = \Phi(X,Y) .$$

In a similar way one proves that $\eta = \pi^*\eta'$.

Remark 4.4 a) The characteristic vector field ξ is horizontal. Indeed, writing $\xi = v\xi + h\xi$, we have $\eta(h\xi) = g'(\pi_*\xi, \pi_*h\xi) = g'(\xi', \xi') = 1$. On the other hand $\eta(v\xi) + \eta(h\xi) = 1$, and then $v\xi = 0$.
b) Since ξ is horizontal, we have $\eta(U) = 0$ for any vertical vector field U and this implies $\mathcal{V}_p \subset \ker \eta_p$, for any $p \in M$.
c) Let F be a fibre of a contact Riemannian submersion and denote by $\widehat{J}$ the restriction of φ to the vertical distribution $\mathcal{V}$. Then, $\widehat{J}^2(U) = \varphi^2(U) = -U$ so that any fibre $(F, \widehat{J}, g_{|F})$ is an almost Hermitian manifold.

The following result can be proved in a standard way.

Proposition 4.3 *Let $\pi : M \to B$ be a contact Riemannian submersion. If X, Y are basic vector fields on M, π-related to X', Y' on B, then we have:*

1) *φX is the basic vector field π-related to $\varphi' X'$,*
2) *$h(S(X,Y))$ is the basic vector field π-related to $S'(X',Y')$,*
3) *$h((\nabla_X \varphi)Y)$ is the basic vector field π-related to $(\nabla'_{X'}\varphi')Y'$,*

where S and S' are the normality tensor fields of M and B, respectively.

Proposition 4.4 *Let $\pi : M \to B$ be a contact Riemannian submersion. If the almost contact structure of M is normal, then the almost contact structure of B is normal and the fibres are Hermitian manifolds.*

Proof. Let X', Y' be vector fields on B and X, Y their horizontal lifts. Since $S(X,Y) = 0$, then 2) in the previous proposition implies $S'(X',Y') = 0$ and B is normal. On the other hand the Nijenhuis tensor $N_{\widehat{J}}$ is the restriction of N_φ to any fibre F and $\widehat{J}$ is integrable.

The following theorem collects results due to B. Watson ([330]), D. Chinea ([64]) and T. Tshikuna-Matamba ([297; 299]).

Theorem 4.1 *Let $\pi : M \to B$ be a contact Riemannian submersion. If M belongs to any of the classes QS, AC, NKC, C, KN, then the base space B belongs to the same class.*

Proof. If M is a quasi-Sasakian manifold, we have $d\Phi = 0$ and by Proposition 4.2 we obtain $d\Phi' = 0$. Then, applying Proposition 4.4, B is a quasi-Sasakian manifold. Now, we suppose that M is an NKC-manifold. By Proposition 4.3, we have $(\nabla'_{X'}\varphi')X' = 0$ for any vector field X' on B. Moreover, since ξ is the lift of ξ' and ξ is ∇-parallel, ξ' is ∇'-parallel also, so B is an NKC-manifold. Finally, assuming that M is a Kenmotsu manifold and using Proposition 4.3, we obtain that B is a Kenmotsu manifold.

Theorem 4.2 *Let $\pi : M \to B$ be a contact Riemannian submersion. If the total space belongs to any of the classes S, AS, KC, NKS, then π is a Riemannian covering map.*

Proof. By Remark 4.4, $d\eta(U, V) = 0$ for any vector fields U, V tangent to a fibre F. Since $\Phi = d\eta$ for the given classes, we obtain $g(U, \varphi V) = 0$. Putting $V = \varphi U$, we get $g(U, U) = 0$ on F, so that F is discrete.

Theorem 4.3 *Let $\pi : M \to B$ be a contact Riemannian submersion. Then,*

1) *If M is a QS, C or KN-manifold, the fibres are K-manifolds.*
2) *If M is an AC, NKC-manifold, then the fibres are AK, NK-manifolds, respectively.*

Moreover, the fibres are minimal in the above cases, except for the KN case where the mean curvature vector field is $H = -\xi$.

Proof. We assume that M is a KN-manifold. Using the Gauss's formula for a fibre submanifold F, and the condition defining a Kenmotsu manifold, we obtain:

$$\widehat{\nabla}_U \widehat{J}V = \widehat{J}\widehat{\nabla}_U V \ , \tag{4.12}$$

$$T_U \widehat{J}V = \varphi(T_U V) + g(\varphi U, V)\xi \ , \tag{4.13}$$

for any U, V vector fields tangent to F. The first formula implies that F is a K-manifold. Now, fixing a local Hermitian frame $\{e_i, \widehat{J}e_i\}$, $i \in \{1, \dots, s\}$

on F, $2s$ being the dimension of F, applying twice (4.13) we get:

$$T_{\widehat{J}e_i}\widehat{J}e_i = -T_{e_i}e_i - 2g(e_i, e_i)\xi ,$$

and then $H = -\xi$. Finally, in the other cases, a straightforward computation gives the statement.

We end this section with a result due to Chinea, Marrero and Rocha on contact Riemannian submersions whose total space is an l.c.C. manifold. For further properties we refer to [70], where the concept of $D(\sigma)$-conformal cosymplectic submersion from a g.c.C. manifold with $\omega = d\sigma$ is introduced and investigated; this can be regarded as the ACM-counterpart to the subject treated in Sec. 3.8.

Theorem 4.4 *Let $\pi : M \to M'$ be a contact Riemannian submersion with M an l.c.C. manifold. Then, we have:*

i) *the Lee vector field of M is horizontal;*
ii) *the horizontal distribution is integrable;*
iii) *the fibres of π are K-manifolds; they are minimal in M if and only if M is cosymplectic;*
iv) *M' is l.c.C. and its Lee form pulls back, via π, to the Lee form of M.*

Proof. Since ξ is basic, via Proposition 4.1, for any vertical vector field U we get

$$0 = d\eta(U, \xi) = (\eta \wedge \omega)(U, \xi) = -\frac{1}{2}\omega(U) ,$$

and then i). Considering now basic vector fields X, Y and U vertical, applying again Proposition 4.1, we have

$$\begin{aligned} 2g(A_X Y, \varphi U) &= \Phi([X, Y], U) = -3d\Phi(X, Y, U) = 6(\Phi \wedge \omega)(X, Y, U) \\ &= 2\Phi(X, Y)\omega(U) = 0 , \end{aligned}$$

which implies the vanishing of the invariant A and then ii). Moreover, via ii) in Proposition 4.1, one has

$$(\nabla_U \varphi)V = g(U, \varphi V)B - g(U, V)\varphi B ,$$

so, the vertical component of $(\nabla_U \varphi)V$ vanishes and the almost complex structure on any fibre of π is Kähler. Since the invariant T satisfies

$$T_U \varphi V - \varphi(T_U V) = g(U, \varphi V)B - g(U, V)\varphi B ,$$

we have $T_U V + T_{\varphi U}\varphi V = 2g(U,V)B$, so the restriction of B to a fibre F is just the mean curvature vector field of F. Thus ω vanishes, *i.e.* M is cosymplectic, if and only if any fibre is minimal. Finally, we put $\dim M = 2m+1$, $\dim M' = 2n+1$, $s = m-n$, and denote by $(\varphi', \xi', \eta', g')$ the ACM-structure on M' and by ω' its Lee form. Firstly, we prove that $\omega(\xi) = \omega'(\xi') \circ \pi$. In fact, considering a π-adapted local orthonormal frame $\{U_1, \dots, U_s, \varphi U_1, \dots, \varphi U_s, X_1, \dots, X_n, \varphi X_1, \dots, \varphi X_n, \xi\}$ of M, where any X_i is basic, one has:

$$\begin{aligned}\delta\eta &= \delta\eta' \circ \pi + \textstyle\sum_j \{(\nabla_{U_j}\eta)U_j + (\nabla_{\varphi U_j}\eta)\varphi U_j\} \\ &= \delta\eta' \circ \pi - \textstyle\sum_j g(T_{U_j}U_j + T_{\varphi U_j}\varphi U_j, \xi) \\ &= \delta\eta' \circ \pi - 2s\omega(\xi)\ ,\end{aligned}$$

and then

$$\omega'(\xi') \circ \pi = -\frac{1}{2n}\delta\eta' \circ \pi = -\frac{1}{2n}(\delta\eta + 2s\omega(\xi)) = \omega(\xi)\ .$$

Now, if $n > 1$, we consider a basic vector field X orthogonal to ξ, which projects to X', and we have:

$$\begin{aligned}2(m-1)\omega(X) &= 2(n-1)\omega'(X') \circ \pi \\ &\quad + \textstyle\sum_j \{(\nabla_{U_j}\Phi)(U_j, \varphi X) + (\nabla_{\varphi U_j}\Phi)(\varphi U_j, \varphi X)\} \\ &= 2(n-1)\omega'(X') \circ \pi + \textstyle\sum_j g(\varphi(T_{U_j}U_j + T_{\varphi U_j}\varphi U_j), \varphi X) \\ &= 2(n-1)\omega'(X') \circ \pi + 2s\omega(X)\ .\end{aligned}$$

This implies $\omega(X) = \omega'(X') \circ \pi$ for any X basic and orthogonal to ξ; since $\omega(\xi) = \omega'(\xi') \circ \pi$ and ω is vertical, we have $\pi^*\omega' = \omega$ and, via iii) in Proposition 4.1, one easily obtains that M' is also locally conformal cosymplectic. Finally, when $n = 1$, we consider $\{X', \varphi' X', \xi'\}$ as a local orthonormal φ-frame on M' and denote by X the horizontal lift of X'. Then, we have:

$$\begin{aligned}\omega(X) &= -2(\eta \wedge \omega)(X, \xi) = -2d\eta(X, \xi) = -2d\eta'(X', \xi') \circ \pi \\ &= ((\nabla_{\xi'}\eta')X') \circ \pi = \omega'(X') \circ \pi\ .\end{aligned}$$

Analogously, one gets $\omega(\varphi X) = \omega'(\varphi' X') \circ \pi$; since moreover $\omega(\xi) = \omega'(\xi') \circ \pi$ and ω is vertical, one has $\pi^*\omega' = \omega$. Combining with Proposition 4.1, one obtains that M' is l.c.C.

Corollary 4.1 *The horizontal distribution of a contact Riemannian submersion with total space a cosymplectic manifold is integrable.*

The previous results have been extended in [92], where contact Riemannian submersions with total space a locally conformal almost quasi-Sasakian (l.c.AQS) manifold are considered. As stated in [308] an ACM-manifold, with dim $M \geq 5$, is l.c.AQS if and only if its Lee form ω is closed and $d\Phi = -2\Phi\wedge\omega$. These manifolds set up a wide class of ACM-manifolds, which contains many of the just considered classes, such as quasi-Sasakian, Kenmotsu, l.c. cosymplectic manifolds. In particular, the action of the invariant A of a contact Riemannian submersion $\pi : M \to M'$, with M l.c.AQS is determined by the Lee vector field B of M. In fact, on the horizontal distribution, one has $A_X Y = \Phi(X,Y)v(\varphi B)$. This implies the integrability of the horizontal distribution, provided that M is quasi-Sasakian, Kenmotsu or, possibly, l.c. cosymplectic.

4.3 Contact-complex Riemannian Submersions

In this section we consider Riemannian submersions having an almost contact metric manifold $(M^{2m+1}, \varphi, \xi, g)$ and an almost Hermitian manifold (B^{2n}, J, g') as total and base space, respectively.

Definition 4.6 A Riemannian submersion $\pi : M \to B$ from an almost contact metric manifold M onto an almost Hermitian manifold B is called a *contact-complex Riemannian submersion* if:

$$J \circ \pi_* = \pi_* \circ \varphi \,.$$

Remark 4.5 As in Proposition 4.2 and Remark 4.4, one verifies that the vertical and horizontal distributions $\mathcal{V}$ and $\mathcal{H}$ are φ-invariant, *i.e.* $\varphi(\mathcal{V}) \subseteq \mathcal{V}$ and $\varphi(\mathcal{H}) \subseteq \mathcal{H}$. Moreover, the characteristic vector field ξ is vertical and $\mathcal{H} \subseteq \ker\eta$; *i.e.* $\eta(X) = 0$ for any horizontal vector field X.

Example 4.1 The Hopf fibration $\pi : \mathbf{S^{2n+1}} \to \mathbf{P_n(C)}$ described in Sec. 1.2 is the basic example of a contact-complex Riemannian submersion.

More generally, if M is a compact regular K-contact manifold, then the Boothby–Wang fibration is a contact-complex Riemannian submersion from M onto an almost Kähler manifold ([33]). Furthermore, in a similar way A. Morimoto ([206]) obtained a fibration of a compact N-manifold with ξ regular.

Example 4.2 Let (B, J, g') be an almost Hermitian manifold and G a 1-dimensional Lie group. We consider a principal G-bundle P over B with a connection 1-form η and denote by ξ the fundamental vector field such that $\eta(\xi) = 1$. We define the tensor field φ on P putting:

$$\varphi_p X = (J_{p'}(\pi_* X))^{\#} \ , \quad X \in T_p P \ ,$$

where $p' = \pi(p)$ and $\#$ means the horizontal lift with respect to the connection η. It is easy to prove that (φ, ξ, η) is an almost contact structure on P, and φ and η are G-invariant. Finally, putting $g = \pi^* g' + \eta \otimes \eta$, we obtain a Riemannian metric associated with the almost contact structure (φ, ξ, η). Then, $\pi : P \to B$ is a Riemannian submersion and we get $\pi_* \circ \varphi = J \circ \pi_*$.

With the same technique used for the proofs of Theorems 4.1, 4.3, one obtains the following result.

Theorem 4.5 *Let $\pi : M \to B$ be a contact-complex Riemannian submersion. If M belongs to any of the classes N, S, AS, KC, NKS, QS, C, AC, NKC, KN, then the base space belongs to H, K, AK, AK, NK, K, K, AK, NK, K, respectively. Furthermore, in any of the above cases, the fibres inherit from the total space a structure of the same type.*

Now, we recall a formula involving the Levi-Civita connection for any almost contact metric manifold ([33; 34]).

$$\begin{aligned} 2g((\nabla_E \varphi)F, G) = {} & 3d\Phi(E, \varphi F, \varphi G) - 3d\Phi(E, F, G) \\ & + g(S(F, G), \varphi E) + S^{(2)}(F, G)\eta(E) \\ & + 2d\eta(\varphi F, E)\eta(G) - 2d\eta(\varphi G, E)\eta(F) \ , \end{aligned} \tag{4.14}$$

where S is the Sasaki–Hatakeyama's tensor and $S^{(2)}$ is the tensor field of type $(2, 0)$ defined by:

$$S^{(2)}(E, F) = d\eta(\varphi E, F) + d\eta(E, \varphi F) \ ,$$

for any E, F, G vector fields on M. Note that $S = 0$ implies $S^{(2)} = 0$.

Theorem 4.6 *Let $\pi : M \to B$ be a contact-complex Riemannian submersion whose total space belongs to any of the classes in Table 4.1, with the only exception for the classes N and AS. Then the fibres are minimal submanifolds.*

Proof. We discuss only the case of a quasi-Sasakian manifold, the other ones can be proved with the same technique, see also [296]. Then, (4.14) becomes:

$$g((\nabla_E\varphi)F,G) = d\eta(\varphi F,E)\eta(G) - d\eta(\varphi G,E)\eta(F) .$$

Considering a vertical vector field U, orthogonal to ξ, for any X horizontal, we get:

$$g(\nabla_U\varphi U - \varphi\nabla_U U, X) = d\eta(U,\varphi X)\eta(U) = 0 .$$

This implies $T_U\varphi U = \varphi T_U U$ and using the properties of the O'Neill tensor T we have: $T_U U + T_{\varphi U}\varphi U = 0$. Moreover $T_\xi\xi = 0$ and then $H = 0$, *i.e.* any fibre is minimal.

4.4 Complex-contact Riemannian Submersions

The attempt to consider a Riemannian submersion from an almost Hermitian manifold to an almost contact metric manifold such that the projection map commutes with the fundamental tensors fails. Indeed, let $\pi : M \to B$ be a Riemannian submersion from an AH-manifold (M,J,g) to an ACM-manifold (B,φ,ξ,η,g'), and assume that $\pi_*\circ J = \varphi\circ\pi_*$. If we denote by ξ^* the horizontal lift of ξ, then, pointwise, we have $\pi_*J\xi^* = \varphi\pi_*\xi^* = \varphi\xi = 0$ and this implies: $-\xi = \pi_*(J^2\xi^*) = \varphi(\pi_*J\xi^*) = 0$, giving a contradiction. To avoid this contradiction, according to Chinea ([67]), one can use an idea closely related to the concept of semi-invariance due to Blair, Ludden and Yano ([36]) and state the following definition.

Definition 4.7 Let (M,J,g) and (B,φ,ξ,η,g') be an almost Hermitian manifold and an almost contact metric manifold, respectively. A Riemannian submersion $\pi : M \to B$ is called a *complex-contact* Riemannian submersion if there exists a 1-form $\tilde\eta$ on M such that

$$\pi_*\circ J = \varphi\circ\pi_* - \tilde\eta\otimes\xi .$$

Denoting by ξ^* the horizontal lift of ξ, it is easy to verify that $J\xi^*$ is a vertical vector field and $\tilde\eta$ is its dual 1-form. Moreover, the following result can be easily proved.

Proposition 4.5 *Let $\pi : M \to B$ be a complex-contact Riemannian submersion. Then one has:*

1) $J(\mathcal{V}) \subset \mathcal{V}\oplus$ span $\{\xi^*\}$,
2) $J(\mathcal{H}) \subset \mathcal{H}\oplus$ span $\{J\xi^*\}$,

3) *If X is the basic vector field on M, π-related to the vector field X' on B, then $JX - g(X,\xi^*)J\xi^*$ is the basic vector field π-related to $\varphi X'$.*

For any vertical vector V we write:

$$JV = \widehat{\varphi}V - \widehat{\eta}(V)\xi^* ,$$

obtaining a field $\widehat{\varphi}$ of endomorphisms on the vertical distribution $\mathcal{V}$ and a 1-form $\widehat{\eta}$ on $\mathcal{V}$ as the restriction of $\tilde{\eta}$. Considering a fibre F of M, putting $\widehat{\xi} = (J\xi^*)_{|F}$, $(\widehat{\varphi}, \widehat{\eta}, \widehat{\xi})$ is an almost contact structure on F with the restriction of g as an associated metric. Therefore, F is a semi-invariant submanifold of M with respect to the vector field ξ^*, ([36]). Furthermore, the fundamental 2-forms on M and B are related by $\Omega = \pi^*\Phi$ and $d\Omega = \pi^* d\Phi$, on the horizontal distribution. By a long calculation, via Proposition 4.5, we obtain the following formulas:

$$\begin{aligned}(\nabla'_{X'}\Phi)(Y',Z')\circ\pi = {}& (\nabla_X\Omega)(Y,Z) + g(Y,\xi^*)g(\nabla_X J\xi^*, Z) \\ & -g(Z,\xi^*)g(Y,\nabla_X J\xi^*) ,\end{aligned} \tag{4.15}$$

$$\begin{aligned}(\nabla_{J\xi^*}\Omega)(X,Y) = {}& g(X,\nabla_{JY}J\xi^*) + g(JX,\nabla_Y J\xi^*) \\ & -g(X,\xi^*)g(\nabla_{J\xi^*}J\xi^*, Y) ,\end{aligned} \tag{4.16}$$

$$(\nabla_{J\xi^*}\Omega)(J\xi^*, X) = g(\nabla_{\xi^*}J\xi^*, X) - g(\nabla_{J\xi^*}J\xi^*, JX) , \tag{4.17}$$

for any X, Y, Z basic vector fields on M π-related to X', Y', Z', respectively.

Theorem 4.7 *Let $\pi : M \to B$ be a complex-contact Riemannian submersion. Then, we have:*

1) *If M is a K-manifold and ξ^* is Killing, then any fibre is a C-manifold.*
2) *If M is an NK-manifold and ξ^* is Killing and analytic, then any fibre is an NKC-manifold.*
3) *If M is an H-manifold and ξ^* is analytic, then any fibre is an N-manifold.*

Proof. We suppose M an NK-manifold and consider a fibre F with the induced ACM-structure. By calculating the Levi-Civita connection $\widehat{\nabla}$ on F, we have:

$$2(\widehat{\nabla}_V\widehat{\Phi})(V,W) = \widehat{\eta}(W)(L_{\xi^*}g)(V,V) - \widehat{\eta}(V)(L_{\xi^*}g)(V,W) , \tag{4.18}$$

$$
\begin{aligned}
2(\widehat{\nabla}_V\widehat{\eta})(W) = {} & g((L_{\xi^*}J)W,V) - g((L_{\xi^*}J)V,W) \\
& +2\widehat{\eta}(V)(L_{\xi^*}g)(\xi^*,W) - 2\widehat{\eta}(W)(L_{\xi^*}g)(\xi^*,V) \\
& -(L_{\xi^*}g)(V,\widehat{\varphi}W) ,
\end{aligned}
\tag{4.19}
$$

for any vertical vector fields V, W on M. Since ξ^* is Killing and analytic we obtain $(\widehat{\nabla}_V\widehat{\Phi})(V,W) = 0$ and $\widehat{\nabla}\widehat{\eta} = 0$, which are equivalent to $(\widehat{\nabla}_V\widehat{\varphi})V = 0$ and $\widehat{\nabla}_V\widehat{\xi} = 0$, so that any fibre is an NKC-manifold. In the case 1), a formula analogous to (4.18) implies $(\widehat{\nabla}_V\widehat{\varphi})W = 0$ for any V, W vertical vector fields, and any fibre is a C-manifold. If M is an H-manifold, by a long calculation, we have:

$$
\begin{aligned}
& (\widehat{\nabla}_U\widehat{\Phi})(V,W) - (\widehat{\nabla}_{\widehat{\varphi}U}\widehat{\Phi})(\widehat{\varphi}V,W) - \widehat{\eta}(V)(\widehat{\nabla}_{\widehat{\varphi}U}\widehat{\eta})W \\
& = \tfrac{1}{2}\{\widehat{\eta}(U)(g((L_{\xi^*}J)JV,W) - g((L_{\xi^*}J)W,JV)) \\
& \quad +\widehat{\eta}(W)(g(L_{\xi^*}J)U,JV) + g(JU,(L_{\xi^*}J)V)) \\
& \quad -\widehat{\eta}(V)(g((L_{\xi^*}J)W,JU) + g(JW,(L_{\xi^*}J)U))\} ,
\end{aligned}
\tag{4.20}
$$

for any U, V, W vertical vector fields. Since ξ^* is analytic, we get the normality.

Theorem 4.8 *Let $\pi : M \to B$ be a complex-contact Riemannian submersion such that the total space M belongs to any of the classes H, K, AK and $J\xi^*$ is parallel. Then, the base space B belongs to the classes N, C, AC, respectively. Furthermore, if M is an NK-manifold and ξ^* is parallel, then B is NKC.*

Proof. Let X, Y, Z be basic vector fields on M, π-related to X', Y', Z' on B. From (4.15), since $J\xi^*$ is parallel, we obtain:

$$
(\nabla'_{X'}\Phi)(Y',Z') \circ \pi = (\nabla_X\Omega)(Y,Z) ,
$$

so that, if M is Kähler, then B is cosymplectic. From (4.15) – (4.17), using that ξ^* is parallel, we have:

$$
\begin{aligned}
& ((\nabla'_{X'}\Phi)(Y',Z') - (\nabla_{\varphi'X'}\Phi)(\varphi Y',Z') - \eta(Y')(\nabla_{\varphi'X'}\Phi)(\xi,\varphi Z')) \circ \pi \\
& = (\nabla_X\Omega)(Y,Z) - (\nabla_{JX}\Omega)(JY,Z) .
\end{aligned}
$$

Thus, if M is an H-manifold, then B is an N-manifold. The other cases can be proved in a similar way.

Finally, we consider on the total space (M, J, g) a conformally related metric $g_c = e^{2\sigma}g$, where σ is a non-constant differentiable function on M satisfying the conditions: $J\xi^*(\sigma) = 1$ and $X(\sigma) = 0$ for any horizontal vector field on M. Then, we have the following result ([67]).

Theorem 4.9 *Let $\pi : M \to B$ be a complex-contact Riemannian submersion. If g_c is a (almost) Kähler metric on M, then B is a (almost) Sasakian manifold.*

Example 4.3 Let $(M, \varphi, \xi, \eta, g)$ be a $(2n+1)$-dimensional almost contact metric manifold and $\pi : TM \to M$ its tangent bundle. We consider the Sasakian metric g^D on TM and the tensor field J on TM defined by:

$$J = \varphi^H + \eta^V \otimes \xi^V - \eta^H \otimes \xi^H ,$$

where the indexes H and V mean horizontal and vertical lifts, respectively ([284; 347]). Then, the following relations hold:

$$JX^V = (\varphi X)^V - \eta(X)^V \xi^H , \qquad JX^H = (\varphi X)^H + \eta(X)^V \xi^V .$$

Moreover, it is well known that $\pi : (TM, g^D) \to (M, g)$ is a Riemannian submersion and, using the above relations, one has $\pi_* \circ J = \varphi \circ \pi_* - \eta^H \otimes \xi$, so that π is a complex-contact Riemannian submersion whose fibres are totally geodesic and semi-invariant submanifolds of (TM, J, g^D).

Example 4.4 Let $(N, \varphi, \xi, \eta, g)$ and $(N', \varphi', \xi', \eta', g')$ be almost contact metric manifolds. On the product manifold $N \times N'$ we define an almost Hermitian structure $(\tilde{J}, \tilde{g})$ setting $\tilde{g} = g + g'$ and

$$\tilde{J}(X, X') = (\varphi X + \eta'(X')\xi, \varphi' X' - \eta(X)\xi') ,$$

for any vector fields X, X' on N, N' respectively ([206]). Then, the second projection $\pi_2 : N \times N' \to N'$ is a complex-contact Riemannian submersion and N is a semi-invariant submanifold of $N \times N'$ with respect to ξ'.

Example 4.5 The canonical projection

$$\pi : \mathbf{S^6} \times \mathbf{R^{2n+1}} \times \mathbf{R} \to \mathbf{S^6} \times \mathbf{R}$$

is a complex-contact Riemannian submersion from a nearly Kähler manifold to a nearly-K-cosymplectic manifold.

Example 4.6 The canonical projection

$$\pi : \mathbf{S^{2m+1}} \times \mathbf{S^{2n+1}} \times \mathbf{R^{2k+1}} \times \mathbf{R} \to \mathbf{S^{2m+1}} \times \mathbf{S^{2n+1}} \times \mathbf{R}$$

is a complex-contact Riemannian submersion from a Hermitian manifold to a normal manifold ([68]).

4.5 Curvature Properties

Let $\pi : M \to M'$ be a contact Riemannian submersion, where M and M' are almost contact metric manifolds with structures (φ, ξ, η, g) and $(\varphi', \xi', \eta', g')$, respectively. We denote by B the *φ-holomorphic bisectional curvature*, defined for any pair of nonzero vectors E and F on M orthogonal to ξ by the formula:

$$B(E,F) = \frac{R(E,\varphi E, F, \varphi F)}{||E||^2||F||^2} .$$

The *φ-holomorphic sectional curvature* is $H(E) = B(E,E)$ for any nonzero vector E orthogonal to ξ.

The first properties of these curvatures have been studied by E. Moskal in his unpublished thesis, K. Ogiue ([224]), S. Tanno ([284]) and others. D. Chinea ([63; 64]) and B. Watson ([330]) studied the curvature relations for some remarkable classes of Riemannian submersions when the total space or/and the base space is an almost contact metric manifold. We denote by B' and H' the φ-holomorphic bisectional and φ-holomorphic sectional curvatures of M'. Similarly, $\widehat{B}$ and $\widehat{H}$ denote the bisectional and the sectional holomorphic curvatures of a fibre.

Proposition 4.6 *Let $\pi : M \to M'$ be a contact Riemannian submersion. Let V and U be unit vertical vectors, and X and Y unit horizontal vectors orthogonal to ξ. Then, one has:*

1) $B(V,U) = \widehat{B}(V,U) + g(T_V \widehat{J}U, T_{\widehat{J}V}U) - g(T_V U, T_{\widehat{J}V}\widehat{J}U)$,

2) $$\begin{aligned} B(X,V) = {} & g((\nabla_V A)(X,\varphi X), \widehat{J}V) - g(A_X \widehat{J}V, A_{\varphi X} V) \\ & + g(A_X V, A_{\varphi X}\widehat{J}V) - g((\nabla_{\widehat{J}V} A)(X,\varphi X), V) \\ & - g(T_V X, T_{\widehat{J}V}\varphi X) + g(T_{\widehat{J}V} X, T_V \varphi X) , \end{aligned}$$

3) $$\begin{aligned} B(X,Y) = {} & B'(\pi_* X, \pi_* Y) - 2g(A_X \varphi X, A_Y \varphi Y) \\ & + g(A_{\varphi X} Y, A_X \varphi Y) - g(A_X Y, A_{\varphi X}\varphi Y) , \end{aligned}$$

4) $H(V) = \widehat{H}(V) + \| T_V \widehat{J}V \|^2 - g(T_V V, T_{\widehat{J}V}\widehat{J}V)$,

5) $H(X) = H'(\pi_* X) - 3 \| A_X \varphi X \|^2$.

Proof. These formulas are a direct application of the O'Neill formulas stated in Sec. 1.3.

The above formulas simplify when the total space belongs to suitable classes ([63; 64; 330]).

Theorem 4.10 *Let $\pi : M \to M'$ be a contact Riemannian submersion with M quasi-Sasakian. Then, we have:*

$$\begin{aligned} &\text{a)}\ B(V,U) = \widehat{B}(V,U) + 2\|T_V U\|^2\ ,\\ &\text{b)}\ B(X,Y) = -2(\|T_V X\|^2 + \|A_X V\|^2)\ ,\\ &\text{c)}\ B(X,Y) = B'(\pi_* X, \pi_* Y) - 2\|A_X Y\|^2\ , \end{aligned}$$

where X, Y are unit horizontal vectors orthogonal to ξ and V, U are unit vertical vectors.

Proof. We have $T_V \widehat{J}U = T_{\widehat{J}V} U = \varphi T_V U$ and then $T_{\widehat{J}V} \widehat{J}U = -T_V U$ and this implies a). Now, for any basic vector field X, we have:

$$\begin{aligned} (\nabla_V A)(X, \varphi X) &= \nabla_V (A_X \varphi X) - A_{\nabla_V X}(\varphi X) - A_X(\nabla_V \varphi X)\\ &= -A_{h\nabla_V X}(\varphi X) - A_X(h\nabla_V \varphi X) - A_X(v\nabla_V \varphi X)\\ &= -A_{A_X V}(\varphi X) - A_X A_{\varphi X} V - A_X T_V(\varphi X)\\ &= A_{\varphi X} A_X V - A_X A_{\varphi X} V - \varphi A_X T_V X\\ &= 2\widehat{J}(A_X A_X V) - \varphi A_X T_V X\ . \end{aligned}$$

Similarly, $(\nabla_{\widehat{J}V} A)(X, \varphi X) = 2\widehat{J}(A_X A_X \widehat{J}V) - \varphi(A_X T_{\widehat{J}V} X)$, so that the sum of the first and the fourth terms on the right in relation 2) of Proposition 4.6 gives $-4\|A_X V\|^2$. On the other hand, it is easy to prove that $A_X \varphi Y = \varphi A_X Y$ and $T_V \varphi X = \widehat{J} T_V X$, and b) follows. Finally c) follows from 3) and the properties of A.

Corollary 4.2 *Let $\pi : M \to M'$ be a contact Riemannian submersion, where M is a quasi-Sasakian manifold. Then:*

$$B(X,Y) \le B'(\pi_* X, \pi_* Y)\ ,$$

for any horizontal vectors X, Y orthogonal to ξ and the equality holds if and only if the horizontal distribution is integrable. In particular, the equality holds if M is cosymplectic.

Theorem 4.11 *Let $\pi : M \to M'$ be a contact Riemannian submersion with M quasi-Sasakian. Then,*

$$\begin{aligned} &\text{a)}\ H(V) = \widehat{H}(V) + 2\|T_V V\|^2\ ,\\ &\text{b)}\ H(X) = H'(\pi_* X)\ , \end{aligned}$$

for any unit vertical vector V and any horizontal vector X orthogonal to ξ. Consequently, if M is a quasi-Sasakian manifold of constant φ-holomorphic sectional curvature c, then M' has the same property.

Proposition 4.7 *Let $\pi : M \to B$ be a contact-complex Riemannian submersion. Then we have:*

$$1)\ B(V,U) = \widehat{B}(V,U) + g(T_V\varphi U, T_U\varphi V) - g(T_VU, T_{\varphi V}\varphi U)\ ,$$

$$\begin{aligned}2)\ B(X,V) = {} & g((\nabla_V A)(X,\varphi X),\varphi V) - g(A_X\varphi V, A_{\varphi X}V) \\ & + g(A_XV, A_{\varphi X}\varphi V) - g((\nabla_{\varphi V}A)(X,\varphi X),V) \\ & + g(T_{\varphi V}X, T_V\varphi X) - g(T_VX, T_{\varphi V}\varphi X)\ ,\end{aligned}$$

$$\begin{aligned}3)\ B(X,Y) = {} & B'(\pi_*X,\pi_*Y) - \{2g(A_X\varphi X, A_Y\varphi Y) \\ & - g(A_{\varphi X}Y, A_X\varphi Y) + g(A_XY, A_{\varphi X}\varphi Y)\}\ ,\end{aligned}$$

$$4)\ H(V) = \widehat{H}(V) + ||T_V\varphi V||^2 - g(T_VV, T_{\varphi V}\varphi V)\ ,$$

$$5)\ H(X) = H'(\pi_*X) - 3||A_X\varphi X||^2\ .$$

Theorem 4.12 *Let $\pi : M \to B$ be a contact-complex Riemannian submersion whose total space is a Sasakian manifold. Then, for any unit horizontal vector X and unit vertical vectors U, V orthogonal to ξ, one has:*

$$\begin{aligned}&1)\ B(U,V) = \widehat{B}(U,V) + 2||T_UV||^2\ ,\\ &2)\ B(X,V) = -2(1 + ||T_VX||^2 + ||A_XV||^2)\ ,\\ &3)\ H(V) = \widehat{H}(V) + 2||T_VV||^2\ ,\\ &4)\ H(X) = H'(\pi_*X) - 3\ .\end{aligned}$$

Corollary 4.3 *Let $\pi : M \to B$ be a contact-complex Riemannian submersion. If the total space M is a Sasakian manifold of constant φ-holomorphic sectional curvature c, then B has constant holomorphic sectional curvature $c + 3$.*

4.6 Riemannian Submersions Involving Quaternionic Kähler and 3-Sasakian Manifolds

We begin with some basic definitions about the differential geometry of $4n$-dimensional manifolds whose tangent spaces have a quaternionic structure. Such manifolds often appear unexpectedly, but naturally, in many different areas of mathematics and mathematical physics. This confirms the conviction of Hamilton (who introduced them) that the quaternions will play an important role in mathematical physics ([7; 143]). We also consider $(4n+3)$-dimensional manifolds which carry an almost contact 3-structure. These manifolds are linked to the quaternionic ones as the almost contact manifolds are linked to the almost complex ones.

Definition 4.8 An *almost hypercomplex structure* on a $4n$-dimensional manifold M is a triplet $\mathcal{J} = (J_i)_{i\in\{1,2,3\}}$ of almost complex structures satisfying the quaternionic-like relations, *i.e.*:

$$J_i^2 = -I \ , \quad J_i \circ J_j = -J_j \circ J_i = J_k \ , \tag{4.21}$$

for any cyclic permutation (i,j,k) of $(1,2,3)$. If each J_i is integrable, we say that $\mathcal{J}$ is a *hypercomplex structure* on M. Given a Riemannian metric g compatible with each J_i, we say that M is endowed with a (almost) *hyper-Hermitian* structure. An almost hyper-Hermitian manifold is called *hyper-Kähler* if its reduced holonomy group is a subgroup of $Sp(n)$, or, equivalently, if each J_i is parallel with respect to the Levi-Civita connection of g.

Definition 4.9 An *almost quaternionic structure* on a $4n$-dimensional manifold is a subbundle $\mathcal{Q} \subset End(T(M))$ of rank 3 which is locally spanned by almost hypercomplex structures $(J_i)_{1\leq i\leq 3}$. In this case the bundle $\mathcal{Q}$ has the structure group $SO(3)$.

Definition 4.10 A Riemannian manifold (M,g) of dimension $4n$ is an *almost quaternionic metric* manifold if $End(T(M))$ admits a subblunde $\mathcal{Q}$ of rank 3, locally spanned by almost Hermitian structures $(J_i)_{i\in\{1,2,3\}}$ such that:

a) $J_i^2 = -I \ , \quad J_i \circ J_j = -J_j \circ J_i = J_k$ for any cyclic permutation (i,j,k) of $(1,2,3)$.

(M,g) is said to be *quaternionic Kähler* if the subbundle $\mathcal{Q}$ also satisfies:

b) if ψ is a local section of $\mathcal{Q}$ and X is a tangent vector field on M, then $\nabla_X\psi$ is a local section of $\mathcal{Q}$, ∇ denoting the Levi-Civita connection of g .

We remark that condition b) is equivalent to:

b') In any coordinate neighborhood U of M there exist local 1-forms $\{\alpha_{ij}\}$, $i,j\in\{1,2,3\}$, such that $\nabla_X J_i = \alpha_{ij}(X)J_j + \alpha_{ik}J_k$ for any cyclic permutation (i,j,k) of $(1,2,3)$ and $\alpha_{ij}+\alpha_{ji}=0$.

Moreover, if U and U' are coordinate neighborhoods with $U \cap U' \neq \emptyset$, then in $U \cap U'$ one has $J'_i = \sum_{j=1}^{3} c_{ij} J_j$, where the matrix (c_{ij}) belongs to $SO(3)$ and $\{J'_i\}$ is a local basis for $\mathcal{Q}$ in U'. As proved by D. Alekseevski in [1], the holonomy group of a quaternionic Kähler manifold is $Sp(n)Sp(1) = Sp(n) \times Sp(1)/\{\pm I\}$.

Definition 4.11 A $(4n+3)$-dimensional Riemannian manifold (M, g), $n \geq 1$, is an *almost 3-contact metric manifold* if M admits three almost contact structures $(\varphi_i, \xi_i, \eta^i)_{1 \leq i \leq 3}$, each compatible with the metric g and satisfying the relations:

$$\begin{aligned} &\varphi_i \circ \varphi_j - \eta^j \otimes \xi_i = -\varphi_j \circ \varphi_i + \eta^i \otimes \xi_j = \varphi_k \,, \\ &\eta^i(\xi_j) = \delta^i_j \,, \varphi_i(\xi_j) = -\varphi_j(\xi_i) = \xi_k \,, \\ &\eta^i \circ \varphi_j = -\eta^j \circ \varphi_i = \eta^k \,, \end{aligned} \tag{4.22}$$

for any cyclic permutation (i, j, k) of $(1, 2, 3)$.

The structure group of an almost 3-contact metric manifold is $Sp(n) \times \{I_3\}$, where I_3 denotes the identity matrix of order 3 ([178]).

Definition 4.12 A set of three almost contact structures $(\varphi_i, \xi_i, \eta^i)_{1 \leq i \leq 3}$ satisfying (4.22) is called an *almost contact 3-structure*. An almost contact 3-structure is said to be *hypernormal* if the three almost contact structures are normal, *i.e.* $N_{\varphi_i} + 2d\eta^i \otimes \xi_i = 0$, $i \in \{1, 2, 3\}$. Let Φ_i be the 2-form given by $\Phi_i(X, Y) = g(X, \varphi_i Y)$, $i \in \{1, 2, 3\}$. When each $(\varphi_i, \xi_i, \eta^i)$ is a (almost) cosymplectic structure, then we say that $(\varphi_i, \xi_i, \eta^i)_{1 \leq i \leq 3}$ is a *(almost) cosymplectic 3-structure*. An almost contact metric 3-structure is called a *contact metric 3-structure* if $\Phi_i = d\eta^i$ for each $i \in \{1, 2, 3\}$. Moreover, a contact metric 3-structure is called a *K-contact 3-structure* if each ξ_i is Killing, whereas it is called a *Sasakian 3-structure* if it is hypernormal and in this case M is said to be a *3-Sasakian manifold*. Finally, an almost contact metric 3-structure is said to be a *quasi-Sasakian 3-structure* if it is hypernormal and the three fundamental 2-forms Φ_i, $1 \leq i \leq 3$ are closed.

Every almost 3-contact metric manifold possesses a local orthonormal basis:

$$(e_1, \dots, e_n, \varphi_i(e_1), \dots, \varphi_i(e_n), \xi_1, \xi_2, \xi_3) \,, \quad i \in \{1, 2, 3\} \,.$$

Example 4.7 Let M^{4n+4} be an almost quaternionic metric manifold. Then, any orientable hypersurface N^{4n+3} of M^{4n+4} admits an almost contact 3-structure ([303]). In particular $\mathbf{S^{4m+3}}$ and $\mathbf{P_{4m+3}(R)} \simeq \mathbf{S^{4m+3}/Z_2}$ are 3-Sasakian manifolds.

Example 4.8 Any parallelizable $(4n+3)$-dimensional manifold has a natural almost contact metric 3-structure. In particular the Lie groups $SU(n)$ and $SL(n, \mathbf{R})$ admit such structures for any even number n. The group $SO(n)$ admits such structure for $n = 3 + 8k$ and $n = 6 + 8k$, $k \in \mathbf{N}$.

Example 4.9 Let $H(p, r)$ be the generalized Heisenberg group. It is a nilpotent, connected Lie group of dimension $p(r+1)+r$. For $p = 3$, $H(3, r)$ is a $(4r+3)$-dimensional Lie group. Some invariant almost contact metric 3-structures have been studied by Monar ([205]).

Example 4.10 H. Yanamoto ([344]) defined an almost contact metric 3-structure on a 7-dimensional Riemannian manifold using the vector cross product ([127]). In the particular case of the 7-dimensional standard sphere, D. Monar proved that the canonical almost contact metric 3-structure does not belong to any of the classes above defined ([205]). A classification of 7-dimensional Riemannian manifolds with vector cross product is given in [98].

A 2-*fold vector cross product* on a Riemannian manifold (M, g) is a $(1,2)$-tensor field $P : \mathcal{X}(M) \times \mathcal{X}(M) \to \mathcal{X}(M)$ satisfying:

1) $P(X,Y) = -P(Y,X)$,
2) $g(P(X,Y),Z) = g(X,P(Y,Z))$,
3) $P(P(X,Y),Z) + P(X,P(Y,Z)) = 2g(X,Z)Y - g(Y,Z)X - g(X,Y)Z$,

for any $X, Y, Z \in \mathcal{X}(M)$. An equivalent definition is given in [127]. Now, we briefly describe the Yanamoto construction. Let P be a vector cross product on a 7-dimensional Riemannian manifold (M^7, g) and suppose that there exist two orthonormal vector fields ξ_1, ξ_2. We construct two almost contact metric structures $(\varphi_i, \xi_i, \eta^i, g)$, $i \in \{1, 2\}$, putting, for any vector field X,

$$\varphi_i(X) = P(\xi_i, X) \ , \quad \eta^i(X) = g(X, \xi_i) \ . \tag{4.23}$$

Then, we obtain an almost contact metric 3-structure on (M^7, g) defining $(\varphi_3, \xi_3, \eta^3)$ by:

$$\varphi_3 = \varphi_1 \circ \varphi_2 - \eta^2 \otimes \xi_1 \ , \quad \xi_3 = \varphi_1(\xi_2) \ , \quad \eta^3 = \eta^1 \circ \varphi_2 \ . \tag{4.24}$$

Conversely, let $(\varphi_i, \xi_i, \eta^i, g)_{1 \le i \le 3}$, be an almost contact metric 3-structure on M^7. We can define a vector cross product in each tangent space T_pM, $p \in M$. Indeed, let $X \in T_pM$ be a unit vector and consider an orthonormal basis $\{X, \varphi_1 X, \varphi_2 X, \varphi_3 X, \xi_1, \xi_2, \xi_3\}$ in T_pM. It is easy to check that the action of P defined by the following table:

P	X	$\varphi_1 X$	$\varphi_2 X$	$\varphi_3 X$	ξ_1	ξ_2	ξ_3
X	0	ξ_1	ξ_2	ξ_3	$-\varphi_1 X$	$-\varphi_2 X$	$\varphi_3 X$
$\varphi_1 X$	$-\xi_1$	0	$-\xi_3$	$-\xi_2$	X	$\varphi_3 X$	$\varphi_2 X$
$\varphi_2 X$	$-\xi_2$	ξ_3	0	ξ_1	$-\varphi_3 X$	X	$-\varphi_1 X$
$\varphi_3 X$	$-\xi_3$	ξ_2	$-\xi_1$	0	$\varphi_2 X$	$-\varphi_1 X$	$-X$
ξ_1	$\varphi_1 X$	$-X$	$\varphi_3 X$	$-\varphi_2 X$	0	ξ_3	$-\xi_2$
ξ_2	$\varphi_2 X$	$-\varphi_3 X$	$-X$	$\varphi_1 X$	$-\xi_3$	0	ξ_1
ξ_3	$-\varphi_3 X$	$-\varphi_2 X$	$\varphi_1 X$	X	ξ_2	$-\xi_1$	0

is a vector cross product on T_pM^7 and one obtains a vector cross product on the Riemannian manifold (M^7, g). Moreover, if P is a parallel tensor field (or nearly parallel, almost parallel, semi-parallel ([127])), then the corresponding structures are cosymplectic, nearly cosymplectic, semi-cosymplectic, respectively ([205]).

Example 4.11 The following homogeneous spaces:

$$Sp(n)/Sp(n-1) \cong \mathbf{S^{4n-1}} \; ; \quad Sp(n)/(Sp(n-1)\times \mathbf{Z_2}) \cong \mathbf{P_{4n-1}(R)} \; ;$$

$$SU(m)/S(U(m-2)\times U(1)) \; ; \quad SO(k)/(SO(k-4)\times Sp(1)) \; ;$$

$$G_2/Sp(1) \; ; \; F_4/Sp(3) \; ; \; E_6/SU(6) \; ; \; E_7/Spin(12) \; ; \; E_8/E_7 \; ,$$

for $n \geq 1$, $m \geq 3$ and $k \geq 7$ are the only 3-Sasakian homogeneous spaces, *i.e.* the only 3-Sasakian manifolds with transitive action of the group of automorphisms of the Sasakian 3-structure ([53]).

We report the following Lemma.

Lemma 4.1 **(Hitchin)** *Let (M, g) be a $4n$-dimensional Riemannian manifold and $\{J_i\}_{1\leq i\leq 3}$ an almost hyper-Hermitian structure on M. If each fundamental 2-form Ω_i, $1 \leq i \leq 3$ is closed, then each J_i is integrable.*

Theorem 4.13 *Let $\mathcal{F} = (\varphi_i, \xi_i, \eta^i)$, $i \in \{1, 2, 3\}$ be an almost contact metric 3-structure on a $(4n+3)$-dimensional Riemannian manifold (M, g).*

1) *If $\mathcal{F}$ is a contact metric 3-structure, then it is a Sasakian 3-structure;*
2) *if $\mathcal{F}$ is an almost cosymplectic 3-structure, then each $(\varphi_i, \xi_i, \eta^i)$ is a cosymplectic structure, i.e. $\mathcal{F}$ is a cosymplectic 3-structure.*

Proof. On the product manifold $\tilde{M} = M \times \mathbf{R}$ we consider the three almost Hermitian structures $(J_i, \tilde{g})$ associated with the contact metric structures $(\varphi_i, \xi_i, \eta^i, g)$ for any $i \in \{1, 2, 3\}$ as in Sec. 4.1, formulas (4.4), (4.8). Using (4.22) we can easily check that $(J_i)_{1 \leq i \leq 3}$ satisfies (4.21). Now, take the metric $\tilde{g}_c = e^{2\sigma}\tilde{g}$, defined in (4.9), which is also compatible with each J_i, as we mentioned in Sec. 4.1. By a simple computation, one can prove that the fundamental 2-forms $\tilde{\Omega}_i$ associated with $(J_i, \tilde{g}_c)$ are closed. Consequently, the three contact metric structures $(\varphi_i, \xi_i, \eta^i)$, $1 \leq i \leq 3$, are normal, *i.e.* $\mathcal{F}$ is a Sasakian 3-structure. In the almost cosymplectic case, it is easy to verify that the three fundamental 2-forms Ω_i associated with $(J_i, \tilde{g})$ are closed, so that the three almost cosymplectic structures are normal.

We remark that part 1) in the previous theorem is a result of T. Kashiwada ([164]).

Proposition 4.8 *Let $\pi : (M, g) \to (B, g')$ be a Riemannian submersion and suppose that $\mathcal{F} = (\varphi_i, \xi_i, \eta^i)_{1 \leq i \leq 3}$ is a 3-structure on M compatible with g and that $(J_i)_{1 \leq i \leq 3}$ is an almost quaternionic structure on B compatible with g'. Assume also that $\pi_* \circ \varphi_i = J_i \circ \pi_*$, $1 \leq i \leq 3$. Then:*

a) *the vertical and horizontal distributions are invariant under each φ_i ;*
b) *ξ_1, ξ_2, ξ_3 are vertical vector fields;*
c) *$\eta^i(X) = 0$ for any horizontal vector field X and $i \in \{1, 2, 3\}$;*
d) *$d\eta^i(X, Y) = -\frac{1}{2}\eta^i(v[X, Y])$ for any horizontal vector fields X, Y and for any $i \in \{1, 2, 3\}$.*

Proof. A standard computation gives the statement.

Theorem 4.14 **(Watson).** *Let $\pi : M \to B$ be a Riemannian submersion from a 3-quasi Sasakian manifold to an almost quaternionic metric manifold. If π_* commutes with each tensor field φ_i, then B is a quaternionic Kähler manifold and the fibres are minimal.*

Proof. Proposition 4.8 easily implies $h(\nabla_X \varphi_i)Y = 0$ for any basic vector fields X, Y and $i \in \{1, 2, 3\}$. Then, if X, Y are π-related to X', Y', we have $(\nabla'_{X'} J_i)Y' = 0$. Now, the statement follows from a result of A. Gray ([126]) stating that the quaternionic Kähler property for B can be characterized by the vanishing of $\sigma(\nabla'_{X'} J_1)Y' \wedge J_2 Y' \wedge J_3 Y'$, σ denoting the cyclic sum over $\{1, 2, 3\}$. Finally, the technique used in proving Theorem 4.6 gives the minimality of the fibres.

One of the first important results in 3-Sasakian geometry is the following theorem of Tanno ([285]), where $\mathbf{S^3}(1)$ denotes the unit sphere with the Sasakian structure of constant φ-holomorphic sectional curvature 1 and $\mathbf{P_3}(\mathbf{R})(1) = \mathbf{S^3}(1)/\{I, -I\}$.

Theorem 4.15 *Let (M, g) be a complete Riemannian manifold admitting a K-contact 3-structure $(\varphi_i, \xi_i, \eta^i, g)_{1\leq i\leq 3}$. If one of the characteristic vector fields ξ_i is regular, then M is a principal G-bundle, ($G = \mathbf{S^3}(1)$ or $G = \mathbf{P_3}(\mathbf{R})(1)$), over a Riemannian manifold (M', g') such that g and g' are related by:*

$$g(X,Y) = g'(\pi_* X, \pi_* Y) \circ \pi + \sum_{i=1}^{3} g(\xi_i, X) g(\xi_i, Y) ,$$

for any $X, Y \in \mathcal{X}(M)$. The $\mathfrak{g}$-valued 1-form ω defined by:

$$\omega(X) = \sum_{i=1}^{3} g(\xi_i, X) \xi_i ,$$

is an infinitesimal connection form on M. Here each ξ_i is the left invariant vector field on G corresponding in $\mathfrak{g}$ to the characteristic vector field ξ_i. Moreover, g' is an Einstein metric on M'.

On the other hand, in [257] K. Sakamoto showed that there exists a principal $SO(3)$-bundle over any quaternionic Kähler manifold. Then, S. Ishihara and M. Konishi proved that there exists a principal $SO(3)$-bundle M with a K-contact 3-structure (*i.e.* a 3-Sasakian structure) over any quaternionic Kähler manifold with positive scalar curvature ([152; 174]). Recently, in [286; 155], S. Tanno and W. Jelonek extended these results to the case in which the base space of the principal bundle is a quaternionic Kähler manifold with negative scalar curvature. In such a case, they showed that the total space of the principal bundle carries a triplet of K-contact structures satisfying similar relations as those characterizing K-contact 3-structures (*i.e.* 3-Sasakian structures). Of course, the projection is a Riemannian submersion.

A good recent survey on 3-Sasakian geometry is the paper of C. Boyer and K. Galicki ([51]). We quote, now, one of the important results of Boyer, Galicki and Mann ([53]).

Theorem 4.16 *Let $(M, \varphi_i, \xi_i, \eta^i, g)$ be a 3-Sasakian manifold of dimension $4n + 3$ such that the Killing vector fields ξ_i are complete. Then:*

i) *M is an Einstein manifold of positive scalar curvature* $2(2n+1)(4n+3)$ *;*

ii) *M admits a second Einstein metric g', of positive scalar curvature, which is not homothetic to g ;*

iii) *the metric g is bundle-like with respect to the foliation $\mathcal{L}$ spanned by $\{\xi_1, \xi_2, \xi_3\}$;*

iv) *each leaf of the foliation $\mathcal{L}$ is a 3-dimensional homogeneous spherical space form;*

v) *the space of the leaves $M/\mathcal{L}$ is a quaternionic Kähler orbifold of dimension $4n$ with positive scalar curvature* $16n(n+2)$.

Hence, every complete 3-Sasakian manifold is compact with finite fundamental group and diameter $\delta \leq \pi$.

Results from the topological point of view have been given by K. Galicki and S. Salamon, who proved that the odd Betti numbers b_{2k+1} of a compact 3-Sasakian manifold of dimension $4n+3$ vanish for $0 \leq k \leq n$ ([103]). Furthermore, R. Bielawski studied the Betti numbers of 3-Sasakian quotients of spheres by tori ([29]).

4.7 Regular f-structures and Riemannian Submersions

As we saw in the previous section, under regularity conditions, 3-Sasakian spaces fiber over quaternionic Kähler manifolds ([53; 185]), and a quaternionic Kähler manifold with positive scalar curvature is the base space of a principal $SO(3)$-bundle whose total space admits a metric with an associate 3-Sasakian structure ([154]).

Now, we present some results, due to G. Hernandez ([141]), motivated by the problem of finding suitable structures on manifolds to obtain fibrations on hypercomplex or hyperKähler manifolds, possibly with totally geodesic fibres. We begin reviewing the basic data on the f-structures introduced by Yano ([346]).

Definition 4.13 An *f-structure* on an m-dimensional manifold is a non-vanishing $(1,1)$-tensor field f having constant rank and satisfying the relation $f^3 + f = 0$.

Obviously, $T(M)$ splits as the direct sum of the subbundles $\operatorname{im} f$ and $\ker f$ and f acts as an almost complex structure on $\operatorname{im} f$. Thus, the rank of f is even, say $2n$, and $m = 2n + s$, $s \geq 0$ being the dimension of $\ker f$

at any point in M. The existence of an f-structure is also equivalent to a reduction of the structural group of $T(M)$ to $U(n) \times O(s)$ ([346]). If $s = 1$ and $T(M)$ reduces to $U(n) \times \{1\}$, we obtain an almost contact structure ([131]), and it is well known that this is equivalent to the existence of a never vanishing vector field ξ such that $f^2 = -I + \eta \otimes \xi$, η being the 1-form dual to ξ ([33]).

The analogous situation, for $s \geq 2$, occurs when one supposes that the subbundle $\ker f$ is parallelizable, giving a *pk.f-structure*, also called a *globally framed f-structure* ([123]). In this case there exist global vector fields $\xi_1, \ldots, \xi_s$ linearly independent at any point, and dual 1-forms $\eta^1, \ldots, \eta^s$ such that:

$$f^2 = -I + \sum_{i=1}^{s} \eta^i \otimes \xi_i \ ,$$

which easily implies $\eta^i \circ f = 0$, for any $i \in \{1, \ldots, s\}$. Moreover, $T(M)$ reduces to $U(n) \times \{I_s\}$. A standard argument, as for almost contact structures, allows to prove the existence of a Riemannian metric g satisfying:

$$g(f(X), f(Y)) = g(X, Y) - \sum_{i=1}^{s} \eta^i(X)\eta^i(Y) \ ,$$

for any $X, Y \in \mathcal{X}(M)$. Such a metric is called a metric *compatible* with the structure f and $(M^{2n+s}, f, \xi_i, \eta^i, g)$ is called a *metric pk.f-manifold* or a *globally framed metric f-manifold*. One also has $\eta^i(X) = g(X, \xi_i)$, $\|\xi_i\| = 1$ and the splitting $T(M) = \ker f \oplus \operatorname{im} f$ becomes orthogonal. As usual, the fundamental 2-form F of the structure is defined by $F(X, Y) = g(X, fY)$, for any $X, Y \in \mathcal{X}(M)$. Furthermore, the Levi-Civita connection satisfies the following formula, which can be easily proved as for an almost contact metric structure ([33]):

$$\begin{aligned} 2g((\nabla_X f)Y, Z) = {} & 3dF(X, fY, fZ) - 3dF(X, Y, Z) \\ & + g(S(Y, Z), fX) + \textstyle\sum_{j=1}^{s} S_j^{(2)}(Y, Z)\eta^j(X) \\ & + 2\textstyle\sum_{j=1}^{s} d\eta^j(fY, X)\eta^j(Z) \\ & - 2\textstyle\sum_{j=1}^{s} d\eta^j(fZ, X)\eta^j(Y) \ . \end{aligned} \tag{4.25}$$

Here, the tensor fields $S_j^{(2)}$ are defined by:

$$\begin{aligned} S_j^{(2)}(X, Y) &= (L_{f(X)}\eta^j)(Y) - (L_{f(Y)}\eta^j)(X) \\ &= 2d\eta^j(fX, Y) - 2d\eta^j(fY, X) \ , \end{aligned}$$

the tensor field S is defined as $[f,f] + \sum_{i=1}^{s} 2d\eta^i \otimes \xi_i$, $[f,f]$ being the Nijenhuis tensor of f, and its vanishing represents the so-called *normality* condition for the structure f.

The geometric meaning is the following. Consider the product manifold $\tilde{M} = M^{2n+s} \times \mathbf{R}^s$ and define a tensor field J on $\tilde{M}$ putting:

$$J(X, \alpha^i \frac{\partial}{\partial x^i}) = (fX - \alpha^i \xi_i, \eta^j(X) \frac{\partial}{\partial x^j}),$$

where $(x^1, \dots, x^s)$ are coordinates in $\mathbf{R}^s$, α^i are differentiable functions on $\tilde{M}$ for any $i \in \{1, \dots, s\}$ and the Einstein convention is assumed. Then, it is easy to verify that $J^2 = -I$ and J is integrable (*i.e.* $\tilde{M}$ is a complex manifold) if and only if $S = 0$. Moreover, in the metric case $(\tilde{M}, J)$ is a Hermitian manifold with respect to the metric $\tilde{g} = g + g_0$, where g is the metric on M associated with f and g_0 the Euclidean metric on $\mathbf{R}^s$.

The following two theorems are essentially due to Goldberg and Yano ([123]).

Theorem 4.17 *Let $(M^{2n+s}, f, \xi_i, \eta^i)$ be a pk.f-manifold of even dimension and put $s = 2p$. Then, the tensor field*

$$J = f + \sum_{i=1}^{p} (\eta^i \otimes \xi_{p+i} - \eta^{p+i} \otimes \xi_i)$$

makes M^{2n+s} an almost complex manifold. Moreover, J is integrable if f is normal.

Remark 4.6 Assume that $M^{2n+s}(f, \xi_i, \eta^i, g)$ is a $\mathcal{K}$-*manifold* in the sense of Blair ([32]) *i.e.* normal and metric with $dF = 0$. Then, (M^{2n+s}, J, g) is a Hermitian manifold and its Kähler form is $\Omega = F - \sum_{i=1}^{p} \eta^i \wedge \eta^{p+i}$. It follows that any $\mathcal{C}$-manifold ($\mathcal{K}$-manifold with all $d\eta^i$ vanishing) of even dimension admits a Kähler structure ([121]).

Theorem 4.18 *Let $(M^{2n+s}, f, \xi_i, \eta^i)$ be a pk.f-manifold of odd dimension and put $s = 2p+1$. Then, the tensor field*

$$\tilde{f} = f + \sum_{i=1}^{p} (\eta^i \otimes \xi_{p+i} - \eta^{p+i} \otimes \xi_i)$$

defines an almost contact structure on M^{2n+s} with ξ_{2p+1} as characteristic vector field. Moreover, the normality of f implies the normality of $\tilde{f}$.

Remark 4.7 Obviously, rearranging the ξ_i's, with the above construction one can make any vector field ξ_i the characteristic vector field of a suitable almost contact structure. Moreover, assuming that M^{2n+s} is a $pk.f$-manifold with associated metric g, the induced almost contact structures are also metric with respect to g.

Now, we examine some other consequences of the normality condition. It is easy to prove that in a normal metric $pk.f$-manifold the distribution $\mathcal{D}'$ spanned by the ξ_i's is completely integrable, since $[\xi_i, \xi_j] = -S(\xi_i, \xi_j) = 0$. Moreover, if each ξ_i is a Killing vector field, then the metric is bundle-like, there exists a local action of $\mathbf{R}^s$ on M^{2n+s} by isometries and we can look at the quotient space $M^{2n+s}/\mathcal{D}'$. Even in the compact case, unless one requires a regularity condition for the distribution $\mathcal{D}'$, this topological space does not carry a structure of manifold ([235]). Nevertheless, simply requiring that there exists a free action of a torus T^s, $M^{2n+s}/\mathcal{D}'$ turns out to be an orbifold. For more details see ([204; 53]).

Definition 4.14 Let $(M^{2n+s}, f, \xi_i, \eta^i)$ be a $pk.f$-manifold. The f-structure is said to be *regular* if the distribution $\mathcal{D}'$ is regular, *i.e.* for any $p \in M^{2n+s}$ there exists a cubical coordinate system $(U, x^1, \ldots, x^{2n+s})$ centered at p such that $\{\frac{\partial}{\partial x^{2n+i}}\}_{i=1,\ldots,s}$ span $\mathcal{D}'_{|U}$ and the leaf through p intersects U in precisely one s-dimensional slice.

A result of Palais, ([235]), allows us to conclude that in a compact regular $pk.f$-manifold with $\mathcal{D}'$ integrable, each leaf of the foliation defined by $\mathcal{D}'$ is compact and $M^{2n+s}/\mathcal{D}'$ is a compact manifold. Furthermore, if M^{2n+s} is connected, then the projection $\pi : M^{2n+s} \to M^{2n+s}/\mathcal{D}'$ is a C^∞-fibration having the leaves as fibres. In particular, when each ξ_i is regular, then any leaf is a torus T^s. Namely, the distribution $\mathcal{D}'$ consists of s 1-dimensional distributions each of which is spanned by a ξ_i. M^{2n+s} being compact, the integral curves of any ξ_i are closed and homeomorphic to a circle $\mathbf{S^1}$, and the regularity and the linear independence of the ξ_i's imply that the fibres are homeomorphic to a torus T^s.

Now, we state a theorem of Blair, Ludden and Yano which gives a link between f-structures and Riemannian submersions ([35]).

Firstly, we recall that the period function of a closed regular vector field X on M is defined putting $\lambda_X(p) = \inf\{t > 0 \mid (\exp tX)(p) = p\}$ and a result of Morimoto ([206]) states that, given a complex manifold M with complex structure J and an analytic vector field X on M such that X and JX are both closed and regular, then the function $l = \lambda_X + \mathrm{i}\lambda_{JX}$ is holomorphic on M.

Lemma 4.2 *Let $(M^{2n+s}, f, \xi_i, \eta^i)$ be a normal, compact, connected $pk.f$-*

manifold with each ξ_i regular and closed. Then, the period functions $\lambda_i = \lambda_{\xi_i}$ are constant.

Proof. The differentiability of the λ_i's was proved by Boothby and Wang ([42]). Assuming that s is even, M^{2n+s} admits a complex structure, via Theorem 4.17, and any ξ_i is analytic. So, putting $\xi_{i*} = J\xi_i$, the functions $l = \lambda_i + \mathrm{i}\lambda_{i^*}$ are holomorphic and then any λ_i is constant. If s is odd, by Theorem 4.18, fix any of the ξ_i's, say ξ'. Then M^{2n+s} admits a normal almost contact structure with characteristic vector field ξ'. Therefore, we can construct a complex structure J on $\tilde{M} = M^{2n+s} \times \mathbf{S^1}$ and ξ' is analytic on $\tilde{M}$. Then, $l = \lambda_{\xi'} + \mathrm{i}\lambda_{J\xi'}$ is holomorphic and so constant on $\tilde{M}$. It follows that any λ_i is constant on $\tilde{M}$ and on M^{2n+s}.

Theorem 4.19 (Blair, Ludden, Yano). *Let $(M^{2n+s}, f, \xi_i, \eta^i)$ be a compact, connected manifold with a regular normal pk.f-structure such that the ξ_i's are regular. Then, M^{2n+s} is the total space of a principal torus bundle over a complex manifold $N^{2n} = M^{2n+s}/\mathcal{D}'$. Moreover, if the fundamental 2-form associated with a compatible metric is closed (i.e. M^{2n+s} is a $\mathcal{K}$-manifold), then N^{2n} is a Kähler manifold.*

Proof. By the above discussion we know that there exists a C^∞-fibration $\pi : M^{2n+s} \to N^{2n} = M^{2n+s}/\mathcal{D}'$ with fibres homeomorphic to a torus T^s, and Lemma 4.2 implies that the period function of any ξ_i is a constant, say c_i. Denote by S^1_i the group of real numbers modulo c_i. S^1_i acts on M^{2n+s} by:

$$(t, p) \mapsto (\exp t\xi_i)(p), \ t \in \mathbf{R} ,$$

and the corresponding action of $T^s = S^1_1 \times \cdots \times S^1_s$ is free and without fixed points, so N^{2n} is a manifold. To prove that M^{2n+s} is the total space of a T^s-bundle, we need to show that it is locally trivial. To this aim, one considers a cover $\{U_\alpha\}$ of N^{2n} such that any U_α is obtained as the projection by π of a regular neighborhood on M^{2n+s} and denotes by $\sigma_\alpha : U_\alpha \to M^{2n+s}$ the section corresponding to $x^{2n+1} = \text{const.}, \dots, x^{2n+s} = \text{const.}$ Thus, we obtain coordinate maps for M^{2n+s} considering the maps $\psi_\alpha : U_\alpha \times T^s \to M^{2n+s}$ such that:

$$\psi_\alpha(x, t_1, \dots, t_s) = \left(\exp \left(\sum_{i=1}^{s} t_i \xi_i \right) \right) (\sigma_\alpha(x)) .$$

Now, on the principal T^s-bundle we consider the Lie algebra valued connection form $\eta = (\eta^1, \dots, \eta^s)$, whose horizontal distribution $\mathcal{D}$ is complementary to $\mathcal{D}'$ and denote by $\tilde{\pi}$ the horizontal lift with respect to η. Putting $JX = \pi_*(f(\tilde{\pi}X))$, for any $X \in \mathcal{X}(N^{2n})$, we get an almost complex structure on N^{2n}, whose Nijenhuis tensor can be written as:

$$N(X,Y) = \pi_*([f,f](\tilde{\pi}X,\tilde{\pi}Y) + 2\sum_{i=1}^{s} d\eta^i(\tilde{\pi}X,\tilde{\pi}Y)\xi_i) ,$$

so N^{2n} is a complex manifold, since M^{2n+s} is normal. Finally, assuming that g is a compatible metric on M^{2n+s}, we define a metric g' on N^{2n} putting $g'(X,Y)\circ\pi = g(\tilde{\pi}X,\tilde{\pi}Y)$. It is easy to verify that g' is J-Hermitian, π is a Riemannian submersion and the Kähler form Ω is related to F by $\pi^*\Omega = F$. In particular, if M^{2n+s} is a $\mathcal{K}$-manifold, then $dF = 0$ implies $d\Omega = 0$ and N^{2n} is Kähler.

Remark 4.8 Two interesting subclasses of $\mathcal{K}$-manifolds are the so-called *$\mathcal{C}$-manifolds* (with each $d\eta^i = 0$) and *$\mathcal{S}$-manifolds* (with each $d\eta^i = F$), ([32]). The previous theorem applies to them; in particular, for $\mathcal{C}$-manifolds, we get $[f,f] = 0$, the distribution im$f$ is integrable and M^{2n+s} is locally the product of a Kähler manifold and a flat abelian Lie group. This result is obtained by Blair even in the non-compact case ([32]). In the case of $\mathcal{S}$-manifolds, we have the following result, also due to Blair, Ludden and Yano ([35]).

Theorem 4.20 *Let $(M^{2n+s}, f, \xi_i, \eta^i, g)$ be a compact, connected regular $\mathcal{S}$-manifold such that each ξ_i is regular. Then, there exists a commutative diagram*

$$\begin{array}{ccc} M^{2n+s} & \xrightarrow{\tau} & M^{2n+1} \\ \pi \searrow & & \swarrow \pi_1 \\ & N^{2n} & \end{array}$$

where N^{2n} is a compact Kähler manifold, M^{2n+1} is a compact regular Sasakian manifold and is the total space of an $\mathbf{S}^1$-bundle over N^{2n}. All the maps are Riemannian submersions with totally geodesic fibres.

Proof. The existence of π comes from Theorem 4.19. Applying again the first part of the proof of the same theorem to the vector fields $\xi_1, \dots, \xi_{s-1}$ we get that M^{2n+s} is a principal T^{s-1}-bundle over a manifold M^{2n+1} with projection τ. The normality condition implies that f, ξ_s, η^s are projectable and we define on M^{2n+1} a $(1,1)$-tensor field φ, by $\varphi(X) = \tau_*(f(\tilde{\tau}(X)))$, $\tilde{\tau}$ denoting the horizontal lift with respect to the connection on M^{2n+s}

defined by the 1-form $(\eta^1, \dots, \eta^{s-1})$. Putting $\xi = \tau_* \xi_s$, $\eta(X) \circ \tau = \eta^s(\tilde{\tau} X)$, it is easy to check that:

$$\eta(\xi) = 1\,,\ \varphi(\xi) = 0\,,\ \eta \circ \varphi = 0\,, \varphi^2 = -I + \eta \otimes \xi\,, [\varphi, \varphi] + 2d\eta \otimes \xi = 0\,,$$

so obtaining a normal almost contact structure on M^{2n+1}. Now, we define a metric $\tilde{g}$ on M^{2n+1} by $\tilde{g}(X,Y) \circ \tau = g(\tilde{\tau} X, \tilde{\tau} Y)$ and τ becomes a Riemannian submersion. Moreover, $\tilde{g}(X, \xi) = \eta(X)$, $\tilde{g}$ is compatible with the structure φ and the 2-form Φ satisfies $\tau^* \Phi = F$. Thus, since $F = d\eta^s$, we obtain:

$$\begin{aligned} \Phi(X,Y) \circ \tau &= \Phi(\tau_* \tilde{\tau} X, \tau_* \tilde{\tau} Y) \circ \tau = d\eta^s(\tilde{\tau} X, \tilde{\tau} Y) \\ &= \tfrac{1}{2}((\tilde{\tau} X)(\eta^s(\tilde{\tau} Y)) - (\tilde{\tau} Y)(\eta^s(\tilde{\tau} X)) - \eta^s([\tilde{\tau} X, \tilde{\tau} Y])) \\ &= \tfrac{1}{2}(X(\eta(Y)) - Y(\eta(X)) - \eta([X,Y])) \circ \tau \\ &= (d\eta)(X,Y) \circ \tau\,, \end{aligned}$$

since $\eta([X,Y]) \circ \tau = \eta^s(\tilde{\tau}[X,Y])$, and $\tilde{\tau}([X,Y]) = h([\tilde{\tau} X, \tilde{\tau} Y])$. Hence, M^{2n+1} is compact, Sasakian, with ξ regular and then is a principal $\mathbf{S}^1$-bundle over N^{2n} with projection a Riemannian submersion π_1 such that $\pi_1 \circ \tau = \pi$.

Definition 4.15 A manifold M admits an *almost quaternionic f-structure of corank m* if for any open set U there exist three local sections f_1, f_2, f_3 of the bundle $End(T(M))$ and m global vector fields $\xi_1, \dots, \xi_m$ satisfying:

a) the quaternionic-like relations $f_i \circ f_j = -f_j \circ f_i = f_k$, for $(i,j,k) = (1,2,3)$ and cyclic permutations of the indexes,
b) the relations:

$$\begin{aligned} &f_i(\xi_t) = 0\,,\ \eta^t \circ f_i = 0\,,\ \ t \in \{1, \dots, m\}, i \in \{1,2,3\}\,, \\ &f_i^2 = -I + \textstyle\sum_{t=1}^m \eta^t \otimes \xi_t\,, \end{aligned}$$

where the η^t are the 1-forms dual to ξ_t.

Moreover, when the f_i's are global, any of them is a $pk.f$-structure and we say that M admits a *hyper f-structure of corank m*.

Furthermore, the existence of an almost quaternionic f-structure is equivalent to a reduction of the structural group of $T(M)$ to $Sp(n)Sp(1) \times \{I_m\}$, and to $Sp(n) \times \{I_m\}$ in the case of a hyper f-structure. Consequently, M must be of dimension $4n+m$. For $m = 0$, we obtain almost quaternionic structures in the first case and hypercomplex structures in the second one. An interesting situation occurs when the hyper f-structure has corank 3. In this case M admits three $pk.f$-structures whose kernels define the same subbundle of $T(M)$, of rank 3. Such structures are strictly related to the almost contact 3-structures introduced in Definition 4.11.

Proposition 4.9 *A hyper f-structure of corank 3 determines an almost contact 3-structure and vice versa.*

Proof. Let $M(f_1, f_2, f_3, \xi_1, \xi_2, \xi_3, \eta^1, \eta^2, \eta^3)$ be a hyper f-structure of corank 3. Then, the three tensor fields defined by:

$$\varphi_i = f_i + \eta^j \otimes \xi_k - \eta^k \otimes \xi_j \ , \tag{4.26}$$

with $(i,j,k) = (1,2,3)$ and cyclic permutations, define an almost contact 3-structure. Namely, any of the f-structures f_1, f_2, f_3, determines three almost contact structures according to Remark 4.7; so $\varphi_1, \varphi_2, \varphi_3$ are almost contact structures on M having ξ_1, ξ_2, ξ_3 as characteristic vector field, respectively. It is easy to verify that, taking account of the cyclic permutations, we have

$$\varphi_i(\xi_j) = -\varphi_j(\xi_i) = \xi_k \ , \ \eta^i \circ \varphi_j = -\eta^j \circ \varphi_i = \eta^k \ ,$$
$$\varphi_i \circ \varphi_j - \eta^j \otimes \xi_i = -\varphi_j \circ \varphi_i + \eta^i \otimes \xi_j = \varphi_k \ ,$$

obtaining an almost contact 3-structure. Vice versa, given an almost contact 3-structure, (4.26) allows us to define the f_i's which give the hyper f-structure.

It is easy to check that any metric compatible with the hyper f-structure (*i.e.* with any f_i) is also compatible with the almost contact 3-structure (*i.e.* with any φ_i) and vice versa. Fixing any compatible metric g, we consider the corresponding fundamental 2-forms:

$$F_i(X,Y) = g(X, f_i(Y)), \ \ \Phi_i(X,Y) = g(X, \varphi_i(Y)), \ i \in \{1,2,3\}$$

which are related by:

$$\Phi_i = F_i - \eta^j \wedge \eta^k \ , \tag{4.27}$$

$(i,j,k) = (1,2,3)$ and cyclic permutations.

From now on we will consider hyper f-structures of corank 3.

Definition 4.16 Let M be a $(4n+3)$-dimensional manifold with a hyper f-structure.
a) M is said to be a *contact 3-manifold* if for any $i \in \{1,2,3\}$ $d\eta^i = \Phi_i$, *i.e.* any η^i is a contact structure on M.
b) M is said to be a *3-$\mathcal{S}$-manifold* if, for any $i \in \{1,2,3\}$, $d\eta^i = F_i$.

Remark 4.9 In the case a) of the above definition, the f_i's are far from being $\mathcal{K}$-structures since the condition $dF_i = 0$ would imply $F_i = 0$ and $f_i = 0$. Namely, (4.27) implies $dF_i = \Phi_j \wedge \eta^k - \eta^j \wedge \Phi_k$, so supposing $dF_i = 0$ and computing in (X, Y, ξ_k) we get $\Phi_j = \eta^i \wedge \eta^k$, $(i,j,k) = (1,2,3)$ and cyclic permutations. So, using again (4.27), one has $F_i = 0$. Moreover, we get a contradiction even if we simply require the f_i's to be normal *pk.f*-structures, since this would imply $[\xi_i, \xi_j] = 0$, whereas the induced almost contact 3-structure would be 3-Sasakian and then $[\xi_i, \xi_j] = 2\xi_k$. Now, we consider the case b) of Definition 4.16. As pointed out by Hernandez ([141]), the normality of the f_i's contradicts the quaternionic-like relations; thus the f_i's cannot be $\mathcal{S}$-structures in the sense of Blair. We observe that the f_i's cannot be almost $\mathcal{S}$-structures, either ([79]). Namely, an almost $\mathcal{S}$-structure is a metric *pk.f*-structure (f, ξ_i, η^i, g), $i \in \{1, \dots, s\}$ such that $d\eta^i = F$ for any $1 \leq i \leq s$. In this case we would have $d\eta^1 = d\eta^2 = d\eta^3 = F_i$ for any $i \in \{1,2,3\}$ and then $f_1 = f_2 = f_3$. Another definition of normality for hyper f-structures of corank 3 was suggested by Hernandez ([141]).

Definition 4.17 A hyper f-structure of corank 3 is said to be *normal* if

$$[f_i, f_i] + 2d\eta^i \otimes \xi_i - 2d\eta^j \otimes \xi_j - 2d\eta^k \otimes \xi_k = 0 ,$$

for $(i,j,k) = (1,2,3)$ and cyclic permutations. A normal hyper f-structure satisfying $d\eta^i = F_i$ is called a *hyper $\mathcal{PS}$-structure.*

As proved in [141] the almost contact 3-structure associated with a normal hyper f-structure is normal. In a hyper $\mathcal{PS}$-manifold each f_i shares some properties with an $\mathcal{S}$-structure ([141]). For example, any ξ_i is Killing and

$$\nabla_{\xi_i}\xi_j = 0 \, , \quad d\eta^i(f_iX, Y) + d\eta^i(X, f_iY) = 0 \, , \quad L_{\xi_j} f_i = 0.$$

On the other hand, in an $\mathcal{S}$-manifold the last condition implies $\nabla_{\xi_j} f_i = 0$, whereas for hyper $\mathcal{PS}$-manifolds one obtains $\nabla_{\xi_i} f_i = 0$ and $\nabla_{\xi_i} f_j = -2f_k$, $(i,j,k) = (1,2,3)$ and cyclic permutations, as well as $d\eta^t(f_iX, Y) - d\eta^t(X, f_iY) = 0$, for $t \neq i$. Moreover, computing the Levi-Civita connection for a hyper $\mathcal{PS}$-manifold, using the normality condition and $d\eta^i = F_i$, from (4.25) one easily obtains:

$$\begin{aligned}(\nabla_X f_i)(Y) = & -d\eta^i(Y, f_iX)\xi_i + d\eta^j(Y, f_iX)\xi_j \\ & + d\eta^k(Y, f_iX)\xi_k + \eta^i(Y) f_i^2(X) \\ & + \eta^j(Y) f_k(X) - \eta^k(Y) f_j(X) \\ & + 2\eta^j(X) f_k(Y) - 2\eta^k(X) f_j(Y) \, .\end{aligned} \tag{4.28}$$

Following the proof of the Blair, Ludden and Yano theorem, we have:

Theorem 4.21 (Hernandez). *Let M^{4n+3} be a compact connected manifold, with a regular normal hyper f-structure. Then M^{4n+3} is the total space of a principal T^3-bundle over a hypercomplex manifold N^{4n}. Moreover, if M^{4n+3} is a hyper $\mathcal{PS}$-manifold then N^{4n} is a hyperKähler manifold and the projection defines a Riemannian submersion with totally geodesic fibres.*

Proof. The normality implies $[\xi_i, \xi_j] = 0$ for any $i, j \in \{1, 2, 3\}$ and the compactness, together with the regularity, ensures that the leaves are tori and the space of leaves, $N^{4n} = M^{4n+3}/T^3$, is a manifold. Consider the connection form on M^{4n+3} given by $\eta = (\eta^1, \eta^2, \eta^3)$, the corresponding horizontal distribution and denote by $\tilde{\pi}$ the horizontal lift with respect to the projection $\pi : M^{4n+3} \to N^{4n}$. Since the f_i's are projectable, for any $i \in \{1, 2, 3\}$ and $X \in \mathcal{X}(N^{4n})$ we can put:

$$J_i(X) = \pi_*(f_i(\tilde{\pi}(X))) ,$$

obtaining three tensor fields of type $(1, 1)$ on N^{4n}. It is easy to verify that $J_i^2 = -I$, $J_i J_j = -J_j J_i = J_k$, for any cyclic permutation (i, j, k) of $(1, 2, 3)$, and

$$\begin{aligned}[J_i, J_i](X, Y) = \pi_*([f_i, f_i](\tilde{\pi}(X), \tilde{\pi}(Y)) + d\eta^i(\tilde{\pi}(X), \tilde{\pi}(Y))\xi_i \\ -d\eta^j(\tilde{\pi}(X), \tilde{\pi}(Y))\xi_j - d\eta^k(\tilde{\pi}(X), \tilde{\pi}(Y))\xi_k) = 0 ,\end{aligned}$$

for any $X, Y \in \mathcal{X}(N^{4n})$. Hence, N^{4n} is a hypercomplex manifold. Now, assuming that M^{4n+3} is a hyper $\mathcal{PS}$-manifold, we fix a compatible metric g and define on N^{4n} a metric $\tilde{g}$ by $\tilde{g}(X, Y) \circ \pi = g(\tilde{\pi}X, \tilde{\pi}Y)$. Then $\tilde{g}$ is Hermitian with respect to each J_i and, for the corresponding Kähler 2-forms Ω_i, we have $\pi^*(\Omega_i) = F_i$. Thus, $\pi^*(d\Omega_i) = dF_i = 0$ and N^{4n} is hyperKähler. Finally, since $\nabla_{\xi_i}\xi_j = 0$ and $\nabla_{\xi_i}X$ is a horizontal vector field for X horizontal, we obtain $T = 0$ and the Riemannian submersion π has totally geodesic fibres.

Remark 4.10 Replacing the regularity condition with a locally free action of T^3 on M^{4n+3}, the metric remains bundle-like since the ξ_i's are Killing. The hyper $\mathcal{PS}$-manifold fibers over a hyperKähler orbifold, whereas, as proved in [53], 3-Sasakian manifolds fiber over quaternionic Kähler orbifolds.

Now, we prove a converse of the above theorem, using a result of Lutz ([188]). In the attempt to generalize contact structures, Lutz defined a p-

contact structure as a subbundle of $T^*(M)$, locally spanned by p 1-forms $\eta^1, \dots, \eta^p$ such that $\eta^1 \wedge \dots \wedge \eta^p \wedge (\sum_{i=1}^p \lambda_i d\eta^i)^{2n}$ never vanishes, for any $(\lambda_1, \dots, \lambda_p) \in (\mathbf{R^p})^*$.

Theorem 4.22 *Any compact hyperKähler manifold $(N^{4n}, \tilde{g}, J_i)_{1 \le i \le 3}$, having the three Kähler 2-forms Ω_i with integral period is the base manifold of a principal T^3-bundle whose total space is a compact manifold carrying a regular hyper $\mathcal{PS}$-structure.*

Proof. The Ω_i are linearly independent and of maximal rank, so they satisfy $(\sum_{i=1}^3 \lambda_i \Omega_i)^{2n} \neq 0$, with $\sum_{i=1}^3 \lambda_i = 1$. Then, we can apply a Theorem of Lutz ([188] Prop. 2), obtaining a principal bundle $\pi : M^{4n+3} \to N^{4n}$ with structural group T^3 such that the connection form $\eta = (\eta^1, \eta^2, \eta^3)$ having curvature $F = (F_1, F_2, F_3)$, $F_i = \pi^*(\Omega_i)$, defines a regular 3-contact structure on M^{4n+3} and moreover $\eta^1 \wedge \eta^2 \wedge \eta^3 \wedge (\sum_{i=1}^3 \lambda_i d\eta^i)^{2n} \neq 0$. Let ξ_i's be the vector fields dual to the η^i's and define a metric g by:

$$g = \pi^* \tilde{g} + \sum_{i=1}^{3} \eta^i \otimes \eta^i .$$

We set $\Phi_i = F_i - \eta^j \wedge \eta^k$, $(i, j, k) = (1, 2, 3)$ and cyclic permutations. Since these 2-forms are of maximal rank and $\eta^i \wedge (\Phi_i)^{2n+1} \neq 0$, any (η^i, Φ_i) defines an almost contact structure in the sense of Libermann or, equivalently, in the sense of Sasaki ([33]). Each Φ_i is nondegenerate on $\ker \eta^i$ and the tensor field φ_i defined by $\Phi_i(X, Y) = g(X, \varphi_i Y)$ satisfies $\varphi_i^2 = -I$ on $\ker \eta^i$. We extend φ_i to $T(M^{4n+3})$ requiring $\varphi_i(\xi_i) = 0$, and we easily obtain $\varphi_i^2 = -I + \eta^i \otimes \xi_i$, $\eta^i(\xi_i) = 1$, $\eta^i \circ \varphi_i = 0$. Furthermore, for $i \neq j$ and for any vector field X, we have:

$$\begin{aligned} g(\varphi_i(\xi_j), X) &= -\Phi_i(\xi_j, X) = -F_i(\xi_j, X) + (\eta^j \wedge \eta^k)(\xi_j, X) \\ &= \eta^k(X) = g(\xi_k, X) , \end{aligned}$$

which implies $\varphi_i(\xi_j) = \xi_k$ (and cyclic permutations). It is easy to check that $\eta^i \circ \varphi_j = \eta^k = -\eta^j \circ \varphi_i$. Clearly, any φ_i is an almost contact structure on M^{4n+3} and setting

$$f_i = \varphi_i - \eta^j \otimes \xi_k - \eta^k \otimes \xi_j ,$$

as in Proposition 4.9, we have $\pi_* \circ f_i = \pi_* \circ \varphi_i = J_i \circ \pi_*$. Thus, pointwise we get, for any $X, Y \in \mathcal{X}(M^{4n+3})$,

$$\begin{aligned}\Phi_i(\varphi_j X, Y) &= (\pi^*\Omega_i)(\varphi_j X, Y) - \eta^j(\varphi_j X)\eta^k(Y) + \eta^j(Y)\eta^k(\varphi_j X)\\ &= \tilde{g}(\pi_*\varphi_j X, J_i\pi_* Y)\circ\pi - \eta^j(Y)\eta^i(X)\\ &= g(X, \varphi_k Y) - \eta^j(X)\eta^i(Y)\\ &= \Phi_k(X, Y) - \eta^j(X)\eta^i(Y)\,.\end{aligned}$$

Hence, $\Phi_i(\varphi_j X, Y) + \eta^j(X)\eta^i(Y) = \Phi_k(X, Y)$, $(i, j, k) = (1, 2, 3)$ and cyclic permutations, which gives $-\varphi_j \circ \varphi_i + \eta^i \otimes \xi_j = \varphi_k = \varphi_i \circ \varphi_j - \eta^j \otimes \xi_i$. So $(\varphi_i, \xi_i, \eta^i)$, $i \in \{1, 2, 3\}$ defines an almost contact 3-structure on M^{4n+3} and the corresponding hyper$f$-structure (f_i, ξ_i, η^i) as in Proposition 4.9 turns out to be a 3-$\mathcal{S}$-structure, since $d\eta^i = F_i$. Finally, to prove the normality, it is convenient to use the relations $\pi_* \circ f_i = J_i \circ \pi_*$. We observe that for a fixed $i \in \{1, 2, 3\}$, we have:

$$d\eta^i(f_i X, f_i Y) = F_i(X, Y) = d\eta^i(X, Y)\,, \tag{4.29}$$

whereas, for $i \neq j$, we have:

$$d\eta^j(f_i X, f_i Y) = -F_j(X, Y) = -d\eta^j(X, Y)\,. \tag{4.30}$$

Thus, for $(i, j, k) = (1, 2, 3)$ and cyclic permutations, denoting by N_i the normality tensors in Definition 4.17, since N^{4n} is hyperKähler, pointwise we have:

$$\pi_*(N_i(X, Y)) = \pi_*([f_i, f_i](X, Y)) = [J_i, J_i](\pi_* X, \pi_* Y) = 0\,,$$

so $N_i(X, Y)$ is a vertical vector field. On the other hand, with a direct computation, using (4.29) and (4.30), for any $t \in \{1, 2, 3\}$ one obtains $g(\xi_t, N_i(X, Y)) = 0$.

Theorem 4.23 *Hyper $\mathcal{PS}$-manifolds do not admit any Einstein metric in the canonical variation of the metric g above described.*

Proof. Let $M^{4n+3}(f_i, \xi_i, \eta^i, g)$ be a hyper $\mathcal{PS}$-manifold. Then, for any vector field Y and for any $h \in \{1, 2, 3\}$, one has:

(1) $R(\xi_h, Y)\xi_h = f_h^2(Y) = -Y + \sum_{r=1}^{3} \eta^r(Y)\xi_r$,

(2) $R(\xi_i, Y)\xi_j = f_k(Y)$, for $(i, j, k) = (1, 2, 3)$ and cyclic permutations,

(3) $R(X, Y)\xi_h = -(\nabla_X f_h)(Y) + (\nabla_Y f_h)(X)$ for X, Y horizontal vector fields.

It follows that the sectional curvatures are given by $K(\xi_s, \xi_h) = 0$ and $K(X, \xi_i) = 1$ for X unitary horizontal vector, and computing the Ricci tensor one has:

$$\rho(\xi_i, \xi_j) = 0\,,\ i \neq j\,;\ \ \rho(\xi_i, \xi_i) = 4n\,;\ \ \rho(X, \xi_i) = 0\,,$$

for X a horizontal vector field, since, by (4.28), $R(X,Y)\xi_h$ in (3) is a vertical vector field. Finally, considering X, Y horizontal vector fields, (1.36) and (1.35) give:

$$\rho(X,Y) = -2g(A_X, A_Y) = -2\sum_{h=1}^{3} g(f_h X, f_h Y) = -6g(X,Y)\,,$$

since the fibres are totally geodesic, N^{4n} is hyperKähler and then Ricci-flat, and $A_X(\xi_h) = \nabla_X \xi_h = -f_h(X)$. Hence, with respect to a local orthonormal frame $\{X_1, \ldots, X_{4n}, \xi_1, \xi_2, \xi_3\}$, the Ricci tensor is represented by the matrix

$$\begin{pmatrix} -6I_{4n} & 0 \\ 0 & 4nI_3 \end{pmatrix}.$$

Obviously, the scalar curvature is constant equal to $-12n$. To complete the proof, we consider the canonical variation of the metric on M^{4n+3} given by $g_t(\xi_r, \xi_s) = tg(\xi_r, \xi_s)$, $t > 0$, $g_t = g$ on the horizontal distribution, and requiring the orthogonality between the vertical and the horizontal distributions. Then, applying 9.71 and 9.72 in [26] (see also Sec. 5.2), Einstein metrics in the canonical variation correspond to the critical points of the function:

$$\phi(t) = t^{3/(4n+3)}\left(\frac{1}{t}\widehat{\tau} + \tau' - t\|A\|^2\right).$$

Since, in this case $\tau' = \widehat{\tau} = 0$, then $\phi'(t) \neq 0$, so that none of the g_t's is Einstein.

Remark 4.11 The basic example of a hyper $\mathcal{PS}$-structure is given in [141], where it is proved that the quaternionic analogue of the Heisenberg group admits a natural hyper $\mathcal{PS}$-structure representing also an example of a fibration of a torus bundle over a torus, which is a fat fibration since the mixed sectional curvatures are positive ([337]).

Theorem 4.24 *Let M^{4n+3} be a compact, connected, regular hyper $\mathcal{PS}$-manifold such that each ξ_i is regular. Then, there exists a commutative* *diagram*

$$\begin{array}{ccc} & \tau & \\ M^{4n+3} & \longrightarrow & M^{4n+1} \\ \pi \searrow & & \swarrow \pi_1 \\ & N^{4n} & \end{array}$$

where M^{4n+1} is a compact regular Sasakian manifold, N^{4n} is a compact hyperKähler manifold and all the maps are Riemannian submersions with totally geodesic fibres which are tori.

Proof. The existence of the submersion π comes from Theorem 4.21. We fix any of the three regular almost contact structures on M^{4n+3} associated with the $\mathcal{PS}$-structure (f_i, ξ_i, η^i, g), $i \in \{1,2,3\}$, say $(\varphi_1, \xi_1, \eta^1)$. The distribution spanned by the regular vector fields ξ_2, ξ_3 is integrable and M^{4n+3} fibers, with fibre T^2 and projection τ, over a compact manifold M^{4n+1}, which is a principal bundle on N^{4n} with fibre T^1 and projection π_1. Since $[\xi_r, \xi_1] = 0, L_{\xi_r} f_1 = 0, L_{\xi_r} \eta^1 = 0$, for any $r \in \{1,2,3\}$, we have that ξ_1, f_1 and η^1 are τ-projectable. Now, denoting by $\tilde{\tau}$ the corresponding horizontal lift, we put:

$$\varphi(X) = \tau_*(\varphi_1(\tilde{\tau}(X))), \quad \xi = \tau_*(\xi_1), \quad \eta(X) \circ \tau = \eta^1(\tilde{\tau}(X)) \,,$$

for any $X \in \mathcal{X}(M^{4n+1})$, obtaining a normal almost contact structure. It is also easy to verify that the metric $\tilde{g}$ on M^{4n+1} given by $\tilde{g}(X,Y) \circ \tau = g(\tilde{\tau}X, \tilde{\tau}Y)$ is a compatible metric and for the related fundamental 2-form F, using (4.27) one has:

$$\begin{aligned} F(X,Y) \circ \tau = \tilde{g}(X, \varphi Y) \circ \tau &= g(\tilde{\tau}X, \varphi_1(\tilde{\tau}Y)) = \Phi_1(\tilde{\tau}X, \tilde{\tau}Y) \\ &= F_1(\tilde{\tau}X, \tilde{\tau}Y) = d\eta^1(\tilde{\tau}X, \tilde{\tau}Y) = d\eta(X,Y) \circ \tau \,, \end{aligned}$$

and M^{4n+1} is a Sasakian manifold.

Chapter 5

Einstein Spaces and Riemannian Submersions

This chapter consists in a systematic exposition of several results concerning the existence of Einstein metrics and of Einstein–Weyl structures on manifolds which can be considered as the total space of a Riemannian submersion with totally geodesic fibres.

Firstly, given a compact Lie group G, we study the principal G-bundles such that the canonical projection is a Riemannian submersion whose horizontal distribution satisfies a Yang–Mills condition. Then, we introduce the concept of canonical variation of the metric on the total space M of a Riemannian submersion and derive existence conditions for Einstein metrics in such canonical variation, provided that M is compact.

This is the subject of Secs. 5.1, 5.2, while in Sec. 5.3 we explain the method of W. Ziller which allows to describe the invariant metrics on the spheres and on the projective spaces and the Einstein ones among them. Examples of homogeneous spaces which do not admit invariant Einstein metrics are also given. The results of Secs. 5.1, 5.2 are also useful in proving, under suitable conditions, the existence of an Einstein metric on the total space of a principal torus bundle over the product of Einstein Kähler manifolds.

Then, in Sec. 5.4, using the technique of Y. Sakane, we prove the existence of Einstein metrics on suitable homogeneous manifolds which are the total space of sphere bundles over a **C**-space.

Another interesting problem, connected with the ones just mentioned, is the existence of Einstein–Weyl structures on a Riemannian manifold. In particular, in Sec. 5.5, considering the projection of a principal circle bundle over a compact Einstein manifold, we state the interrelation between Einstein metrics and Einstein–Weyl structures in the canonical variation of the metric on the total space. A sufficient condition for the existence of

Einstein–Weyl structures on the total space of principal torus bundles is also stated.

Finally, in Sec. 5.6, we point out the interrelation between Riemannian submersions and Hermitian or almost contact Einstein–Weyl structures. We show how, in the compact case, locally conformal Kähler manifolds with parallel Lee form describe Hermitian Einstein–Weyl spaces; this makes the subject of Secs. 3.5, 3.6, 3.7 relevant for Hermitian–Weyl theory, too. In particular, we state the existence of a Kähler structure on the leaves space of a compact generalized Hopf manifold determined by the vertical foliation. This theorem, originally due to I. Vaisman, has been improved by H. Pedersen, Y. Poon, A. Swann ([241]) and by K. Tsukada ([301]). Finally, we prove the results of F. Narita on the existence of Sasakian structures on Einstein–Weyl manifolds and of Einstein–Weyl structures on Sasakian manifolds, provided that the considered manifolds are the total space of a Riemannian submersion.

5.1 Einstein Metrics on the Total Space of a Riemannian Submersion

The aim of this section is the investigation of the Riemannian submersions whose total space is an Einstein manifold. In the case of totally geodesic fibres, they are characterized as follows.

Proposition 5.1 *Let $\pi : (M, g) \to (B, g')$ be a Riemannian submersion with totally geodesic fibres. Then, (M, g) is Einstein if and only if the following relations hold:*

$$\hat{\rho}(U, V) + \sum_{i=1}^{n} g(A_{X_i}U, A_{X_i}V) = \frac{\tau}{m} g(U, V) , \tag{5.1}$$

$$\rho'(X', Y') \circ \pi - 2\sum_{j=1}^{r} g(A_X U_j, A_Y U_j) = \frac{\tau}{m} g(X, Y) , \tag{5.2}$$

$$\sum_{i=1}^{n} (\nabla_{X_i} A)_{X_i} = 0 , \tag{5.3}$$

where $\dim M = m$, $\dim B = n$, $r = m - n$, *U, V are vertical, X, Y are basic vector fields π-related to X', Y' and $\{X_i\}_{1\leq i\leq n}$, $\{U_j\}_{1\leq j\leq r}$ respectively denote local orthonormal frames of the distributions $\mathcal{H}$, $\mathcal{V}$.*

Proof. Since the totally geodesicity of the fibres implies the vanishing of the invariant T and of the mean curvature vector field, via (1.33), formulas (1.36) reduce to:

$$\rho(U,V) = \hat{\rho}(U,V) + \sum_{i=1}^{n} g(A_{X_i}U, A_{X_i}V) \ ,$$

$$\rho(X,Y) = \rho'(X',Y') \circ \pi - 2\sum_{j=1}^{r} g(A_X U_j, A_Y U_j) \ ,$$

$$\rho(U,X) = \sum_{i=1}^{n} g((\nabla_{X_i} A)(X_i, X), U) \ .$$

Thus, the Einstein condition for (M,g) directly implies (5.1), (5.2) and also, for any horizontal X and vertical U, $\sum_{i=1}^{n} g((\nabla_{X_i}A)(X_i,X),U) = 0$ or, equivalently, $\sum_{i=1}^{n} g((\nabla_{X_i}A)(X_i,U),X) = 0$. So the vector fields $\sum_{i=1}^{n}(\nabla_{X_i}A)(X_i,X)$, $\sum_{i=1}^{n}(\nabla_{X_i}A)(X_i,U)$ vanish, since they are, respectively, vertical and horizontal; this proves (5.3). The converse in the statement is trivial.

Remark 5.1 Equations (5.1), (5.2) allow to relate the scalar curvatures $\hat{\tau}$ of the fibres and τ' of B to τ and $||A||$ according to:

$$\hat{\tau} + ||A||^2 = \frac{r}{m}\tau \ , \quad \tau' \circ \pi - 2||A||^2 = \frac{n}{m}\tau \ . \tag{5.4}$$

Thus, if M is connected, $\dim M = m \geq 3$, the Einstein condition also implies that $||A||$ and $\hat{\tau}$ are constant functions on any fibre; so, when any two fibres are isometric, they are constant on M and via (5.4) one also obtains the constancy of τ'. This situation occurs when M is complete (Corollary 2.1) or if π acts as the projection of a principal fibre bundle.

We are going to describe this case in detail.
Let G be a compact, simple Lie group, equipped with the standard bi-invariant metric, and $\pi : (M,g) \to (B,g')$ the projection of a principal G-bundle. We assume that π is a Riemannian submersion with totally geodesic fibres isometric to G and denote by θ the 1-form of the principal connection corresponding to the horizontal distribution $\mathcal{H}$. Since at any point $p \in M$, the restriction $\theta_{p|\mathcal{V}_p}$ is a linear isomorphism onto the Lie algebra $\mathfrak{g}$ of G and the curvature form acts as $\Omega(X,Y) = -\frac{1}{2}\theta(v([X,Y]))$ for any $X,Y \in \mathcal{X}^h(M)$, using Proposition 1.5, one gets:

$$(A_X Y)_p = -(\theta_{p|\mathcal{V}_p})^{-1}(\Omega(X,Y)_p) \ . \tag{5.5}$$

Then, (5.5) implies the $Ad(G)$-invariance of the symmetric bilinear form given by:

$$(U,V) \mapsto \sum_i g(A_{X_i}U^*, A_{X_i}V^*) \ ,$$

U^*, V^* being the fundamental vector fields corresponding to U, V and $\{X_i\}$ a local orthonormal frame of $\mathcal{H}$. Consequently, its constancy on each fibre forces it to coincide with the Killing–Cartan form, up to a real number a, and one has:

$$\sum_i g(A_{X_i}U^*, A_{X_i}V^*) = a\, g(U^*, V^*), \quad U, V \in \mathfrak{g} \ .$$

Taking the traces, one obtains $||A||^2 = a \dim G$, so $||A||$ is constant. Furthermore, for any $W, W' \in \mathcal{X}^v(M)$, we have: $\sum_i g(A_{X_i}W, A_{X_i}W') = a\, g(W, W')$. The fibres are Einstein manifolds of constant scalar curvature $\frac{1}{4}\dim G$, since they are isometric to G. Combining with (1.37), the scalar curvature τ of M is constant if and only if τ' is a constant function on B. Finally, we say that the distribution $\mathcal{H}$ determined by θ is a Yang–Mills connection if the invariant A satisfies (5.3). We observe that this condition is independent of the metric on the fibres, but only involves the metric g' on B and the distribution $\mathcal{H}$, *i.e.* the 1-form θ. In fact, we can consider a local orthonormal frame $\{X_i\}_{1 \le i \le n}$ of basic vector fields such that $\pi_* X_i = X'_i$. Then, the vanishing of $\sum_i (\nabla_{X_i} A)_{X_i}$ is equivalent to the vanishing of the vertical component of $\sum_i (\nabla_{X_i} A)(X_i, Y)$, for any basic vector field Y. Since $T = 0$, using the Schouten connection considered in Sec. 1.3, one easily obtains:

$$\begin{aligned} v\Big(\sum_i (\nabla_{X_i} A)(X_i, Y)\Big) = \frac{1}{2} \sum_i v([X_i, v[X_i, Y]] &- [h(\nabla_{X_i} X_i), Y] \\ &-[X_i, h(\nabla_{X_i} Y)]) \ . \end{aligned}$$

Note that each term in the last expression involves the 1-form θ and the Levi-Civita connection on B, since $h(\nabla_{X_i} X_i)$, $h(\nabla_{X_i} Y)$ are π-related to $(\nabla'_{X'_i} X'_i)$, $(\nabla'_{X'_i} Y')$.

This allows to reformulate Proposition 5.1 in the following way.

Corollary 5.1 *Let $\pi : (M, g) \to (B, g')$ be the projection of a principal fibre bundle with structure group a compact, simple Lie group G. Suppose that π is a Riemannian submersion with horizontal distribution $\mathcal{H}$ determined by a principal connection and totally geodesic fibres isometric to G.*

Then (M, g) is Einstein if and only if $\mathcal{H}$ is a Yang–Mills connection, τ' is constant and

$$\rho'(X', Y') \circ \pi - 2 \sum_j g(A_X U_j, A_Y U_j) = \frac{\hat{\tau} + \|A\|^2}{\dim G} g(X, Y) \, ,$$

for any $X, Y \in \mathcal{X}^b(M)$, π-related to X', Y'.

This corollary applies to the projection of an $\mathbf{S^1}$-bundle $\pi : M \to B$ satisfying the previous hypotheses. Let θ be the 1-form of the principal connection on M which determines the π-horizontal distribution and denote by U^* the unit fundamental vector field such that $\theta(U^*) = 1$. Since the adjoint representation of $\mathbf{S^1}$ on $\mathbf{R}$ is trivial, there exists a closed 2-form α on B such that the curvature form is given by $\Omega = \pi^* \alpha$ and, via (5.5), the invariant A acts on the horizontal vector fields as $(A_X Y)_p = -\alpha_{\pi(p)}(\pi_{*p} X_p, \pi_{*p} Y_p) U^*_p$. This makes (5.3) equivalent to the condition that α is coclosed. In fact, with respect to a local orthonormal frame $\{X_i\}_{1 \le i \le n}$ of basic vector fields, π-related to X'_i, for any basic Y, π-related to Y', one has:

$$\begin{aligned} \textstyle\sum_i (\nabla_{X_i} A)(X_i, Y) = &- \textstyle\sum_i ((\nabla'_{X'_i} \alpha)(X'_i, Y') \circ \pi) U^* \\ &- \textstyle\sum_i (\alpha(X'_i, Y') \circ \pi)(\nabla_{X_i} U^*) \, , \end{aligned}$$

and taking the vertical component, since $\|U^*\| = 1$, we get $v(\nabla_{X_i} U^*) = 0$ and

$$v(\sum_i (\nabla_{X_i} A)(X_i, Y)) = -((\delta' \alpha)(Y') \circ \pi) U^* \, ,$$

$\delta'\alpha$ denoting the codifferential of α. Thus, $\delta'\alpha = 0$ if and only if $\sum_i (\nabla_{X_i} A)(X_i, Y) = 0$ for any horizontal Y, *i.e.* if and only if $\sum_i (\nabla_{X_i} A)_{X_i} = 0$, since $\sum_i (\nabla_{X_i} A)_{X_i}$ is g-skew-symmetric. Finally, via the skew-symmetry of A, for any basic vector field X, π-related to X', we have $A_X U^* = \sum_{i=1}^n \alpha(X', X'_i) X_i$. Thus, $\|A\|^2 = \|\alpha\|^2$ and the Ricci tensor of M satisfies:

$$\begin{aligned} &\rho(V, W) = \|\alpha\|^2 g(V, W) \, , \\ &\rho(X, Y) = (\rho'(X', Y') - 2 \textstyle\sum_{i=1}^n \alpha(X', X'_i) \alpha(Y', X'_i)) \circ \pi \, , \end{aligned}$$

for any V, W vertical vector fields and X, Y basic vector fields π-related to X', Y'. So, we have proved the following result.

Corollary 5.2 *Let $\pi : (M, g) \to (B, g')$ be the projection of a principal $\mathbf{S^1}$-bundle, which is also a Riemannian submersion with totally geodesic fibres of length 2π and invariant horizontal distribution. Let α denote the*

2-form on B which lifts to the curvature form of the connection associated with $\mathcal{H}$. Then, (M,g) is Einstein if and only if α is closed, coclosed, $||\alpha||$ is constant and

$$\rho'(X',Y') = 2\sum_{i=1}^{n} \alpha(X',X'_i)\alpha(Y',X'_i) + ||\alpha||^2 g'(X',Y') ,$$

for any $X',Y' \in \mathcal{X}(B)$, $\{X'_i\}_{1\leq i\leq n}$ being a local orthonormal frame on B.

5.2 The Canonical Variation of the Metric in the Total Space

We present the proof of an unpublished result, due to Bérard-Bergery, concerning the existence of Einstein metrics on the total space (M,g) of a Riemannian submersion. Such metrics are realized as a suitable deformation of g.

Definition 5.1 Let $\pi : (M,g) \to (B,g')$ be a Riemannian submersion. The canonical variation of g is the family of metrics $\{g_t\}_{t\in \mathbf{R}^*_+}$ on M such that:

$$\begin{aligned} g_t(U,V) &= tg(U,V) , && U,V \in \mathcal{X}^v(M) \\ g_t(X,Y) &= g(X,Y) , && X,Y \in \mathcal{X}^h(M) \\ g_t(X,U) &= 0 , && U \in \mathcal{X}^v(M), X \in \mathcal{X}^h(M) . \end{aligned} \tag{5.6}$$

Remark 5.2 Any metric in the canonical variation makes π a Riemannian submersion with the same horizontal distribution $\mathcal{H}$. The invariants of π with respect to g_t are denoted by A^t, T^t, as well as ∇^t stands for the Levi-Civita connection of (M,g_t). Via a straightforward computation, one gets:

$$\begin{aligned} &v(\nabla^t_U V) = v(\nabla_U V) , \quad h(\nabla^t_U V) = th(\nabla_U V) , \\ &\nabla^t_X U = \nabla_X U , \quad \nabla^t_U X = \nabla_U X , \quad \nabla^t_X Y = \nabla_X Y , \end{aligned} \tag{5.7}$$

U,V being vertical vector fields and X,Y horizontal ones. Thus, combining with (1.19) one has:

$$\begin{aligned} &T^t_U V = tT_U V , \quad T^t_U X = T_U X , \\ &A^t_X Y = A_X Y , \quad A^t_X U = tA_X U . \end{aligned} \tag{5.8}$$

Now, considering the local g_t-orthonormal vertical frame $\{t^{-\frac{1}{2}}U_j\}_{1\leq j\leq r}$, $\{U_j\}_{1\leq j\leq r}$ being a g-orthonormal one, the first equation in (5.8) implies

$N = \sum_j T_{U_j} U_j = N^t$. Thus, the mean curvature vector field of any fibre is independent of t and this leads to the following result.

Lemma 5.1 *The Riemannian submersion $\pi : (M, g) \to (B, g')$ has minimal fibres if and only if $\pi : (M, g_t) \to (B, g')$ has minimal fibres for any t. Moreover, the fibres of $\pi : (M, g) \to (B, g')$ are totally geodesic if and only if for any t the fibres of $\pi : (M, g_t) \to (B, g')$ are totally geodesic.*

Proposition 5.2 *Let $\pi : (M, g) \to (B, g')$ be a Riemannian submersion with totally geodesic fibres. For any $X, Y \in \mathcal{X}^b(M)$, π-related to X', Y' and $U, V \in \mathcal{X}^v(M)$, the Ricci tensor ρ_t of the metric g_t in the canonical variation of g satisfies:*

$$\begin{aligned} \rho_t(U, V) &= \hat{\rho}(U, V) + t^2 \textstyle\sum_{i=1}^n g(A_{X_i} U, A_{X_i} V) \,, \\ \rho_t(X, Y) &= \rho'(X', Y') \circ \pi - 2t \textstyle\sum_{i=1}^n g(A_X X_i, A_Y X_i) \,, \\ \rho_t(U, X) &= t \textstyle\sum_{i=1}^n g((\nabla_{X_i} A)(X_i, X), U) = t\rho(U, X) \,, \end{aligned} \tag{5.9}$$

where $\{X_i\}_{1 \le i \le n}$ is a local horizontal orthonormal frame. Moreover, the scalar curvatures are related by:

$$\tau_t = -t||A||^2 + \tau' \circ \pi + \frac{1}{t} \hat{\tau} \,. \tag{5.10}$$

Proof. Firstly, we compute the Ricci tensor $\hat{\rho}_t$ of any fibre of the submersion $\pi : (M, g_t) \to (B, g')$. Formula (5.7) implies $R^t(U, V, W) = R(U, V, W)$. Thus, considering $U, V \in \mathcal{X}^v(M)$ and a local g-orthonormal vertical frame $\{U_j\}_{1 \le j \le r}$, since the fibres are totally geodesic, we have:

$$\hat{\rho}_t(U, V) = -\textstyle\sum_{j=1}^r g_t(R^t(U, t^{-1/2} U_j, V), t^{-1/2} U_j) = \hat{\rho}(U, V) \,.$$

Given a local horizontal orthonormal frame $\{X_i\}_{1 \le i \le n}$, via (5.8) one has:

$$\sum_{i=1}^n g_t(A^t_{X_i} U, A^t_{X_i} V) = t^2 \sum_{i=1}^n g(A_{X_i} U, A_{X_i} V) \,,$$

so the first formula follows from (1.36). The second relation is again an application of (1.36) and (5.8). Via (5.7), (5.8) we have:

$$\sum_{i=1}^n g_t((\nabla^t_{X_i} A)(X_i, X), U) = t \sum_{i=1}^n g((\nabla_{X_i} A)(X_i, X), U) \,,$$

and the last formula in (5.9) follows from (1.36). Finally, considering $\{X_i, t^{-1/2}U_j\}$ as a g_t-orthonormal, π-adapted frame, we get:

$$\tau_t = \sum_{i=1}^{n} \rho_t(X_i, X_i) + \sum_{j=1}^{r} \frac{1}{t}\rho_t(U_j, U_j) = \tau' \circ \pi - 2t\|A\|^2 + \frac{1}{t}(\hat{\tau} + t^2\|A\|^2) \, ,$$

and (5.10) follows.

Let M be a compact manifold with $\dim M = m \geq 3$ and denote by $\mathcal{M}_1(M)$ the (infinite dimensional) manifold consisting of the metrics g on M with fixed volume $Vol(g) = 1$. The total scalar curvature is the functional S defined on $\mathcal{M}_1(M)$ by $S(g) = \int_M \tau_g dv_g$, τ_g denoting the scalar curvature of g. It is well known that a metric g is Einstein if and only if it is a critical point of $S \,/\, Vol^{(m-2)/m}$. Now, given a Riemannian submersion $\pi : (M, g) \to (B, g')$, with M compact, connected and $\dim M = m \geq 3$, one looks for the Einstein metrics in the canonical variation of g. Any of such metrics has constant scalar curvature and so, according to Remark 5.1, the functions $\hat{\tau}, \tau', \|A\|$ are constant. Moreover, such metrics arise as critical points of the restriction of the above functional to the canonical variation, which is the function defined by:

$$\begin{aligned} S(t)\, Vol(M, g_t)^{(2-m)/m} &= Vol(M, g_t)^{(2-m)/m} \int_M \tau_t dv_g \\ &= (-t\|A\|^2 + \tau' + \tfrac{1}{t}\hat{\tau}) t^{r/m} Vol(M, g)^{2/m} \, , \end{aligned}$$

since $Vol(M, g_t) = t^{r/2} Vol(M, g)$. Thus, we are interested in the critical points of the function $\phi : \mathbf{R}_+^* \to \mathbf{R}$ defined by:

$$\phi(t) = \left(-t\|A\|^2 + \tau' + \frac{1}{t}\hat{\tau}\right) t^{r/m} \, . \tag{5.11}$$

Example 5.1 Let $\pi : (\mathbf{S^{2n+1}}, g) \to (\mathbf{P_n}(\mathbf{C}), g')$ be the Hopf fibration considered in Sec. 1.2. We have $\hat{\tau} = 0, \tau' = 4n(n+1)$, since the fibres are isometric to $\mathbf{S^1}$ and g' is the Fubini–Study metric of holomorphic sectional curvature 4. From Remark 1.6, one has $\|A\|^2 = 2n$ and then

$$\phi(t) = 2n(-t + 2n + 2) t^{1/(2n+1)}, \; t > 0 \, .$$

The unique critical point of ϕ is $t = 1$, since

$$\phi'(t) = -\frac{4n(n+1)}{2n+1}(t-1) t^{\frac{2n+1}{2n}} \, .$$

Thus, g is the unique Einstein metric in the canonical variation.

Example 5.2 Let $\pi : (\mathbf{S^{2n+1}}(1,k), g_a) \to (\mathbf{P_n}(\mathbf{C})(k+3), G)$ be the submersion with totally geodesic fibres considered in Remarks 1.3, 1.6. The metric $g_a = \pi^*G + a^2\eta\otimes\eta$ is the canonical variation of $g = \pi^*G + \eta\otimes\eta$. Since $(\mathbf{S^{2n+1}}(1,k), g)$ is a Sasakian manifold of constant φ-sectional curvature k, the Ricci tensor acts as:

$$\rho(X,Y) = \frac{1}{2}(k(n+1)+3n-1)g(X,Y) - \frac{1}{2}(n+1)(k-1)\eta(X)\eta(Y)\ ,$$

for any $X, Y \in \mathcal{X}(\mathbf{S^{2n+1}})$. In this case $||A||^2 = 2n, \hat{\tau} = 0, \tau' = n(n+1)(k+3)$ and the function ϕ is given by $\phi(a^2) = (-2a^2 + n(n+1)(k+3))a^{2/(2n+1)}$, its derivative vanishing at $a^2 = (k+3)/4$. Via (5.9) and the above formula, the Ricci tensor of g_a is given by:

$$\begin{aligned} &\rho_a(\xi,\xi) = 2na^2 g_a(\xi,\xi),\ \ \rho_a(\xi, X) = 0\ , \\ &\rho_a(X,Y) = (\tfrac{1}{2}(k+3)(n+1) - 2a^2)g(X,Y)\ , \end{aligned}$$

where $X, Y \in \mathcal{X}^h(\mathbf{S^{2n+1}})$. Thus, for $a = \frac{1}{2}(k+3)^{1/2}$, one obtains an Einstein metric on $\mathbf{S^{2n+1}}$. It is isometric to the canonical metric on the sphere of radius $1/a$, since $a^2 = (k+3)/4$ is the constant sectional curvature of g_a. Indeed we recall that the corresponding invariant $\tilde{A}$ acts as $\tilde{A}_X Y = a g_a(\varphi X, Y)\tilde{\xi}$, $\tilde{A}_X\tilde{\xi} = -a\varphi(X)$, where $a\tilde{\xi} = \xi$ and X, Y are horizontal. Via (1.31), for any orthonormal basis $\{X, Y\}$ of a horizontal 2-plane α, one obtains:

$$K(\alpha) = \frac{k+3}{4}(1 + 3g(\varphi X, Y)^2) - 3a^2 g(\varphi X, Y)^2 = \frac{k+3}{4}\ .$$

Moreover, considering $\{X, \tilde{\xi}_p\}$ as an orthonormal basis of a mixed 2-plane α at a point $p \in \mathbf{S^{2n+1}}$, we have

$$K(\alpha) = g_a(\tilde{A}_X\tilde{\xi}_p, \tilde{A}_X\tilde{\xi}_p) = \frac{k+3}{4}\ .$$

Now, we state conditions for the existence of two Einstein metrics in the canonical variation of g.

Lemma 5.2 *The function ϕ has two critical points if and only if*

$$\hat{\tau} > 0\ ,\ \tau' > 0\ ,\ r^2\tau'^2 - 4n(n+2r)||A||^2\hat{\tau} > 0\ .$$

The scalar curvature corresponding to a critical point is strictly positive.

Proof. The discussion of the equation $\phi'(t) = 0$, or equivalently,

$$(2r+n)||A||^2t^2 - r\tau' t + n\hat{\tau} = 0\ , \tag{5.12}$$

implies the existence of two positive solutions only when

$$\hat{\tau} > 0\,, \quad \tau' > 0\,, \quad \Delta = r^2\tau'^2 - 4n(n+2r)\|A\|^2\hat{\tau} > 0\,.$$

Moreover, via (5.10), the scalar curvature corresponding to a solution t_0 of (5.12) is

$$\tau_0 = \frac{m}{t_0(2r+n)}(\tau' t_0 + 2\hat{\tau}) > 0\,.$$

This Lemma helps in proving the main result of this section (see also [26] 9.73).

Theorem 5.1 (Bérard-Bergery). *Let $\pi : (M,g) \to (B,g')$ be a Riemannian submersion with totally geodesic fibres of dimension $r = m-n$, $m = \dim M$, $n = \dim N$. If M is compact, connected with $m \geq 3$, there are two different Einstein metrics in the canonical variation of g if and only if:*

i) *the fibres and (B,g') are Einstein manifolds with positive scalar curvature;*
ii) *$\|A\|$ is constant and $r^2\tau'^2 - 4n(n+2r)\|A\|^2\hat{\tau} > 0$;*
iii) *with respect to a local orthonormal π-adapted frame $\{X_i, U_j\}$, one has:*

$$\textstyle\sum_{i=1}^n g(A_{X_i}U, A_{X_i}V) = \frac{1}{r}\|A\|^2 g(U,V)\,, \quad U,V \in \mathcal{X}^v(M)\,,$$

$$\textstyle\sum_{j=1}^r g(A_X U_j, A_Y U_j) = \frac{1}{n}\|A\|^2 g(X,Y), \quad X,Y \in \mathcal{X}^h(M)\,;$$

iv) $\sum_{i=1}^n (\nabla_{X_i}A)_{X_i} = 0$.

Proof. Assuming the existence of two Einstein metrics g_{t_1}, g_{t_2}, according to Lemma 5.2, we have $\hat{\tau} > 0, \tau' > 0, r^2\tau'^2 - 4n(n+2r)\|A\|^2\hat{\tau} > 0$, with $\tau', \hat{\tau}$ and $\|A\|$ constant. From (5.9) we obtain iv) and the relations:

$$\textstyle\hat{\rho}(U,V) + t_h^2 \sum_{i=1}^n g(A_{X_i}U, A_{X_i}V) = \frac{1}{m}\tau_h t_h g(U,V)\,,$$

$$\textstyle\rho'(X',Y') \circ \pi - 2t_h \sum_{i=1}^n g(A_X X_i, A_X X_i) = \frac{1}{m}\tau_h g(X,Y)\,,$$

for any $h \in \{1,2\}$, $U,V \in \mathcal{X}^v(M)$, $X,Y \in \mathcal{X}^b(M)$, π-related to X',Y'. Since

$$\frac{1}{m}t_h\tau_h = \frac{1}{r}\hat{\tau} + t^2\|A\|^2\,, \quad \frac{1}{m}n\tau_h = \tau' - 2t_h\|A\|^2$$

and $t_1 \neq t_2$, we have:

$\hat{\rho}(U,V) = \frac{1}{r}\hat{\tau} g(U,V)$,

$\sum_{i=1}^{n} g(A_{X_i}U, A_{X_i}V) = \frac{1}{r}\|A\|^2 g(U,V)$,

$\rho'(X',Y') = \frac{1}{n}\tau' g'(X',Y')$,

$\sum_{j=1}^{r} g(A_X U_j, A_Y U_j) = \sum_{i=1}^{n} g(A_X X_i, A_Y X_i) = \frac{1}{n}\|A\|^2 g(X,Y)$,

and i), iii) hold, too. The converse statement follows immediately from Lemma 5.2 and (5.9).

5.3 Homogeneous Einstein Spaces

Several examples of Riemannian submersions with totally geodesic fibres can be obtained by considering reductive homogeneous spaces.

Let H be a compact subgroup of a Lie group G and $p : G \to G/H$ the natural projection, which acts as the projection of an H-principal bundle. Let $\mathfrak{g}, \mathfrak{h}$ be the Lie algebras of G, H, respectively. Since H is compact, one can consider a splitting

$$\mathfrak{g} = \mathfrak{h} \oplus \mathfrak{m} , \tag{5.13}$$

where $\mathfrak{m}$ is an $Ad(H)$-invariant complement of $\mathfrak{h}$ in $\mathfrak{g}$, *i.e.* $[\mathfrak{h}, \mathfrak{m}] \subseteq \mathfrak{m}$.

It is well known ([172] Vol. II, Chap. X) that any $Ad_G(H)$-invariant inner product $<,>$ on $\mathfrak{m}$ canonically induces a left-invariant metric g' on G/H such that:

$$g'(X,Y)_{eH} = < X, Y >, \quad X, Y \in \mathfrak{m} . \tag{5.14}$$

Identifying the Lie algebra $\mathfrak{g}$ with the set of the Killing vector fields of $(G/H, g')$ generating one parameter subgroups of G, the Lie subalgebra $\mathfrak{h}$ of $\mathfrak{g}$ corresponds to the subalgebra of the Killing vector fields which vanish at eH and $\mathfrak{m}$ is identified with the tangent space $T_{eH}(G/H)$, via the map which associates to any Killing vector field its value at eH. The Levi-Civita connection ∇' on G/H acts on any pair (X,Y) of Killing vector fields according to:

$$(\nabla'_X Y)_{eH} = -\frac{1}{2}[X,Y]_{\mathfrak{m}} + U(X,Y) , \tag{5.15}$$

where $U : \mathfrak{m} \times \mathfrak{m} \to \mathfrak{m}$ is determined by:

$$2 < U(X,Y), Z >=< [Z,X]_{\mathfrak{m}}, Y > + < X, [Z,Y]_{\mathfrak{m}} > , \quad Z \in \mathfrak{m} .$$

Now, let K be a closed subgroup of H, with Lie algebra $\mathfrak{k}$ and consider an $Ad(K)$-invariant complement $\mathfrak{p}$ of $\mathfrak{k}$ in $\mathfrak{h}$. Then $\mathfrak{g}$ splits as $\mathfrak{g} = \mathfrak{k} \oplus \mathfrak{p} \oplus \mathfrak{m}$, and $\mathfrak{p} \oplus \mathfrak{m}$ becomes an $Ad(K)$-invariant complement of $\mathfrak{k}$ in $\mathfrak{g}$. Moreover, considering an $Ad_G(H)$-invariant inner product $<,>$ on $\mathfrak{m}$ and an $Ad_H(K)$-invariant inner product $(,)$ on $\mathfrak{p}$, one obtains the left-invariant metric g on G/K, determined by:

$$g(X+V, Y+W)_{eK} =< X,Y > +(V,W) , \tag{5.16}$$

for any $X, Y \in \mathfrak{m}, \quad V, W \in \mathfrak{p}$.

Let $\pi : G/K \to G/H$ be the map such that $\pi(xK) = xH$, $x \in G$. Then, π acts as the projection of the H/K-bundle associated with $p : G \to G/H$ and it can be viewed as a submersion between the Riemannian manifolds $(G/K, g)$, $(G/H, g')$, g, g' being determined by (5.14), (5.16).

Theorem 5.2 *The map $\pi : (G/K, g) \to (G/H, g')$ is a Riemannian submersion with totally geodesic fibres isometric to H/K with the invariant metric induced by the inner product on $\mathfrak{p}$.*

Proof. With the above notation, $\mathfrak{p} \oplus \mathfrak{m}$ and $\mathfrak{m}$ are respectively identified with the tangent spaces $T_{eK}(G/K), T_{eH}(G/H)$. So we can regard the differential of π at eK as the map:

$$\pi_* : (V, X) \in \mathfrak{p} \oplus \mathfrak{m} \mapsto X \in \mathfrak{m} ,$$

which preserves the length of any vector $X \in \mathfrak{m}$. Thus, the G-equivariance of π implies that π is a Riemannian submersion with $\mathfrak{p}$, $\mathfrak{m}$ as the vertical, horizontal spaces at eK, and the metric induced by the inner product on $\mathfrak{p}$ makes H/K a Riemannian manifold isometric to any fibre. Moreover, according to (5.15), the Levi-Civita connection ∇ of G/K acts on Killing vector fields V, W on $\mathfrak{p}$ as:

$$(\nabla_V W)_{eK} = -\frac{1}{2}[V, W]_{\mathfrak{p}\oplus\mathfrak{m}} + U(V, W) .$$

Since $[V, W]_{\mathfrak{p}\oplus\mathfrak{m}} = [V, W]_{\mathfrak{p}}$ and, for any $Z \in \mathfrak{m}$, $[Z, V], [Z, W]$ are in $\mathfrak{m}$, we have $U(V, W) = 0$ and $(\nabla_V W)_{eK} = -\frac{1}{2}[V, W]_{\mathfrak{p}}$. So, the invariant T vanishes at eK and then $T = 0$.

In [354] W. Ziller describes the invariant metrics on the spheres and on the projective spaces obtained with the following procedure. Looking at each of these spaces as a homogeneous space G/H, H being a compact Lie group, one considers a splitting of the Lie algebra of G as in (5.13). The adjoint action of H on $\mathfrak{m}$ induces a splitting $\mathfrak{m} = \mathfrak{m}_0 \oplus \mathfrak{m}_1 \oplus \cdots \oplus \mathfrak{m}_r$, H acting trivially on $\mathfrak{m}_0$ and irreducibly on any $\mathfrak{m}_i$, $i \in \{1, \dots, r\}$. Assuming that the representations induced by $Ad(H)$ on each $\mathfrak{m}_i$ are inequivalent, then the metrics on G/H are the invariant ones determined by an inner product on $\mathfrak{m}$, $<,>= h \oplus \sum_{i=1}^{r} a_i \beta_{|\mathfrak{m}_i}$, where h is an arbitrary inner product on $\mathfrak{m}_0$, $a_i \in \mathbf{R}_+^*$ and $\beta_{|\mathfrak{m}_i}$ is the restriction to $\mathfrak{m}_i$ of the inner product corresponding to a bi-invariant metric on G.

Moreover, Ziller classifies the Einstein metrics among the ones so obtained. Further results are recently obtained by M. Kerr who classifies the homogeneous Einstein metrics on compact irreducible symmetric spaces of rank greater than 1 which are not simple Lie groups ([167]). Anyway, this construction leaves the context of Riemannian submersions.

Now, we are going to describe the metrics in [354] involving the submersions considered in Chaps. 1, 2.

Example 5.3 Considering $G = U(n+1)$, $H = U(1)U(n)$, $K = U(n)$, the Lie algebra of G splits as $\mathfrak{u}(n+1) = \mathfrak{u}(n) \oplus \mathfrak{u}(1) \oplus \mathfrak{m}$, where

$$\mathfrak{m} = \left\{ \begin{pmatrix} 0 & -{}^t\overline{\eta} \\ \eta & 0 \end{pmatrix} ; \eta \in \mathbf{C^n} \right\}$$

is identified with $\mathbf{C^n}$ and H acts irreducibly on $\mathbf{C^n}$. The inner product β on $\mathfrak{u}(n+1)$ given by:

$$\beta(A, A') = \frac{1}{2} \mathrm{tr}\, A\, {}^t\overline{A'}$$

is $U(n+1)$-invariant and induces the standard inner product on $\mathbf{C^n}$. For any $c > 0$, $\frac{4}{c} <,>$ determines the Fubini–Study metric g_c of holomorphic sectional curvature c on $\mathbf{P_n(C)} = U(n+1)/U(1)U(n)$. As in [354], the $U(n+1)$-invariant metrics on $\mathbf{S^{2n+1}} = U(n+1)/U(n)$ set up a two parameter family of homogeneous metrics. In fact, $U(n)$ acts trivially on $\mathfrak{u}(1)$ and the inner products β' on $\mathfrak{u}(1) \oplus \mathbf{C^n}$ which describe the $U(n+1)$-invariant metrics on $\mathbf{S^{2n+1}}$, via (5.16), are obtained by:

$$\beta'(\lambda + \eta, \lambda' + \eta') = a \,\mathrm{Re}(\lambda \overline{\lambda'}) + \frac{4}{c} < \eta, \eta' > ,$$

for any $\lambda, \lambda' \in \mathfrak{u}(1)$, $\eta, \eta' \in \mathbf{C^n}$, with $a, c \in \mathbf{R}_+^*$. The canonical metric g on $\mathbf{S^{2n+1}}$, which is realized for $a = 1, c = 4$, turns out to be the only Einstein homogeneous metric on $\mathbf{S^{2n+1}}$ ([156]), and the projection π considered in Theorem 5.2 coincides with the Hopf fibration.

Example 5.4 One regards the projective space $\mathbf{P_n(Q)}$ as the irreducible, compact, symmetric space of type I, $Sp(n+1)/Sp(n)Sp(1)$, with the homogeneous metric g' of constant quaternionic sectional curvature 4. Up to homotheties, g' is the only $Sp(n+1)$-invariant metric on $\mathbf{P_n(Q)}$, corresponding to $-B$, B being the Killing–Cartan form on $\mathfrak{sp}(n+1)$. We consider $Sp(n)$ as a compact subgroup of $Sp(n)Sp(1)$ and the splitting $\mathfrak{sp}(n+1) = \mathfrak{sp}(n) \oplus \mathfrak{sp}(1) \oplus \mathfrak{m}$, where

$$\mathfrak{m} = \left\{ \begin{pmatrix} 0 & \eta \\ -{}^t\overline{\eta} & 0 \end{pmatrix} ; \eta \in \mathbf{Q^n} \right\}$$

is identified with $\mathbf{Q^n}$ and $Sp(n)$ acts trivially on $Sp(1)$ and irreducibly on $\mathbf{Q^n}$, via the standard action. Then, one obtains the natural inner products on $\mathbf{Q^n}$ and $\mathfrak{sp}(1)$, *i.e.*

$$< \eta, \eta' >= \mathrm{Re}({}^t\overline{\eta}\eta') \ , \ \eta, \eta' \in \mathbf{Q^n} \ , \quad (\lambda, \lambda') = \mathrm{Re}(\overline{\lambda}\lambda') \ , \ \lambda, \lambda' \in \mathfrak{sp}(1) \ .$$

Via (5.16), the corresponding metric on $\mathbf{S^{4n+3}} = Sp(n+1)/Sp(n)$ is just the metric g of sectional curvature $k = 1$ and the Riemannian submersion $\pi : (\mathbf{S^{4n+3}}, g) \to (\mathbf{P_n(Q)}, g')$ in Theorem 5.2 is just the generalized Hopf fibration described in Example 1.4. Now, we evaluate the function ϕ determined by the canonical variation of g. Since in this case the invariant A acts on the horizontal vector fields as $A_X Y = -g(IX,Y)IN - g(JX,Y)JN - g(KX,Y)KN$, one directly obtains $||A||^2 = 12n$. The scalar curvature of $\mathbf{P_n(Q)}$ is given by $\tau' = 16n(n+2)$ and applying (5.11) we have:

$$\phi(t) = -2\left(6nt - 8n(n+2) - \frac{3}{t}\right) t^{-4n}, \ t > 0 \ .$$

Therefore $\phi' = 0$ yields $(3n+2)t^2 - 2(n+2)t + 1 = 0$, with solutions 1 and $t_0 = 1/(2n+3)$. Via Theorem 5.1, t_0 really determines an Einstein metric on $\mathbf{S^{4n+3}}$ if and only if A satisfies, with respect to horizontal and vertical orthonormal frames $\{X_i\}_{1 \le i \le 4n}$, $\{U_j\}_{1 \le j \le 3}$, the conditions:

$$\textstyle\sum_{i=1}^{4n} g(A_{X_i}U, A_{X_i}V) = 4ng(U,V) \ , \quad U, V \in \mathcal{X}^v(\mathbf{S^{4n+3}}) \ ,$$

$$\textstyle\sum_{j=1}^{3} g(A_X U_j, A_Y U_j) = 3g(X,Y) \ , \quad X, Y \in \mathcal{X}^h(\mathbf{S^{4n+3}}) \ .$$

Considering the vertical orthonormal frame $\{IN, JN, KN\}$, for any horizontal X we have $A_X IN = IX$, $A_X JN = JX$, $A_X KN = KX$ and these formulas imply the previous conditions. As proved in [354], g and g_{t_0} are the only $Sp(n+1)$-invariant metrics on $\mathbf{S^{4n+3}}$ in a four-parameter family of homogeneous metrics.

Example 5.5 The complex projective space $\mathbf{P_{2n+1}(C)}$ can be realized as the homogeneous space $Sp(n+1)/Sp(n)U(1)$, considering the group $K = Sp(n)U(1)$ as a closed subgroup of $Sp(n)Sp(1)$. Then, one obtains the splitting $\mathfrak{sp}(n+1) = \mathfrak{sp}(n) \oplus \mathfrak{u}(1) \oplus \mathfrak{p} \oplus \mathfrak{m}$, where $\mathfrak{p}, \mathfrak{m}$ are identified with $\mathbf{C}$, $\mathbf{Q^n}$, respectively. The action of $K = Sp(n)U(1)$ is irreducible on $\mathbf{C}$, $\mathbf{Q^n}$. In fact, $U(1)$ acts trivially on $\mathbf{Q^n}$ and as a rotation on $\mathbf{C}$, while the action of $Sp(n)$ is trivial on $\mathbf{C}$ and the standard one on $\mathbf{Q^n}$. According to [354], one obtains a two-parameter family of $Sp(n+1)$-invariant metrics on $\mathbf{P_{2n+1}(C)}$ determined by the invariant products on $\mathbf{C} \oplus \mathbf{Q^n}$ such that:

$$< \lambda + \eta, \lambda' + \eta' >= a_1(a_2 \mathrm{Re}(\overline{\lambda}\lambda') + \mathrm{Re}({}^t\overline{\eta}\eta')) \, , \tag{5.17}$$

for any λ, $\lambda' \in \mathbf{C}$, η, $\eta' \in \mathbf{Q^n}$, $a_1, a_2 \in \mathbf{R}^*_+$. The Fubini–Study metric g_4 on $\mathbf{P_{2n+1}(C)}$ is realized for $a_1 = a_2 = 1$. According to Theorem 5.2, the projection $\rho : (\mathbf{P_{2n+1}(C)}, g_4) \to (\mathbf{P_n(Q)}, g')$ is a Riemannian submersion with fibres isometric to $\mathbf{P_1(C)}$; it coincides with the one considered in Example 2.1. Moreover, (5.17) implies that the Einstein $Sp(n+1)$-invariant metrics on $\mathbf{P_{2n+1}(C)}$ are in the canonical variation of g_4, up to homotheties. So, we evaluate the function ϕ corresponding to g. Since in this case $\hat{\tau} = 8$, $\tau' = 16n(n+2)$, $||A||^2 = 8n$, one has:

$$\phi(t) = 8(-nt^2 + 2n(n+2)t + 1)t^{-2n/(2n+1)}, \ \ t > 0 \, ,$$

and the derivative ϕ' vanishes at 1, and at $t_0 = 1/(n+1)$. Now, we prove that t_0 is a critical value of ϕ applying Theorem 5.1. To this aim, we need to prove that the invariant A of ρ satisfies, with respect to local horizontal and vertical frames $\{X_i\}_{1 \le i \le 4n}$, $\{U_j\}_{1 \le j \le 2}$, the conditions:

$$\textstyle\sum_{i=1}^{4n} g_4(A_{X_i}U, A_{X_i}V) = 4n g_4(U,V), \ \ U, V \in \mathcal{X}^v(\mathbf{P_{2n+1}(C)}) \, ,$$

$$\textstyle\sum_{j=1}^{2} g_4(A_X U_j, A_Y U_j) = 2 g_4(X,Y), \ \ X, Y \in \mathcal{X}^h(\mathbf{P_{2n+1}(C)}) \, .$$

As in Example 2.1, we get $\rho \circ \pi = \pi'$, where $\pi : (\mathbf{S^{4n+3}}, g) \to (\mathbf{P_{2n+1}(C)}, g_4)$ and $\pi' : (\mathbf{S^{4n+3}}, g) \to (\mathbf{P_n(Q)}, g')$ are the Hopf submersion and the generalized Hopf submersion, respectively. The invariant A' of π' acts on the vertical distribution by

$$A'_{X'} IN = IX' \, , \ A'_{X'} JN = JX' \, , \ A'_{X'} KN = KX' \, , \quad X' \in \mathcal{X}^h(\mathbf{S^{4n+3}}) \, .$$

Since the vertical distribution of ρ is spanned, pointwise, by $V_1 = \pi_*(JN)$, $I'V_1 = \pi_*(KN)$, I' being the complex structure on $\mathbf{P_{2n+1}}(\mathbf{C})$, we obtain, for any $X \in \mathcal{X}^h(\mathbf{P_{2n+1}}(\mathbf{C}))$:

$$A_X V_1 = \pi_*(A'_{X'} JN) = \pi_*(JX')\ , \quad A_X I'V_1 = I'(\pi_*(JX'))\ ,$$

where X' is the π-horizontal lift of X. Then,

$$\begin{aligned}\textstyle\sum_{i=1}^{4n} g_4(A_{X_i}V_1, A_{X_i}V_1) &= \textstyle\sum_{i=1}^{4n} g_4(\pi_*(JX_i'), \pi_*(JX_i'))\\ &= \textstyle\sum_{i=1}^{4n} g(JX_i', JX_i') = 4n\ ,\end{aligned}$$

and, analogously,

$$\sum_{i=1}^{4n} g_4(A_{X_i}I'V_1, A_{X_i}I'V_1) = 4n\ ,$$

thus implying the first condition. Finally, for any $X \in \mathcal{X}^h(\mathbf{P_{2n+1}}(\mathbf{C}))$, one has:

$$\begin{aligned} g_4(A_X V_1, A_X V_1) + g_4(A_X I'V_1, A_X I'V_1) &= g(JX', JX') + g(KX', KX')\\ &= 2g(X', X') = 2g_4(X, X)\ .\end{aligned}$$

This proves the existence of two invariant Einstein metrics g_4, and g_{t_0} on $\mathbf{P_{2n+1}}(\mathbf{C})$; both of them are Hermitian with respect to the standard complex structure on $\mathbf{P_{2n+1}}(\mathbf{C})$, but, contrary to g_4, g_{t_0} has non-constant holomorphic sectional curvatures.

Example 5.6 Putting $G = Spin(9)$, $H = Spin(8)$, $K = Spin(7)$, H is regarded as a closed subgroup of G via the canonical inclusion, while the inclusion of K in H is the usual one followed by a triality automorphism, that is an outer automorphism of $Spin(8)$ of order three ([184], Chap. 1). The Lie algebra of H splits as $\mathfrak{spin}(8) \cong \mathfrak{so}(8) \cong \mathfrak{spin}(7) \oplus \mathfrak{p}$, where $\dim \mathfrak{p} = 7$ and the isotropy representation of $Spin(7)$ on $\mathfrak{p}$ is the only seven-dimensional irreducible one. With respect to the splitting $\mathfrak{spin}(9) \cong \mathfrak{so}(9) \cong \mathfrak{spin}(7) \oplus \mathfrak{p} \oplus \mathfrak{m}$, with $\dim \mathfrak{m} = 8$, the action of $Spin(7)$ on $\mathfrak{m}$ is irreducible, too. By a suitable normalization, according to (5.16), one obtains a one-parameter family of $Spin(9)$-invariant metrics on $Spin(9)/Spin(7) \cong \mathbf{S}^{15}$ and, up to isometries, all the homogeneous $Spin(9)$-invariant metrics on $\mathbf{S}^{15}$ are in this family. Except for the metric g of constant sectional curvature, a $Spin(9)$-invariant metric is not isometric to any of the metrics on $\mathbf{S}^{15}$ considered in Example 5.4, for $n = 3$. Since moreover for any $n \geq 3$, $Spin(n)/\mathbf{Z_2} \cong SO(n)$,

$Spin(n)$ being the universal covering of $SO(n)$, one obtains:

$$H/K \cong SO(8)/SO(7) = \mathbf{S^7},\ G/H \cong SO(9)/SO(8) = \mathbf{S^8} \cong \mathbf{S^8}\left(\frac{1}{2}\right),$$

and the submersion $\pi : (\mathbf{S^{15}}, g) \to (\mathbf{S^8}(\frac{1}{2}), g')$ in Theorem 2.6 is just the one defined in Remark 2.2. Via (5.4) the corresponding invariant A satisfies $||A||^2 = 56$, (5.12) reduces to $11t^2 - 14t + 3 = 0$, so $t = 3/11$ determines a new Einstein homogeneous metric on $\mathbf{S^{15}}$.

Further examples of compact homogeneous Einstein manifolds can be obtained going deep in the theory just explained, ([26; 156; 321]).

Considering again a compact connected Lie group G, let q be a bi-invariant metric on G. Given a closed subgroup H of G, one considers a q-orthogonal reductive decomposition $\mathfrak{g} = \mathfrak{h} \oplus \mathfrak{m}$ and an invariant metric g on G/H such that $Vol(G/H, g) = 1$. In [321], Wang and Ziller state necessary and sufficient conditions for the boundedness of the total scalar curvature functional. In particular they state the following result.

Theorem 5.3 *Let H be a maximal connected subgroup of a compact Lie group G. Then G/H has a G-invariant Einstein metric, corresponding to a maximum of the total scalar curvature functional.*

Thus, homogeneous spaces which do not admit invariant Einstein metrics can be determined considering pairs (H, K) of closed subgroups of a compact Lie group G, with $K \subset H \subset G$. We assume that G/H, H/K are isotropy irreducible spaces and consider $(-B)$-orthogonal splittings $\mathfrak{g} = \mathfrak{h} \oplus \mathfrak{m} = \mathfrak{k} \oplus \mathfrak{p} \oplus \mathfrak{m}$, B denoting the Killing–Cartan form of $\mathfrak{g}$. Further, we assume that the isotropy representation of G/H stays irreducible when restricted to $\mathfrak{k}$. In this case, any $Ad(K)$-invariant inner product on $\mathfrak{p} \oplus \mathfrak{m}$ is diagonalized with respect to $-B$, and the irreducibility of $\mathfrak{p}$ and $\mathfrak{m}$ also implies that $\mathfrak{p}$, $\mathfrak{m}$ are also orthogonal and $Ad(K)$-invariant with respect to the given inner product. In this way, one obtains a two-parameter family of invariant metrics on G/K, so that, up to homotheties, the invariant Einstein metrics on G/K are in the canonical variation of the Riemannian submersion $\pi : (G/K, g) \to (G/H, g')$ where g, g' are respectively determined by $-B$, $-B_{|\mathfrak{m}}$. At the point eK, the invariant A of π acts as:

$$A_X U = -\frac{1}{2} h([X, U]_{\mathfrak{p} \oplus \mathfrak{m}}) = -\frac{1}{2}[X, U]_{\mathfrak{m}},\ \ X \in \mathfrak{m},\ U \in \mathfrak{p}\ .$$

With respect to an orthonormal basis $\{X_i\}$ of $\mathfrak{m}$, via (5.16), we have:

$$\sum_{i=1}^n g_{eK}(A_{X_i}U, A_{X_i}U) = \tfrac{1}{4}\sum_{i=1}^n B(ad_U(ad_U X_i), X_i)$$

$$= \tfrac{1}{4}(B_H(U,U) - B(U,U)) \; ,$$

where B_H is the Killing–Cartan form of H. We assume that $\mathfrak{p}$, $\mathfrak{m}$ are nonequivalent, irreducible summands and that the Killing–Cartan form of the normal subgroup H' of H acting effectively on H/K satisfies $B' = cB_{|\mathfrak{p}}$, $c > 0$; so the inner product on $\mathfrak{p}$ is given by $-cB_{|\mathfrak{p}}$ and, from the above relation one has:

$$||A||^2 = \frac{c-1}{4}\sum_{j=1}^r B(U_j, U_j) = \frac{1-c}{4} r \; ,$$

$\{U_j\}$ being a local vertical orthonormal frame. Finally, the scalar curvatures τ' of G/H and $\hat{\tau}$ of H/K are given by $\tau' = \frac{1}{2}n \, , \, \hat{\tau} = \frac{1}{2}cr$, provided that G/H, H/K are symmetric spaces. Then, (5.12) reduces to

$$\left(\frac{r}{n} + \frac{1}{2}\right)(1-c)t^2 - t + c = 0 \; ,$$

without real solutions if and only if $2c(1-c)(1+2r/n) > 1$.

Example 5.7 One considers the compact, simple Lie group $G = SO(2q)$ and its closed subgroup $H = U(q)$ canonically embedded via the map

$$a + ib \in U(q) \mapsto \begin{pmatrix} a & -b \\ b & a \end{pmatrix} \in SO(2q) \; .$$

The Killing–Cartan form acts on $\mathfrak{so}(2q)$ as $B(A, A') = 2(q-1)\,\mathrm{tr}\, AA'$. Let $\mathfrak{m}$ be the $(-B)$-orthogonal complement of $\mathfrak{u}(q)$ in $\mathfrak{so}(2q)$; the inner product $<,>= -B_{|\mathfrak{m}}$ induces on $SO(2q)/U(q)$ the standard metric which makes such manifold a $q(q-1)$-dimensional irreducible Riemannian symmetric space. Now, considering $K = U(1)SO(q)$ as a closed subgroup of H (via the canonical embedding), the Lie algebra $\mathfrak{u}(q)$ splits as $\mathfrak{u}(q) = \mathfrak{u}(1) \oplus \mathfrak{so}(q) \oplus \mathfrak{p}$, where

$$\mathfrak{p} = \{ib \mid b = (b^i_j) \in \mathfrak{gl}(q, \mathbf{R}) \; , \; b = {}^t b \; , \; b^1_1 = 0\} \; .$$

One can prove that the representations of $Ad(H)$ on $\mathfrak{p}$, $\mathfrak{m}$ are inequivalent ([321]). Now, the group $SU(q)$ acts effectively on $U(q)/U(1)SO(q)$ and its

Killing form B' is given by $B'(X,Y) = 2q \operatorname{tr} XY, \quad X,Y \in \mathfrak{su}(q)$ ([139]). Then, on $\mathfrak{p}$, we have:

$$B'(ib, ib') = -2q \operatorname{tr} bb' ,$$

$$B(ib, ib') = 2(q-1) \operatorname{tr} \begin{pmatrix} -bb' & 0 \\ 0 & -bb' \end{pmatrix} = -4(q-1) \operatorname{tr} bb' .$$

According to the above notation, $c = q/(2q-2)$, and the metric on the $\frac{1}{2}(q+2)(q-1)$-dimensional irreducible symmetric space $U(q)/U(1)SO(q)$ is determined by the inner product $(\ ,\ ) = -\frac{q}{2(q-1)} B_{|\mathfrak{p}}$. In this case, we have

$$2c(1-c)\left(1 + \frac{2r}{n}\right) = \frac{(q-2)(q+1)}{(q-1)^2} > 1 \quad \text{if and only if} \quad q > 3 .$$

So, if $q \geq 4$, $SO(2q)/U(1)SO(q)$ does not admit invariant Einstein metrics.

Example 5.8 We consider the compact, simple Lie group $G = SU(p+q)$ and its closed subgroup $H = S(U(p)U(q))$ canonically embedded in G. The homogeneous manifold $\mathbf{G_q}(\mathbf{C^{p+q}}) = G/H$ is the complex Grassmannian of the q-planes in $\mathbf{C^{p+q}}$. It is a $2pq$-dimensional irreducible Riemannian symmetric space with the standard metric. Let $K = S(SO(p)U(1)U(q))$ and consider the splitting

$$\mathfrak{s}(\mathfrak{u}(p) \oplus \mathfrak{u}(q)) = \mathfrak{so}(p) \oplus \mathfrak{u}(1) \oplus \mathfrak{su}(q) \oplus \mathfrak{p} ,$$

where

$$\mathfrak{p} = \{ib \mid b = (b^i_j) \in \mathfrak{sl}(p, \mathbf{R}) \ , \ b = {}^t b \ , \ b^1_1 = 0\} .$$

Since $SU(p)$ acts effectively on H/K, let B' be the Killing form of $SU(p)$, whose action on $\mathfrak{p}$ is given by $B'(ib, ib') = -2p \operatorname{tr} bb'$. The restriction to $\mathfrak{p}$ of the Killing form of $SU(p+q)$ is given by:

$$B(ib, ib') = 2(p+q) \operatorname{tr} \begin{pmatrix} -bb' & 0 \\ 0 & 0 \end{pmatrix} = -2(p+q) \operatorname{tr} bb' ,$$

so $c = p/(p+q)$. We observe that H/K is an irreducible symmetric space, with the metric given by $(\ ,\ ) = -\frac{p}{p+q} B_{|\mathfrak{p}}$, and $\dim H/K = \frac{1}{2}(p-1)(p+2)$. Then, we have

$$2c(1-c)\left(1 + \frac{2r}{n}\right) = \frac{p^2 + 2pq + p - 2}{(p+q)^2}$$

and $SU(p+q)/S(SO(p)U(1)U(q))$ does not admit homogeneous Einstein metrics, if $q^2 < p-2$.

We end this section with a technical lemma, useful in the next section, whose proof is reported in [259]. Indeed, this lemma follows from a result proved by Matsuzawa ([197]) and Bérard-Bergery (9.74 in [26]) and used by Wang and Ziller in [320] to get many new examples of homogeneous Einstein metrics. Let H be a compact, semisimple Lie subgroup of a compact, simple Lie group G, and consider again a closed subgroup K of H, the splitting of the corresponding Lie algebras $\mathfrak{g} = \mathfrak{h} \oplus \mathfrak{m} = \mathfrak{k} \oplus \mathfrak{p} \oplus \mathfrak{m}$, where $\mathfrak{p}$, $\mathfrak{m}$ are $Ad(K)$-invariant, irreducible summands such that $[\mathfrak{p}, \mathfrak{p}] \subseteq \mathfrak{k}$, $[\mathfrak{m}, \mathfrak{m}] \subseteq \mathfrak{h}$. For any $t > 0$, the G-invariant metric g_t on G/K corresponding to the inner product $-B_{|\mathfrak{m}} - tB_{|\mathfrak{p}}$ (B being the Killing–Cartan form of $\mathfrak{g}$) makes the projection $\pi : (G/K, g_t) \to (G/H, g')$ a Riemannian submersion onto the Einstein manifold $(G/H, g')$, g' being determined by $-B_{|\mathfrak{m}}$. Indeed, for $t = 1$, g_1 is the so-called standard metric on G/K and $\{g_t\}_{t>0}$ is its canonical variation. As before, the $Ad(K)$-irreducibility of $\mathfrak{p}$ gives the existence of a constant $c > 0$ such that the Killing–Cartan form of $\mathfrak{h}$ is $B' = cB_{|\mathfrak{p}}$.

Lemma 5.3 *Assume that g_{t_0} is an Einstein metric in the canonical variation of the standard metric on G/K. If $c < t_0$, $c \neq t_0/2$, then $t_1 = ct_0/(t_0 - c)$ determines another G-invariant Einstein metric g_{t_1} on G/K.*

5.4 Einstein Metrics on Principal Bundles

Several results in Sec. 5.1 are useful to describe a relevant class of Einstein metrics. Such metrics occur on the total space of torus bundles over a product of Kähler–Einstein manifolds with positive first Chern class. Here we only sketch the method developed by M. Wang and W. Ziller; a detailed description of the topological properties of the considered manifolds, together with several geometric applications, is given in [322].

Let $\pi : P \to B$ be the projection of a principal T^r-bundle, T^r denoting a real r-dimensional torus acting on P on the right. We fix on T^r the left invariant metric $\hat{g}$ corresponding to an inner product $<,>$ on the Lie algebra $\mathfrak{t}^r$ of T^r; note that $\hat{g}$ is bi-invariant, since T^r is abelian. A principal connection θ on P and a Riemannian metric g' on B determine the metric g on P, pointwise given by:

$$g(X,Y) = <\theta(X), \theta(Y)> + g'(\pi_* X, \pi_* Y) \circ \pi \,, \tag{5.18}$$

for $X, Y \in \mathcal{X}(P)$. This makes $\pi : (P, g) \to (B, g')$ a Riemannian submersion with totally geodesic fibres isometric to $(T^r, \hat{g})$ and horizontal distribution determined by θ; moreover, T^r acts on (P, g) by isometries, since $\hat{g}$ is bi-invariant. According to (5.5), the invariant A of π acts on the basic vector fields X, Y, π-related to X', Y', by:

$$\theta(A_X Y) = -\Omega(X, Y) = -\eta(X', Y') \circ \pi \, , \tag{5.19}$$

η being the $\mathfrak{t}^r$-valued closed 2-form on B which pulls back to the curvature form Ω of θ; the existence of η comes from the triviality of the adjoint representation of T^r on $\mathfrak{t}^r$. The Einstein condition for (P, g) is expressed in terms of the metrics g', $\hat{g}$, the Ricci tensor ρ' on (B, g') and the form η. In fact, the Ricci tensor $\hat{\rho}$ of the fibres vanishes, since the fibres are isometric to the flat manifold $(T^r, \hat{g})$. Let $\{X_i'\}_{1 \leq i \leq n}$ be a local orthonormal frame on B with θ-horizontal lift $\{X_i\}_{1 \leq i \leq n}$; given $U \in \mathcal{X}^v(P)$, $X \in \mathcal{X}^b(P)$, π-related to X', from (5.19) one easily obtains:

$$A_X U = \sum_{i=1}^{n} < \eta(X', X_i'), \theta(U) > X_i \, ,$$

$$\begin{aligned} \textstyle\sum_{i=1}^n g((\nabla_{X_i} A)(X_i, X), U) &= -\textstyle\sum_{i=1}^n < (\nabla'_{X_i'} \eta)(X_i', X'), \theta(U) > \\ &= - < (\delta' \eta)(X'), \theta(U) > \, , \end{aligned}$$

$\delta' \eta$ being the codifferential of η. Via Proposition 5.1, the Einstein condition for g is equivalent to the system:

i) $\sum_{i,j} < \eta(X_j', X_i'), \theta(U) >< \eta(X_j', X_i'), \theta(V) >= E < \theta(U), \theta(V) >$

ii) $\rho'(X', Y') - 2 \sum_i < \eta(X', X_i'), \eta(Y', X_i') >= E g'(X', Y')$

iii) η is harmonic,

where $X', Y' \in \mathcal{X}(B)$, $U, V \in \mathcal{X}^v(P)$ and $E = \tau / m$ is the so-called *Einstein constant*.

Now, we consider a compact n-dimensional Kähler Einstein manifold (M, J, g) and denote by ψ its *Ricci form* such that $\psi(X, Y) = \rho(X, JY)$, which is a harmonic form of bidegree (1,1). We recall that $(1/2\pi)\psi$ represents the first real Chern class $c_1^{\mathbf{R}}(M)$ in the cohomology group $H^2(M, \mathbf{R})$, $c_1^{\mathbf{R}}(M)$ denoting the image of the first Chern class $c_1(M) \in H^2(M, \mathbf{Z})$ by the universal *change of coefficients* morphism $H^2(M, \mathbf{Z}) \to H^2(M, \mathbf{R})$

([26]). Then, the sign of $c_1^{\mathbf{R}}(M)$ is the sign of the scalar curvature τ, since the Einstein condition implies

$$\tau = \sum_{i=1}^{n} \psi(Je_i, e_i) = \mathrm{tr}(\psi) .$$

Assuming $\tau > 0$, a theorem of S. Kobayashi states that M is simply connected and hence the cohomology group $H^2(M, \mathbf{Z})$ has no torsion. This implies the existence of an indivisible cohomology class $\alpha \in H^2(M, \mathbf{Z})$ such that $c_1(M) = q\alpha$, with $q \in \mathbf{N}^*$. Since moreover $\psi = \frac{\tau}{n}\Omega$, Ω being the Kähler form of M, one can normalize the metric g so that $2\pi\alpha = [\Omega]$ and then $q = \tau/n$ coincides with the Einstein constant.

Firstly, we consider a principal $\mathbf{S}^1$-bundle over the product manifold $B = M_1 \times \cdots \times M_h$, where each (M_i, J_i, g_i) is an n_i-dimensional compact Einstein Kähler manifold having positive first Chern class $c_i(M_i) = q_i\alpha_i$, α_i indivisible, and Kähler form Ω_i such that $[\Omega_i] = 2\pi\alpha_i$. Denote by $\pi_i : B \to M_i$ the projection onto the i-th factor. The given bundle, of projection $\pi : P \to B$, is classified by its Euler class $e(P) \in H^2(B, \mathbf{Z})$, expressed as $e(P) = \sum_{i=1}^{h} b_i\pi_i^*(\alpha_i)$, $b_i \in \mathbf{Z}$. Then, the coefficients q_i and the classes α_i allow to determine an Einstein metric on P, according to the following result.

Theorem 5.4 *Let $\pi : P \to B$, $B = M_1 \times \cdots \times M_h$, be the projection of a principal $\mathbf{S}^1$-bundle over the product of n_i-dimensional compact, Einstein Kähler manifolds (M_i, J_i, g_i) with positive first Chern class and Euler class $e(P) = \sum_{i=1}^{h} b_i\pi_i^*(\alpha_i)$, α_i indivisible.*

If $e(P) \neq 0$, then P has an Einstein metric g of positive scalar curvature. In fact there exist positive constants $x_1, \ldots, x_h$ depending on b_i, q_i, making $\pi : (P, g) \to (B, \sum_i x_i g_i)$ a Riemannian submersion with totally geodesic fibres. This requirement characterizes g up to homotheties.

Proof. As remarked above, one needs to define a metric g', a harmonic 2-form η on B, a connection θ on P with curvature form $\pi^*\eta$ and an inner product $<,>$ on $\mathbf{R}$, so that conditions i), ii) are satisfied. We put $g' = \sum_{i=1}^{h} x_i g_i$, $x_1, \ldots, x_h$ being positive constants to be determined and observe that, for any $i \in \{1, \ldots, h\}$, the Kähler form Ω_i is the $x_i g_i$-harmonic representative of $2\pi\alpha_i \in H^2(M_i, \mathbf{R})$. Thus, the 2-form $\eta = \sum_{i=1}^{h} b_i\pi_i^*(\Omega_i)$ is the g'-harmonic representative of $2\pi e(P)$. Then, there exists a connection θ on P with curvature form $d\theta = \pi^*\eta$ ([26], 9.75) and θ is determined up to gauge transformations. In fact, considering another connection θ'

with $d\theta' = \pi^*\eta$, since $\theta - \theta'$ vanishes on the vertical vector fields and the adjoint action of $\mathbf{S^1}$ on $\mathbf{R}$ is trivial, there exists a 1-form β on B such that $\theta - \theta' = \pi^*\beta$. In this case B is simply connected, β is closed ($\pi^*\beta$ being closed), so there exists a differentiable function f such that $\beta = df$; f determines the transformation F of P given by $F(u) = u\exp(\mathrm{i}f(\pi(u)))$ and $F^*\theta = \theta'$. Now, we consider the inner product on $\mathbf{R}$ such that $< 1, 1 >= c^2$, $c > 0$; so the metric g on P given in (5.18) makes any fibre isometric to a circle of length $2\pi c$. Applying condition i) to the fundamental vector field U^* such that $\theta(U^*) = 1$, one obtains $c^2\|\eta\|^2 = E$, E being the Einstein constant of g. Moreover, $\|\eta\|^2 = \sum_{i=1}^h b_i^2 \|\pi_i^*(\Omega_i)\|^2_{g'}$, so one needs to evaluate $\|\Omega_i\|^2$ with respect to the metric $x_i g_i$. Considering a local g_i-orthonormal frame $\{X_j, JX_j\}_{1\le j\le m_i}$, with $2m_i = n_i = \dim M_i$, one obtains:

$$\|\Omega_i\|^2_{x_i g_i} = 2\sum_{j,k=1}^{m_i} \Omega_i\left(x_i^{-\frac{1}{2}}X_j, x_i^{-\frac{1}{2}}JX_k\right)\Omega_i\left(x_i^{-\frac{1}{2}}X_j, x_i^{-\frac{1}{2}}JX_k\right) = x_i^{-2}n_i \, .$$

Then, i) reduces to:

$$E = c^2 \sum_{i=1}^h b_i^2 n_i x_i^{-2} \, . \tag{5.20}$$

In order to apply ii), we observe that, for a fixed i, the Ricci tensor ρ' of (B, g') acts on vector fields tangent to M_i as $\rho'(X', Y') = \rho_i'(X', Y') = q_i g_i(X', Y')$, ρ_i' being the Ricci tensor of (M_i, g_i). Moreover, one considers a local g'-orthonormal frame on B, $\{X_j'\}_{1\le j\le n}$, where $n = \sum_{k=1}^h n_k$, the first n_i vector fields are tangent to M_i, and puts $\Omega_k^* = \pi_k^*\Omega_k$. Then, for X', Y' tangent to M_i one has:

$$\begin{aligned}\sum_{j=1}^n < \eta(X', X_j'), \eta(Y', X_j') > &= c^2 \sum_{j=1}^{n_i}\sum_{k=1}^h b_k^2 \Omega_k^*(X', X_j')\Omega_k^*(Y', X_j') \\ &= c^2 b_i^2 \frac{1}{x_i}\sum_{j=1}^{n_i} \Omega_i(X', x_i^{\frac{1}{2}}X_j')\Omega_i(Y', x_i^{\frac{1}{2}}X_j') \\ &= c^2 b_i^2 x_i^{-1} g_i(X', Y') \, .\end{aligned}$$

Thus, condition ii) gives:

$$\frac{q_i}{x_i} - 2c^2\left(\frac{b_i}{x_i}\right)^2 = E \, , \quad 1 \le i \le h \, .$$

Combining with (5.20), one obtains the quadratic system:

$$\frac{q_i}{x_i} - 2\left(\frac{b_i}{x_i}\right)^2 \left(\sum_{j=1}^{h} n_j \left(\frac{b_j}{x_j}\right)^2\right)^{-1} E = E \,, \quad 1 \leq i \leq h \,.$$

Via the mean value theorem, one proves the existence and the uniqueness of a solution $(x_1, \ldots, x_h)$, with any $x_i > 0$, provided that any b_i is nonzero. If some of the b_i's vanish, up to a permutation of the factors, P splits as $P' \times M_{q+1} \times \cdots \times M_h$, and P' is the total space of an $\mathbf{S^1}$-bundle with Euler class $\sum_{i=1}^{q} b_i \pi_i^* \alpha_i$, with $b_i \neq 0$, $1 \leq i \leq q$. Then, one can consider on P' an Einstein metric of positive scalar curvature, and this allows to construct an Einstein metric on P, too.

Suppose now that $\pi : P \to B$ acts as the projection of a T^r-bundle, $r \geq 2$, over the product $B = M_1 \times \cdots \times M_h$ of $h \geq r$ compact, Kähler Einstein manifolds with positive scalar curvature. Such a bundle is classified by the r characteristic classes $\beta_1, \ldots, \beta_r \in H^2(B, \mathbf{Z})$ which are the Euler classes of the $\mathbf{S^1}$-bundles described as follows. One fixes a canonical decomposition $T^r = \mathbf{S^1} \times \cdots \times \mathbf{S^1}$ in r factors corresponding to a canonical basis $\{e_1, \ldots, e_r\}$ of the Lie algebra $\mathfrak{t}^r$ of T^r. For any $i \in \{1, \ldots, r\}$, T^{r-1} denotes the torus obtained from T^r deleting the i-th factor $\mathbf{S^1}$. Putting $P_i = P/T^{r-1}$ let $\sigma_i : P_i \to B$ be the projection of the quotient bundle. Then β_i is the Euler class of this $\mathbf{S^1}$-bundle. Moreover, any automorphism A of T^r allows to change the action of T^r on P and this determines a new principal T^r-bundle $\pi' : P \to B$. With respect to π' the canonical basis of the Lie algebra of T^r is $\{e'_1, \ldots, e'_r\}$, where $e'_i = \sum_{j=1}^{r} a_{ij} e_j$, $i \in \{1, \ldots, r\}$ and $a = (a_{ij}) \in M_r(\mathbf{Z}), \det a = 1$. Then, π' is classified by the classes $\beta'_1, \ldots, \beta'_r \in H^2(B, \mathbf{Z})$ such that

$$\beta'_i = \sum_{j=1}^{r} a'_{ij} \beta_j \text{ and } a' = (a'_{ij}) = ({}^t a)^{-1} \,.$$

By the Bonnet–Meyers theorem, an Einstein metric on P with positive Einstein constant can exist, provided that $\pi_1(P)$ is finite ([26], 6.52). In our case, this request is equivalent to the linear independence in $H^2(B, \mathbf{Z})$ of the characteristic classes of $\pi : P \to B$. So, assuming that $\beta_1, \ldots, \beta_r$ are linearly independent, we look for an Einstein metric g on P defined as in (5.18). For any $k \in \{1, \ldots, h\}$, let g_k be the Einstein metric on M_k such that the Kähler form Ω_k satisfies $[\Omega_k] = 2\pi\alpha_k$, with α_k indivisible in

$H^2(M_k, \mathbf{Z})$. Denoting by $\pi_i : B \to M_i$ the projection on the i-th factor and putting $\beta_i = \sum_{j=1}^h b_{ij}\pi_j^*\alpha_j$, $i \in \{1, \dots, r\}, b_{ij} \in \mathbf{Z}$, the linear independence of $\beta_1, \dots, \beta_r$ means that the matrix (b_{ij}) has maximal rank, r. As in Theorem 5.4, we choose the metric $g' = \sum_{j=1}^h x_j g_j$ on B, where any x_j is a positive constant to be determined. Observe that g' is the product of Kähler Einstein metrics, too. Since, for any $i \in \{1, \dots, r\}$, $\sigma_i : P_i \to B$ is an $\mathbf{S^1}$-bundle with Euler class β_i, there exists a principal connection θ_i on P_i with curvature form $d\theta_i = \sigma_i^*\eta_i$, where $\eta_i = \sum_{j=1}^h b_{ij}\pi_j^*\Omega_j$ is the g'-harmonic representative of $2\pi\beta_i^{\mathbf{R}}$. Let $\Delta : B \to B \times \cdots \times B$ denote the diagonal map and consider the bundle morphism $F : P \to P_1 \times \cdots \times P_r$ such that $\Delta \circ \pi = (\sigma_1 \times \cdots \times \sigma_r) \circ F$. Putting $\theta = F^*(\sum_{i=1}^r \theta_i e_i)$, θ determines a connection on P with curvature form $d\theta = \pi^*\eta$, where $\eta = \sum_{i=1}^r (\sum_{j=1}^h b_{ij}\pi_j^*\Omega_j)e_i$ is a $\mathfrak{t}^r$-valued g'-harmonic 2-form on B. Now, considering an arbitrary inner product $<,>$ on $\mathfrak{t}^r$ and putting $< e_i, e_l >= h_{il}$, $i, l \in \{1, \dots, r\}$, one can reformulate the Einstein condition for the metric g defined by (5.18) in terms of the matrices $(b_{ij}), (h_{il})$, of the Einstein constants q_i of M_i's and of the constants x_i's. In fact, for any $k \in \{1, \dots, r\}$, let U_k^* be the fundamental vector field such that $\theta(U_k^*) = e_k$ and consider a local g'-orthonormal frame on B, $\{X_i'\}_{1 \le i \le n}$, $n = \dim B$. Then:

$$< \eta(X_i', X_j'), \theta U_k^* >= \sum_{q=1}^r \sum_{s=1}^h b_{qs}(\pi_s^*\Omega_s)(X_i', X_j')h_{kq} \ , \ i, j \in \{1, \dots, n\} \ ,$$

and, since $g'(\pi_s^*\Omega_s, \pi_t^*\Omega_t) = \delta_{ts}x_s^{-1}x_t^{-1}\sqrt{n_s}\sqrt{n_t}$, for any $k, l \in \{1, \dots, r\}$, one has:

$$\sum_{i,j} < \eta(X_i', X_j'), \theta U_k^* >< \eta(X_i', X_j'), \theta U_l^* >= \sum_{q,m} \sum_{s=1}^h b_{qs}b_{ms}h_{kq}h_{lm}x_s^{-2}n_s \ ,$$

where the sum indexes i, j range in $\{1, \dots, n\}$ and q, m in $\{1, \dots, r\}$. Thus, condition i) reduces to:

$$\sum_{q,m=1}^r \sum_{s=1}^h b_{qs}b_{ms}h_{kq}h_{lm}x_s^{-2}n_s = Eh_{kl} \ ,$$

where $k, l \in \{1, \dots, r\}$, or equivalently to:

$$\sum_{s=1}^h b_{ls}b_{ms}x_s^{-2}n_s = Eh^{lm} \ , \quad l, m \in \{1, \dots, r\} \ , \tag{5.21}$$

(h^{lm}) denoting the inverse matrix of (h_{lm}). Furthermore, for a fixed i in $\{1,\dots,h\}$, one considers a local orthonormal frame on B, $\{X'_j\}_{1\leq j\leq n}$, with the first n_i vector fields tangent to M_i. Given X' tangent to M_i, one obtains:

$$\sum_{j=1}^{n} < \eta(X',X'_j),\eta(X',X'_j) >= \sum_{q,l=1}^{r} b_{qi}b_{li}h_{ql}x_i^{-1}g_i(X',X') \ .$$

Thus, condition ii) implies:

$$q_i - 2\sum_{q,l=1}^{r} b_{qi}b_{li}h_{ql}x_i^{-1} = Ex_i \ , \quad i \in \{1,\dots,h\} \ . \tag{5.22}$$

After a detailed investigation, one can prove the existence of at least one solution of the system (5.21), (5.22). This allows to state the following result.

Theorem 5.5 **(Wang, Ziller).** *Let $\pi : P \to M_1 \times \cdots \times M_h$ be the projection of a T^r-bundle, $r \leq h$, over the product of compact, Kähler Einstein manifolds with positive first Chern class and characteristic classes $\beta_i = \sum_{j=1}^{h} b_{ij}\pi_j^*\alpha_j$, $i \in \{1,\dots,r\}$. If the matrix (b_{ij}) has rank r, then there exists an Einstein metric g on P of positive scalar curvature which makes π a Riemannian submersion with totally geodesic fibres and such that the metric on the base is the product of Kähler Einstein metrics.*

Existence results of Einstein metric on the total space of a principal circle bundle are due to Sakane, also ([258; 259]). In this case, the base space is a **C**-space, that is a compact, simply connected, homogeneous complex manifold. Firstly, we recall some basic facts on semisimple Lie algebras, referring to [139; 45; 46; 26] for more details.

Let G be a compact semisimple Lie group with Lie algebra $\mathfrak{g}$. We denote by $\mathfrak{g}^{\mathbf{C}}$ the complexification of $\mathfrak{g}$. Fixed a maximal abelian subalgebra $\mathfrak{h}$ of $\mathfrak{g}$, its complexification $\mathfrak{h}^{\mathbf{C}}$ is a Cartan subalgebra of $\mathfrak{g}^{\mathbf{C}}$, so that $\mathfrak{g}^{\mathbf{C}} = \mathfrak{h}^{\mathbf{C}} \oplus \sum_{\alpha\in\Delta}\mathfrak{g}_\alpha$, where Δ is the root system of $\mathfrak{g}^{\mathbf{C}}$ modulo $\mathfrak{h}^{\mathbf{C}}$ and, for any $\alpha \in \Delta$, $\mathfrak{g}_\alpha$ is the 1-dimensional linear space consisting of the vectors X such that $ad_H X = \alpha(H)X$, for any $H \in \mathfrak{h}^{\mathbf{C}}$. Via the Killing form $B^{\mathbf{C}}$ of $\mathfrak{g}^{\mathbf{C}}$, to any $\alpha \in \Delta$ is univocally associated a nonzero element $H_\alpha \in \mathfrak{h}^{\mathbf{C}}$; then $\alpha(H_\alpha)$ is real and positive and $[\mathfrak{g}_\alpha,\mathfrak{g}_{-\alpha}] = \mathbf{C}H_\alpha$. The restriction of $B^{\mathbf{C}}$ to the linear space $\mathfrak{h}_{\mathbf{R}} = \sum_{\alpha\in\Delta}\mathbf{R}H_\alpha$ is real and positive definite, $\dim\mathfrak{h}_{\mathbf{R}} = \dim\mathfrak{h} = l$, $\mathfrak{h}^{\mathbf{C}} = \mathfrak{h}_{\mathbf{R}}\oplus i\mathfrak{h}_{\mathbf{R}}$ and the restriction of any root to $\mathfrak{h}_{\mathbf{R}}$ is real-valued. To the root system Δ there is associated a Weyl basis $\{Z_\alpha\}_{\alpha\in\Delta}$, where, for any α, $Z_\alpha \in \mathfrak{g}_\alpha$, $B^{\mathbf{C}}(Z_\alpha,Z_{-\alpha}) = -1$, $[Z_\alpha,Z_{-\alpha}] = -H_\alpha$. Putting, for any

$\alpha, \beta \in \Delta$,

$$[Z_\alpha, Z_\beta] = N_{\alpha,\beta} Z_{\alpha+\beta} \text{ , if } \alpha + \beta \in \Delta \text{ ; } \quad [Z_\alpha, Z_\beta] = 0 \text{ , otherwise,}$$

the $N_{\alpha,\beta}$ are real constants and $N_{\alpha,\beta} = -N_{-\alpha,-\beta}$. The Weyl basis also determines

$$\sum_{\alpha\in\Delta} \mathbf{R}(\mathrm{i}H_\alpha) \oplus \sum_{\alpha\in\Delta} \mathbf{R}(Z_\alpha + Z_{-\alpha}) \oplus \sum_{\alpha\in\Delta} \mathbf{R}\mathrm{i}(Z_\alpha - Z_{-\alpha}) \text{ ,}$$

as a compact real form of $\mathfrak{g}^{\mathbf{C}}$, so that $\mathfrak{h} \cong \sum_{\alpha\in\Delta} \mathbf{R}(\mathrm{i}H_\alpha)$ and

$$\mathfrak{g} = \mathfrak{h} \oplus \sum_{\alpha\in\Delta} \mathbf{R}(Z_\alpha + Z_{-\alpha}) \oplus \sum_{\alpha\in\Delta} \mathbf{R}\mathrm{i}(Z_\alpha - Z_{-\alpha}) \text{ .}$$

In particular, any root α, identified with H_α, can be regarded as an element of $\mathrm{i}\mathfrak{h}$. Now, we fix a fundamental system $\Pi = \{\alpha_1, \ldots, \alpha_l\}$ of Δ and denote by Δ_+ the set of all positive roots relative to Π. Let $\{\Lambda_1, \ldots, \Lambda_l\}$ be the family of fundamental weights of $\mathfrak{g}^{\mathbf{C}}$ corresponding to Π. Any root can be expressed as a linear combination of $\{\Lambda_1, \ldots, \Lambda_l\}$ by integer coefficients. Given a subset Π_0 of Π, we put $[\Pi_0] = \Delta \cap \mathrm{span}(\Pi_0)$, $\mathrm{span}(\Pi_0)$ being the subspace of $\mathrm{i}\mathfrak{h}$ generated by Π_0, and we consider the parabolic subalgebra of $\mathfrak{g}^{\mathbf{C}}$

$$\mathfrak{u} = \mathfrak{h}^{\mathbf{C}} \oplus \sum_{\alpha\in[\Pi_0]\cup\Delta_+} \mathfrak{g}_\alpha \text{ .}$$

Denote by $G^{\mathbf{C}}$ a simply connected complex semisimple Lie group with Lie algebra $\mathfrak{g}^{\mathbf{C}}$ and by U the subgroup of $G^{\mathbf{C}}$ corresponding to $\mathfrak{u}$. Then G acts transitively on $G^{\mathbf{C}}/U$ and $G^{\mathbf{C}}/U$ is a $\mathbf{C}$-manifold admitting a G-invariant Kähler metric. Any G-invariant almost complex structure on $G^{\mathbf{C}}/U$ is determined by an endomorphism I of $\sum_{\alpha\in\Delta\setminus[\Pi_0]} \mathfrak{g}_\alpha$ such that $I^2 = -id$. The G-invariance of I entails that, for any $\alpha \in \Delta$, $I(\mathfrak{g}_\alpha) = \mathfrak{g}_\alpha$, so Z_α is an eigenvector of I. Since the eigenvalues of I are $\mathrm{i}, -\mathrm{i}$, one has $IZ_\alpha = \mathrm{i}\varepsilon_\alpha Z_\alpha$, $\varepsilon_\alpha = \pm 1$, and $\varepsilon_{-\alpha} = -\varepsilon_\alpha$. The canonical complex structure on $G^{\mathbf{C}}/U$, denoted by $J^{\mathbf{C}}$, is given by $\varepsilon_\alpha = 1$, if $\alpha < 0$, ([262]). Putting $K = G \cap U$, K is a closed connected subgroup of G and the real homogeneous manifold G/K is diffeomorphic to the manifold $G^{\mathbf{C}}/U$ ([208]). For simplicity, we write $M = G/K = G^{\mathbf{C}}/U$. The Lie algebra $\mathfrak{k}$ of K is given by:

$$\mathfrak{k} = \mathfrak{h} \oplus \sum_{\alpha\in[\Pi_0]} \mathbf{R}(Z_\alpha + Z_{-\alpha}) \oplus \sum_{\alpha\in[\Pi_0]} \mathbf{R}\mathrm{i}(Z_\alpha - Z_{-\alpha}) \text{ .}$$

A result of Murakami states that the manifold M does not admit G-invariant real 1-forms and any closed G-invariant 2-form is harmonic with respect to any G-invariant metric ([208]).

Now, let $\mathfrak{m}$ be the orthogonal complement of $\mathfrak{k}$ in $\mathfrak{g}$ with respect to the inner product $(,) = -B$, B being the Killing–Cartan form of $\mathfrak{g}$. The canonical complex structure on G/K corresponds to the endomorphism I of $\mathfrak{m}$ such that $I(Z_\alpha + Z_{-\alpha}) = \mathrm{i}(Z_\alpha - Z_{-\alpha})$, for any root $\alpha < 0$. Let $\mathfrak{m}_1, \dots, \mathfrak{m}_h$ be $Ad(K)$-invariant, irreducible, orthogonal subspaces of $\mathfrak{m}$, so that $\mathfrak{g} = \mathfrak{k} \oplus \sum_{j=1}^h \mathfrak{m}_j$, $[\mathfrak{k}, \mathfrak{m}_j] \subset \mathfrak{m}_j$. The space of the G-invariant metrics on G/K is described by the family of inner products $\sum_{j=1}^h x_j(,)_{\mathfrak{m}_j}$, $x_1, \dots, x_h \in \mathbf{R}^*_+$, where $(,)_{\mathfrak{m}_j}$ is the restriction of $(,)$ to $\mathfrak{m}_j$. Any of these inner products is Hermitian with respect to I and, for any j, one has $I(\mathfrak{m}_j) = \mathfrak{m}_j$; in particular, $\mathfrak{m}_j$ is even-dimensional.

Principal $\mathbf{S^1}$-bundles on $M = G/K$ are classified by the cohomology group $H^2(M, \mathbf{Z})$ which, in this case, is isomorphic to $\mathbf{Z}\Lambda_{i_1} \oplus \dots \oplus \mathbf{Z}\Lambda_{i_r}$, where $\Lambda_{i_1}, \dots, \Lambda_{i_r}$ are the fundamental weights corresponding to the set $\Pi \setminus \Pi_0 = \{\alpha_{i_1}, \dots, \alpha_{i_r}\}$, with $0 < \alpha_{i_1} < \dots < \alpha_{i_r}$. So, let $\pi : P \to M$ be the projection of an $\mathbf{S^1}$-bundle, determined by $\sum_{q=1}^r a_q \Lambda_{i_q}$, $a_1, \dots, a_r \in \mathbf{Z}$. Given $k_1, \dots, k_r \in \mathbf{N}^*$, we put:

$$\Delta(k_1, \dots, k_r) = \{\textstyle\sum_{q=1}^l m_q \alpha_q \in \Delta^+ \mid m_{i_1} = k_1, \dots, m_{i_r} = k_r\}\,,$$

$$\mathfrak{m}(k_1, \dots, k_r) = \bigoplus_{\alpha \in \Delta(k_1, \dots, k_r)} (\mathbf{R}(Z_\alpha + Z_{-\alpha}) + \mathbf{R}\mathrm{i}(Z_\alpha - Z_{-\alpha}))\,.$$

We remark that $I(\mathfrak{m}(k_1, \dots, k_r)) = \mathfrak{m}(k_1, \dots, k_r)$ and $\mathfrak{m}(k_1, \dots, k_r)$ is $Ad(K)$-invariant. For a fixed $j \in \{1, \dots, h\}$, there exists a unique choice for $(k_1, \dots, k_r)$ such that $\mathfrak{m}_j \subset \mathfrak{m}(k_1, \dots, k_r)$. Since, for any $\alpha \in \Delta(k_1, \dots, k_r)$, the value $(\sum_{q=1}^r a_q \Lambda_{i_q}, \alpha) = \sum_{q=1}^r a_q k_q (\Lambda_{i_q}, \alpha_{i_q})$ is independent of α, with the space $\mathfrak{m}_j$ there is associated the number $b_j = \sum_{q=1}^r a_q k_q (\Lambda_{i_q}, \alpha_{i_q})$.

Denote by $\omega_j : \mathfrak{m}_j \times \mathfrak{m}_j \to \mathbf{R}$ the bilinear defined by $\omega_j(X, Y) = (X, IY)_{\mathfrak{m}_j}$ and put $\eta' = \sum_{j=1}^h b_j \omega_j$. The G-invariant 2-form η on M corresponding to η' is closed and then harmonic with respect to any invariant metric. So, one can consider a connection form θ on P with curvature form $\pi^*\eta$. Let g' be the G-invariant metric on M corresponding to a fixed inner product $\sum_{j=1}^h x_j(,)_{\mathfrak{m}_j}$, choose an inner product $<,>$ on $\mathbf{R}$ and put $< 1, 1 >= c^2$. Then θ, g', c determine a Riemannian metric g on P which makes $\pi : (P, g) \to (M, g')$ a Riemannian submersion. Via a direct computation (compare with the proof of Theorem 5.4), the Einstein condition for

g is equivalent to the system:

$$\begin{aligned} & c^2 \sum_{j=1}^h b_j^2 x_j^{-2} n_j = E \\ & r_j - 2c^2 (b_j)^2 (x_j)^{-2} = E \,, \quad j \in \{1, \dots, h\} \,, \end{aligned} \tag{5.23}$$

where E is the Einstein constant of (P, g), $n_j = \dim \mathfrak{m}_j$, $r_j = \rho'(X, X)$, ρ' being the Ricci tensor of $(G/K, g')$ and X any vector in $\mathfrak{m}_j$, with $(X, X)_{\mathfrak{m}_j} = x_j^{-1}$.

Now, we distinguish the following cases: $h \geq 3$, $h = 2$.

Case A): $h \geq 3$. In [258] Sakane states a sufficient condition for the existence of a solution of the above system. This criterion involves the degree of a map $F_0 : \mathbf{S^{h-2}} \to \mathbf{S^{h-2}}$ induced by the following homotopy $f : [0, 1] \times (\mathbf{R}_+^*)^h \to \mathbf{S^{h-1}}$ such that

$$f(t, x_i, \dots, x_h) = \|T\|^{-1} (T_1, \dots, T_h) \,,$$

where, for any $j \in \{1, \dots, h\}$,

$$\left(1 + 2\frac{c^2}{E} \left(\frac{b_j}{x_j} \right)^2 t \right) T_j = x_j, \ \|T\|^2 = \sum_{j=1}^h T_j^2 \,.$$

Particularly interesting are the following applications.

Example 5.9 The flag manifold $M = SU(n)/S(U(n_1)U(n_2)U(n_3))$, with $n = n_1 + n_2 + n_3$, is the twistor space of the complex Grassmannian $G_{n_2+n_3} = SU(n)/S(U(n_1)U(n_2 + n_3))$ considered in Example 5.8. The Lie algebra $\mathfrak{su}(n)$ of $SU(n)$, consisting of the trace zero, skew-Hermitian complex matrices of order n, admits the Lie algebra $\mathfrak{h}$ of the diagonal matrices as Cartan subalgebra. For any $\lambda_1, \dots, \lambda_n \in \mathbf{R}$ such that $\sum_{j=1}^n \lambda_j = 0$, we put:

$$(\lambda_1, \dots, \lambda_n) = \begin{pmatrix} \mathrm{i}\lambda_1 & 0 & \cdots & 0 \\ 0 & \mathrm{i}\lambda_2 & \cdots & 0 \\ \cdots & \cdots & \cdots & \cdots \\ 0 & 0 & \cdots & \mathrm{i}\lambda_n \end{pmatrix} ,$$

and denote by λ_j the linear form on $\mathfrak{h}$ which associates to any $(\lambda_1, \dots, \lambda_n)$ its j-th component. For the canonical ordering, the positive roots are the $n(n-1)/2$ forms $\lambda_i - \lambda_j$, $i < j$. The $n - 1$ forms $\alpha_j = \lambda_j - \lambda_{j+1}$, with $1 \leq j \leq n - 1$, set up a fundamental system of roots; via the Killing–Cartan form B on $\mathfrak{su}(n)$, α_j is identified with $H_j = (0, \dots, 1, -1, \dots, 0)$, with 1, -1 as j-th, $(j+1)$-th components. Thus, for any $(\lambda_1,\dots,\lambda_n) \in \mathfrak{h}$:

$$B(H_j,(\lambda_1,\dots,\lambda_n)) = -2n(\lambda_j - \lambda_{j+1}) = -2n\alpha_j(\lambda_1,\dots,\lambda_n) .$$

With respect to the adjoint action of $S(U(n_1)U(n_2)U(n_3))$ the orthogonal complement of $\mathfrak{s}(\mathfrak{u}(n_1) \oplus \mathfrak{u}(n_2) \oplus \mathfrak{u}(n_3))$ admits three invariant irreducible subspaces $\mathfrak{m}_{ij}$, $1 \le i < j \le 3$, with $\dim \mathfrak{m}_{ij} = 2n_i n_j$. Hence, any $SU(n)$-invariant metric on M is determined by an inner product

$$\sum_{1 \le i < j \le 3} x_{ij}(\, , \,)_{\mathfrak{m}_{ij}} , \qquad x_{ij} > 0 .$$

The weights Λ_1, Λ_2 corresponding to the roots $\alpha_{n_1}, \alpha_{n_1+n_2}$ generate a group isomorphic to $H^2(M,\mathbf{Z})$, and then any pair $(k,l) \in \mathbf{Z}^{2^*}$ determines a principal $\mathbf{S}^1$-bundle, whose total space is denoted by $P_{k,l}$.

Proposition 5.3 (Sakane). *For any $(k,l) \in (\mathbf{Z}^2)^*$ there exist an $SU(n)$-invariant metric on $SU(n)/S(U(n_1)U(n_2)U(n_3))$ and an Einstein metric on $P_{k,l}$ making the projection a Riemannian submersion with totally geodesic fibres.*

Example 5.10 The homogeneous space $M = SO(2n)/U(n-1)U(1)$ can be considered as the manifold of the $2(n-1)$-dimensional linear subspaces of $\mathbf{R}^{2n}$ endowed with a complex structure compatible with the induced Euclidean structure. Assume $n \ge 3$ and consider the Lie algebra $\mathfrak{so}(2n)$ of $SO(2n)$ consisting of the real skew-symmetric matrices of order $2n$ which is a compact real form of $\mathfrak{su}(2n)$. A Cartan subalgebra of $\mathfrak{so}(2n)$ is the Lie algebra $\mathfrak{h}$ consisting of the matrices $A = (a^i_j) = (\lambda_1,\dots,\lambda_n)$ such that

$$a^{2h}_{2h-1} = -a^{2h-1}_{2h} = \lambda_h \in \mathbf{R} , \quad h \in \{1,\dots,n\} ,$$

the other entries being zero. For any $j \in \{1,\dots,n-1\}$ let α_j be the 1-form on $\mathfrak{h}$ such that $\alpha_j(\lambda_1,\dots,\lambda_n) = \lambda_j - \lambda_{j+1}$ and α_n the 1-form defined by $\alpha_n(\lambda_1,\dots,\lambda_n) = \lambda_{n-1} + \lambda_n$. Then $\{\alpha_1,\dots,\alpha_n\}$ is a fundamental system of positive roots identified, via the Killing–Cartan form B on $\mathfrak{so}(2n)$, with $\{H_1,\dots,H_n\}$, where $H_j = (0,\dots,1,-1,\dots,0)$, with $1,-1$ as j-th, $(j+1)$-th components, $1 \le j \le n-1$ and $H_n = (0,\dots,0,1,1)$. The B-orthogonal complement of $\mathfrak{u}(n-1) \oplus \mathfrak{u}(1)$ in $\mathfrak{so}(2n)$ admits three $Ad(U(n-1) \times U(1))$-invariant irreducible subspaces $\mathfrak{m}_j$, $1 \le j \le 3$, with $\dim \mathfrak{m}_1 = \dim \mathfrak{m}_2 = 2(n-1)$, $\dim \mathfrak{m}_3 = (n-1)(n-3)$; thus, the space of $SO(2n)$-invariant Riemannian metrics on M depends on three parameters.

Moreover, $H^2(M, \mathbf{Z}) \cong \mathbf{Z}\Lambda_{n-1} \oplus \mathbf{Z}\Lambda_n$, where Λ_{n-1}, Λ_n are the weights corresponding to α_{n-1}, α_n. So, any pair $(k,l) \in (\mathbf{Z^2})^*$ classifies a principal $\mathbf{S^1}$-bundle over M, whose total space is denoted by $P_{k,l}$.

Proposition 5.4 *For any $(k,l) \in (\mathbf{Z^2})^*$ there exist an $SO(2n)$-invariant metric on $SO(2n)/(U(n-1)U(1))$ and an Einstein metric on $P_{k,l}$ so that the projection of the considered bundle is a Riemannian submersion with totally geodesic fibres.*

Case B): $h = 2$. In [259] Sakane examines a particular situation. Firstly, we illustrate the decomposition of the Lie algebra $\mathfrak{g}$ when $\mathfrak{g}$ is simple and $\Pi \setminus \Pi_0 = \{\alpha_v\}$, with $v \in \{1, \ldots, l\}$ and $n_v = 2$. Here n_v denotes the coefficient of α_v in the expression $\tilde{\alpha} = \sum_{k=1}^{l} n_k \alpha_k$, $\tilde{\alpha}$ being the highest root of Δ ([139]). Since $n_v = 2$, one obtains the splitting $\mathfrak{g} = \mathfrak{k} \oplus \mathfrak{m}_1 \oplus \mathfrak{m}_2$, where, for $j = 1, 2$,

$$\mathfrak{m}_j = \sum_{\alpha \in \Delta_j} \mathbf{R}(Z_\alpha + Z_{-\alpha}) + \mathbf{R}i(Z_\alpha - Z_{-\alpha}) ,$$

and

$$\Delta_j = \{\alpha \in \Delta^+ \mid \alpha = j\alpha_v + \sum_{k \neq v} m_k \alpha_k\} .$$

Then, $\mathfrak{m}_1, \mathfrak{m}_2$ are $Ad(K)$-invariant, orthogonal, irreducible subspaces such that $[\mathfrak{m}_1, \mathfrak{m}_1] \subset \mathfrak{k} \oplus \mathfrak{m}_2$, $[\mathfrak{m}_2, \mathfrak{m}_2] \subset \mathfrak{k}$, $[\mathfrak{m}_1, \mathfrak{m}_2] \subset \mathfrak{m}_1$ and $\mathfrak{k} \oplus \mathfrak{m}_2$ is a semisimple Lie subalgebra of $\mathfrak{g}$. Then, the Lie subgroup L of G corresponding to $\mathfrak{k} \oplus \mathfrak{m}_2$ is closed. Let B' be the Killing form of $\mathfrak{k} \oplus \mathfrak{m}_2$ and put $B'_{|\mathfrak{m}_2} = cB_{|\mathfrak{m}_2}$. The exact values of c corresponding to the classical simple Lie algebras $\mathfrak{g}^{\mathbf{C}}$ are listed in [259]; in any case, one has $c < 1$. Moreover, a result of Borel and Hirzebruch states that the G-invariant metric g_2 on G/K determined by the inner product $(\, , \,)_{\mathfrak{m}_1} + 2(\, , \,)_{\mathfrak{m}_2}$ is Kähler and Einstein ([43]). As before, $(\, , \,)_{\mathfrak{m}_j}$ is the restriction to $\mathfrak{m}_j$ of the inner product $-B$ on $\mathfrak{g}$. So, one can apply Lemma 5.3, thus obtaining another homogeneous Einstein metric on G/K, denoted by $g_{2c/(2-c)}$; in fact it corresponds to the inner product $(\, , \,)_{\mathfrak{m}_1} - 2c/(2-c)(\, , \,)_{\mathfrak{m}_2}$. The Ricci tensor of G/K with respect to an invariant metric

$$x_1(\, , \,)_{\mathfrak{m}_1} + x_2(\, , \,)_{\mathfrak{m}_2} , \quad x_1, x_2 > 0 ,$$

is then determined by its values $r_j(x_1, x_2)$ on (X_j, X_j), X_j being a unit vector in $\mathfrak{m}_j$; they are given by:

$$r_1(x_1, x_2) = \frac{1}{2}\left(1 - \frac{n_2(2-c)}{2(n_1+2n_2)}\frac{x_2}{x_1}\right)x_1^{-1},$$

$$r_2(x_1, x_2) = \left(\frac{2}{c} + \frac{n_1(2-c)}{8(n_1+2n_2)}\left(\frac{x_2}{x_1}\right)^2\right)x_2^{-1},$$

where $n_j = \dim \mathfrak{m}_j$. Note that $r_j = r_j(x_1, x_2)$ can be considered as homogeneous functions of degree -1 and $(r_1(x_1, x_2), r_2(x_1, x_2)) \neq (0, 0)$, for any x_1, x_2. Since $g_1, g_{2c/(2-c)}$ are Einstein metrics, one has:

$$r_1(1,2) = r_2(1,2) > 0\,, \quad r_1\left(1, \frac{2c}{2-c}\right) = r_2\left(1, \frac{2c}{2-c}\right) > 0\,.$$

Now, let $\pi : P \to G/K$ be the projection of a principal $\mathbf{S^1}$-bundle, classified by $a\Lambda_v$, $a \in \mathbf{Z}^*$, and consider again (5.23). Firstly, we observe that, given $\alpha \in \Delta_j$, $j \in \{1, 2\}$, we have $b_j = \alpha_j(\Lambda_v, \alpha_v)$, so $b_2 = 2b_1$. For any $x_1, x_2 > 0$ one puts

$$q(x_1, x_2) = \frac{n_1}{x_1^2} + \frac{4n_2}{x_2^2},$$

$$t_j(x_1, x_2) = r_j(x_1, x_2)\left(1 + \frac{2}{q(x_1, x_2)}\left(\frac{j}{x_j}\right)^2\right)^{-1}, \quad j = 1, 2\,.$$

This allows to define three functions q, t_1, t_2, which are homogeneous of degree $-2, -1, -1$, respectively and t_1, t_2 never vanish. System (5.23) easily reduces to $t_1(x_1, x_2) = t_2(x_1, x_2) = E > 0$, which is independent of the integer a which classifies π. Since t_1, t_2 are homogeneous functions, non-homothetic Einstein metrics on P can be obtained by distinct positive numbers x_2 such that

$$t_1(1, x_2) = t_2(1, x_2) > 0\,. \tag{5.24}$$

Observe that $x_2 = 2$ is a solution of this equation. In fact,

$$t_1(1,2) = r_1(1,2)\left(1 + \frac{2}{n_1+n_2}\right)^{-1} = t_2(1,2) > 0\,.$$

This gives an Einstein metric g on P making $\pi : (P, g) \to (G/K, g_2)$ a Riemannian submersion with fibres isometric to a circle of length $2\pi\rho$, $\rho > 0$. This corresponds to a result of S. Kobayashi ([170]).

To look for another Einstein metric on P, one considers the positive, continuous functions $f_j :]0,2[\to \mathbf{R}$, $j = 1, 2$, such that

$$f_j(x_2) = \frac{t_j(1, x_2)}{\sqrt{t_1^2(1, x_2) + t_2^2(1, x_2)}}, \quad x_2 \in]0, 2[\; .$$

Note that, since $c < 1$ and $r_1(1, 2c/(2-c)) = r_2(1, 2c/(2-c))$, one has

$$t_1\left(1, \frac{2c}{2-c}\right) > t_2\left(1, \frac{2c}{2-c}\right) ,$$

and then

$$f_1\left(\frac{2c}{2-c}\right) > \frac{1}{\sqrt{2}} > f_2\left(\frac{2c}{2-c}\right) > 0 \; .$$

Moreover:

$$f_1^2 + f_2^2 = 1 \; , \quad \lim_{x_2 \to 0+} f_1(x_2) = 0 \; , \quad \lim_{x_2 \to 0+} f_2(x_2) = 1 \; ,$$

then there exists $x_2^0 \in]0, 2c/(2-c)[$ such that $f_1(x_2^0) = f_2(x_2^0) = 1/\sqrt{2}$, that is $t_1(1, x_2^0) = t_2(1, x_2^0) > 0$. So, x_2^0 determines a non-Einstein metric g_0' on G/K (the one corresponding to the inner product $(\, , \,)_{\mathfrak{m}_1} + x_2^0(\, , \,)_{\mathfrak{m}_2}$) and an Einstein metric g_0 on P, g_0 non-homothetic to g, such that the map $\pi : (P, g_0) \to (G/K, g_0')$ is a Riemannian submersion.

This proves the following theorem.

Theorem 5.6 **(Sakane).** *Let G/K be a Kähler* **C***-space, G simple, obtained by the pair (Π, Π_0), such that $\Pi \setminus \Pi_0 = \{\alpha_v\}$ and $n_v = 2$. The total space P of the principal* $\mathbf{S^1}$*-bundle on G/K corresponding to Λ_v has at least two G-invariant Einstein metrics.*

5.5 Einstein Weyl Structures on Principal Bundles

Another interesting problem, connected with the ones just examined, is the existence of Einstein–Weyl structures on a Riemannian manifold. Firstly, we report the recent results on this topic obtained with a method similar to the one of Wang and Ziller described in the previous section.

Let M be an n-dimensional manifold with a conformal structure $[g]$ determined by a Riemannian metric g and assume that D is a symmetric connection on M. The pair $([g], D)$ defines an *Einstein–Weyl structure* if D

preserves $[g]$ and the symmetrized of the Ricci tensor ρ^D of D is proportional to g. Equivalently, there exist a 1-form ω and a function $\tilde{\Lambda}$ on M such that:

$$Dg = \omega \otimes g \ , \quad \rho^D(X,Y) + \rho^D(Y,X) = \tilde{\Lambda} g(X,Y) \ , \tag{5.25}$$

for any $X, Y \in \mathcal{X}(M)$. Note that, given g and ω, the first condition in the above formula uniquely determines the action of D. So, a Weyl structure is also specified by the pair (g, ω); the action of D, called the *Weyl connection* of the structure, is given by:

$$D_X Y = \nabla_X Y - \frac{1}{2}(\omega(X)Y + \omega(Y)X - g(X,Y)B) \ , \tag{5.26}$$

where $B = \omega^{\#}$ is the vector field g-associated with ω and ∇ is the Levi-Civita connection of g. This allows to relate the Ricci tensors ρ^D, ρ of D, ∇, respectively; in fact one has ([219; 242]):

$$\begin{aligned} \rho^D(X,Y) = \rho(X,Y) &+ \tfrac{1}{2}(n-1)(\nabla_X\omega)Y - \tfrac{1}{2}(\nabla_Y\omega)X \\ &+ \tfrac{1}{4}(n-2)\omega(X)\omega(Y) + \tfrac{1}{2}(\operatorname{div} B - \tfrac{1}{2}(n-2)\|\omega\|^2)g(X,Y) \ . \end{aligned}$$

Putting:

$$\Lambda = \tfrac{1}{2}(\tilde{\Lambda} - \operatorname{div} B + \tfrac{1}{2}(n-2)\|\omega\|^2) \ ,$$

$$(\mathcal{D}\omega)(X,Y) = (\nabla_X\omega)Y + (\nabla_Y\omega)X + \omega(X)\omega(Y) \ , \quad X, Y \in \mathcal{X}(M) \ ,$$

it is easy to prove that (g, ω) defines an Einstein–Weyl structure if and only if there exists a function Λ on M such that:

$$\rho + \frac{1}{4}(n-2)\mathcal{D}\omega = \Lambda g \ . \tag{5.27}$$

Note that Λ, which is called the *Einstein–Weyl function* of the Weyl structure, is not necessarily constant.

In [242] Pedersen and Swann determine conditions for the existence of an Einstein–Weyl structure on the total space of principal torus bundles. Firstly, we consider the projection $\pi : P \to B$ of a principal $\mathbf{S^1}$-bundle over a compact Einstein manifold (B, g') with positive scalar curvature. As in Sec. 5.4, let g be a Riemannian metric on P making π a Riemannian submersion with totally geodesic fibres of length 2π and $\{g_t\}_{t \in \mathbf{R}^*_+}$ its canonical variation.

A standard Einstein–Weyl structure in the canonical variation is a pair (g_t, ω) which defines an Einstein–Weyl structure on (P, g) such that ω is a vertical 1-form, *i.e.* ω vanishes on the horizontal distribution. The last condition is equivalent to the request that the fibres of π are totally geodesic with respect to the Weyl connection. In fact, since π has totally geodesic fibres, via (5.26) one has: $g(D_U U, X) = \frac{1}{2}g(U,U)\omega(X)$, U vertical and X horizontal vector fields. Hence $D_U U$ is vertical if and only if $\omega_{|\mathcal{H}} = 0$.

Theorem 5.7 *Let $\pi : (P, g) \to (B, g')$ be the projection of a non-trivial principal circle bundle over a compact, Einstein manifold of positive scalar curvature, classified by $\alpha \in H^2(B, \mathbf{Z})$. If (g_t, ω) is a standard Einstein–Weyl structure on P in the canonical variation of g, then B admits a local almost complex structure J. When the Einstein–Weyl function is constant, J is globally defined, (B, J, g') is an almost Kähler manifold whose Kähler form is a multiple of α and the canonical variation of g admits a unique Einstein metric. Vice versa, if g_{t_0} is an Einstein metric in the canonical variation, then there exists, uniquely determined up to the orientation of the fibres, a standard Einstein–Weyl structure (g_t, ω), $t \leq t_0$, with constant Einstein–Weyl function.*

Proof. As in Sec. 5.4, we denote by θ a principal connection on P so that g is given by (5.18), by U a unitary vertical vector field and by η the closed 2-form on B such that $\pi^*\eta$ is the curvature form of θ. Then, applying (5.19), the invariant A acts on the horizontal vector fields as: $A_X Y = -(\pi^*\eta)(X, Y)U$, and combining with (5.9), we get:

$$\rho_t(U, U) = t^2 ||\eta||^2 \circ \pi \ ,$$

$$\rho_t(U, X) = -t(\delta'\eta)(X') \circ \pi \ ,$$

$$\rho_t(X, Y) = \rho'(X', Y') \circ \pi - 2t \textstyle\sum_{i=1}^n \eta(X', X_i')\eta(Y', X_i') \circ \pi \ ,$$

for any $X, Y \in \mathcal{X}^b(P)$, π-related to X', Y', ρ_t denoting the Ricci tensor of the metric g_t in the canonical variation of g and $\{X_i'\}$ a local orthonormal frame on B. Let ω be a vertical 1-form on P; then there exists a differentiable function f on P such that $\omega = f\theta$. Now, we want to express the Einstein–Weyl condition for (g_t, ω) in terms of f and θ. Thus, we evaluate the tensor field $\mathcal{D}\omega$ corresponding to g_t. Since $\omega(U) = f$ and $\omega(\nabla_U^t U) = tfg(U, \nabla_U U) = 0$, we have:

$$\mathcal{D}\omega(U, U) = 2(\nabla_U^t \omega)(U) + \omega(U)^2 = 2U(f) + f^2 \ .$$

Considering X basic, via (5.7), since $T = 0$ and ω is vertical, we have:

$$\begin{aligned}\mathcal{D}\omega(X, U) &= (\nabla_X^t \omega)(U) + (\nabla_U^t \omega)(X) = X(f) - \omega[X, U] = 2d\omega(X, U) \\ &= 2(df \wedge \theta + fd\theta)(X, U) = X(f) \ .\end{aligned}$$

Finally, $\mathcal{D}\omega$ vanishes on the pairs of horizontal vector fields; in fact:

$$\mathcal{D}\omega(X,Y) = -\omega(\nabla^t_X Y + \nabla^t_Y X) = -\omega(A_X Y + A_Y X) = 0 .$$

Putting $\rho' = E'g'$, E' being the Einstein constant of (B,g'), from (5.27) we obtain that an Einstein–Weyl structure (g_t,ω) corresponds to a solution (t,f,Λ) of the system:

$$\begin{aligned} & t^2\|\eta\|^2 \circ \pi + \tfrac{1}{4}(n-1)(2U(f)+f^2) = t\Lambda , \\ & t(\pi^*(\delta'\eta))(X)\circ\pi - \tfrac{1}{4}(n-1)X(f) = 0 , \\ & 2t\textstyle\sum_{i=1}^n (\pi^*\eta)(X,X_i)(\pi^*\eta)(Y,X_i) = (E'-\Lambda)g(X,Y) , \end{aligned} \tag{5.28}$$

where $X,Y \in \mathcal{X}^h(P)$ and the X_i are the horizontal lifts of the X'_i. Assume the existence of a solution (t_1,f,Λ) of the system; from the last equation we obtain that, if Λ is constant, $\Lambda = E'\circ\pi$, then $\eta = 0$, *i.e.* the bundle would be trivial. Thus, $E'\circ\pi - \Lambda \geq 0$ does not vanish at some point and we consider an open set W of B such that $E'\circ\pi - \Lambda > 0$ on $\pi^{-1}(W)$. Let μ be a smooth function defined on W such that

$$\mu\circ\pi = \frac{E'\circ\pi - \Lambda}{2t_1} .$$

Putting

$$J'X' = \sum_{i=1}^n \eta(X',X'_i)X'_i ,$$

we have:

$$J'^2X' = \sum_{i,j=1}^n \eta(X'_j,X'_i)\eta(X',X'_j)X'_i = -\mu X' ,$$

so $J = \mu^{-1/2}J'$ is a local almost complex structure on B; it is g'-Hermitian. Furthermore, if Λ is constant, μ is constant and J is globally defined on B. The Kähler form of (J,g') is the closed 2-form $-\mu^{-1/2}\eta$, hence (J,g') is an almost Kähler structure on B and the cohomology class of η is a multiple of α. This implies that η is coclosed, so that, for any $t>0$, the Ricci tensor ρ_t is given by:

$$\begin{aligned} & \rho_t(U,U) = t^2\|\eta\|^2\circ\pi , \\ & \rho_t(U,X) = 0 , \\ & \rho_t(X,Y) = (E'-2t\mu)g(X,Y) , \end{aligned}$$

and the positive solution of the equation $||\eta||^2 t^2 + 2\mu t - E' = 0$ determines the unique Einstein metric in the canonical variation. Conversely, if there exists an Einstein metric g_{t_0} with Einstein constant E, we obtain:

$$t_0||\eta||^2 = E > 0 \ , \ \delta'\eta = 0 \ , \ 2\sum_{i=1}^{n} \eta(X', X'_i)\eta(Y', X'_i) = \frac{E' - E}{t_0} g'(X', Y') \ ,$$

for any X', Y' tangent to B. Then, (5.28) reduces to:

$$(n-1)t_0(2U(f) + f^2) = 4t(t_0\Lambda - tE) \ ,$$

$$df_{|\mathcal{H}} = 0 \ ,$$

$$t(E' - E) = t_0(E' - \Lambda) \ ,$$

so that

$$(n-1)(2U(f) + f^2) = 4t\left(1 - \frac{t}{t_0}\right) E' \ .$$

Since this equation does not admit non-constant never vanishing solutions, the only solution of the above system gives an Einstein–Weyl structure (g_t, ω) with $t < t_0$ and constant Einstein–Weyl function.

The above statement allows to produce several examples of manifolds admitting Einstein–Weyl structures; a detailed description is given in [242] and [142]. Here, we only examine the case of the odd dimensional spheres.

Example 5.11 As in Example 5.3, we consider the Hopf submersion

$$\pi : \mathbf{S^{2n+1}} = U(n+1)/U(n) \to \mathbf{P_n(C)} = U(n+1)/U(1)U(n)$$

and the 1-parameter family $\{g_a\}_{a \in \mathbf{R}^*_+}$ of $U(n+1)$-invariant metrics on $\mathbf{S^{2n+1}}$ determined by the inner products

$$\beta'_a(\lambda + \eta, \lambda' + \eta') = a\mathrm{Re}(\lambda\overline{\lambda'}) + <\eta, \eta'>$$

on $u(1) \oplus \mathbf{C^n}$, where $<,>$ is the canonical Hermitian product on $\mathbf{C^n}$. The canonical metric on $\mathbf{S^{2n+1}}$, which is realized for $a = 1$, is the only homogeneous Einstein metric in $\{g_a\}_{a \in \mathbf{R}^*_+}$. Note that the metric on $\mathbf{P_n(C)}$ corresponding to $<,>$ is the Fubini–Study metric, so the base space is compact, Einstein, with positive scalar curvature. Via the above theorem, one obtains an Einstein–Weyl structure in $\{g_a\}_{a \in \mathbf{R}^*_+}$, corresponding to a value $a < 1$.

As in Sec. 5.4, we consider the projection $\pi : P \to B = M_1 \times \cdots \times M_h$ of a principal $\mathbf{S^1}$-bundle over the product of compact Kähler Einstein manifolds (M_i, J_i, g_i) with positive first Chern class $c_i(M_i) = q_i\alpha_i$, α_i indivisible, and Kähler form Ω_i such that $2\pi\alpha_i = [\Omega_i]$. Then, π is classified by the Euler class $e(P) = \sum_{i=1}^h b_i\pi_i^*(\alpha_i)$, where any $b_i \in \mathbf{Z}$, $\pi_i : B \to M_i$ denoting the projection onto the i-th factor. We seek for Einstein–Weyl structures (g, ω) on P, with ω vertical, and positive constants $x_i, \ldots, x_h$ making the map $\pi : (P, g) \to (B, g')$, $g' = \sum_{i=1}^h x_i g_i$, a Riemannian submersion with totally geodesic fibres. We recall that, given positive real numbers $x_i, \ldots, x_h$ the 2-form $\eta = \sum_{i=1}^h b_i\pi_i^*(\Omega_i)$ is the harmonic representative of $2\pi e(P)$ with respect to the metric g'. As in Theorem 5.4, one considers a connection θ on P with curvature form $\pi^*\eta$ and the metric g on P, defined as in (5.18), by means of g' and the standard inner product $<, >$ on $\mathbf{R}$. This makes $\pi : (P, g) \to (B, g')$ a Riemannian submersion with totally geodesic fibres of length 2π. The Ricci tensor ρ of (P, g) acts as:

$$\rho(U, U) = \sum_{i=1}^h b_i^2 x_i^{-2} n_i \ ,$$

$$\rho(U, X) = -(\pi^*(\delta'\eta))(X) = 0 \ ,$$

$$\rho(X, Y) = (q_i - 2b_i^2 x_i^{-1}) g_i(X', Y') \circ \pi_i \circ \pi \ ,$$

for any basic vector fields X, Y whose projections X', Y' are tangent to M_i; moreover U is the vertical vector field such that $\theta(U) = 1$, and for any i, n_i is the real dimension of M_i. Now, putting $\omega = f\theta$, $n = \sum_{i=1}^h n_i$, the Einstein–Weyl condition for (g, ω) is equivalent to the following system:

$$\sum_{i=1}^h b_i^2 x_i^{-2} n_i + \tfrac{1}{4}(n-1)(2U(f) + f^2) = \Lambda$$

$$df_{|\mathcal{H}} = 0$$

$$q_i x_i^{-1} - 2b_i^2 x_i^{-2} = \Lambda \ , \ i \in \{1, \ldots, h\} \ .$$

From the last equation, we get that Λ is constant and this implies the constancy of a globally defined function f. Moreover, for any i such that $b_i = 0$, one has $x_i = q_i/\Lambda$. Thus, when the Euler class $e(P)$ vanishes, for any $\Lambda > 0$, $(x_1, \ldots, x_h) = (q_1/\Lambda, \ldots, q_h/\Lambda)$, $f^2 = \frac{4}{n-1}\Lambda$ determine an Einstein–Weyl structure on P. Otherwise, for any choice of positive numbers k, Λ, with $4\Lambda > k$, one can prove, like in Theorem 5.4, the existence of a solution $(x_1, \ldots, x_h)$ of the above system, with $f^2 = k/(n-1)$. So, we obtain the following result.

Theorem 5.8 *Let $\pi : P \to M_1 \times \cdots \times M_h$ be the projection of a principal $\mathbf{S^1}$-bundle over the product of compact, Kähler Einstein manifolds (M_i, J_i, g_i) with positive first Chern class. Then, there exist positive real numbers $x_1, \ldots, x_h$ and Einstein–Weyl structures (g, ω) on P, with ω vertical, making $\pi : (M, g) \to (M_1 \times \cdots \times M_h, \sum_{i=1}^h x_i g_i)$ a Riemannian submersion with totally geodesic fibres.*

Finally, one considers the projection $\pi : P \to B = M_1 \times \cdots \times M_h$ of a principal T^r-bundle over the product of h, $h \geq r$, n_i-dimensional compact, Kähler Einstein manifolds (M_i, J_i, g_i) with positive first Chern class, Kähler form Ω_i such that $[\Omega_i] = 2\pi\alpha_i$, $c_1(M_i) = q_i \alpha_i$, α_i indivisible. As in Sec. 5.4, let $\beta_1, \ldots, \beta_r$ be the characteristic classes classifying π; for any $k \in \{1, \ldots, r\}$ one puts $\beta_k = \sum_{i=1}^h b_{ki} \pi_i^* \alpha_i$, with $b_{ki} \in \mathbf{Z}$. We can consider a connection θ on P with curvature form $\pi^* \eta$, where $\eta = \sum_{k=1}^r (\sum_{i=1}^h b_{ki} \pi_i^* \Omega_i) e_k$ is a g'-harmonic 2-form on B with values in the Lie algebra $\mathfrak{t}^r$ of T^r, $g' = \sum_{i=1}^h x_i g_i$, $\{e_1, \ldots, e_r\}$ being a canonical basis of $\mathfrak{t}^r$. We denote by $\{U_1, \ldots, U_r\}$ the vertical frame on P such that $\theta(U_k) = e_k$, $k \in \{1, \ldots, r\}$. Given an inner product $<,>$ on $\mathfrak{t}^r$, we put $< e_k, e_l >= h_{kl}$, $1 \leq k, l \leq r$, and consider the Riemannian metric g on P defined as in (5.18) via the inner product $<,>$ and the metric $g' = \sum_{i=1}^h x_i g_i$. Computing the Ricci tensor ρ of (P, g), one has:

$$\rho(U_k, U_l) = \sum_{q,m=1}^r \sum_{s=1}^h n_s b_{qs} b_{ms} x_s^{-2} h_{kq} h_{lm} \ , \quad k, l \in \{1, \ldots, r\} \ ,$$

$$\rho(U_k, X) = 0 \ , \ k \in \{1, \ldots, r\}, \quad X \in \mathcal{X}^h(P) \ ,$$

$$\rho(X, X) = (q_i g_i(X', X') - 2 \sum_{q,l=1}^r b_{qi} b_{li} x_i^{-1} h_{ql} g_i(X', X')) \circ \pi_i \circ \pi \ ,$$

for any basic vector field X whose projection X' is tangent to M_i. Now, considering a vertical 1-form ω on P, one puts $\omega = \sum_{k=1}^r f_k \theta_k$, where $\theta = \sum_{k=1}^r \theta_k e_k$; then $\mathcal{D}\omega$ is determined by:

$$\mathcal{D}\omega(U_k, U_l) = U_k(f_l) + U_l(f_k) + f_k f_l \ , \ k, l \in \{1, \ldots, r\} \ ,$$

$$\mathcal{D}\omega(X, U_k) = X(f_k) \ , \ k \in \{1, \ldots, l\} \ ,$$

$$\mathcal{D}\omega(X, X) = 0 \ ,$$

for any $X \in \mathcal{X}^h(P)$. Therefore, the Einstein–Weyl equation for (g, ω) is equivalent to the system:

$$\sum_{q,m}\sum_s n_s h_{kq} b_{qs} h_{lm} b_{ms} x_s^{-2} + \tfrac{1}{4}(n-2)(U_k(f_l) + U_l(f_k) + f_k f_l) = \Lambda h_{kl}$$

$$(df_k)_{|\mathcal{H}} = 0$$

$$q_i x_i^{-1} - 2\sum_{q,l=1}^{r} h_{ql} b_{qi} b_{li} x_i^{-2} = \Lambda\ ,$$

where $k, l \in \{1, \dots, r\}$, $i \in \{1, \dots, h\}$, $n = \sum_{i=1}^{h} n_i + r$ and the sum indexes q, m range in $\{1, \dots, r\}$ while s ranges in $\{1, \dots, h\}$.

In [242], Pedersen and Swann prove the existence of solutions $(x_1, \dots, x_h, f_1, \dots, f_r)$ of this system, with each f_k constant, when $\mathrm{rank}(b_{qi}) \in \{r-1, r\}$. Since moreover the first Betti number of P is $b_1(P) = r - \mathrm{rank}(b_{qi})$, they prove the following statement.

Theorem 5.9 *Let $\pi : P \to M_1 \times \cdots \times M_h$ be the projection of a principal T^r-bundle over the product of $h \geq r$ compact Einstein Kähler manifolds (M_i, J_i, g_i), with positive first Chern class.*
If $b_1(P) \leq 1$, then there exist positive real numbers $x_1, \dots, x_h$ and a family of Einstein–Weyl structures (g, ω) on P, with ω a vertical 1-form, such that $\pi : (P, g) \to (M_1 \times \cdots \times M_h, \sum_{i=1}^{h} x_i g_i)$ is a Riemannian submersion with totally geodesic fibres.

5.6 Einstein Weyl Structures on Hermitian and Sasakian Manifolds

The Einstein–Weyl theory on Hermitian or contact metric manifolds has been developed successfully in the last ten years. In particular, existence results of Einstein–Weyl structures on manifolds which are the total space of a Riemannian submersion can be easily obtained via the properties of the invariants of the submersion. In this section, we firstly consider Hermitian Einstein–Weyl spaces, then we'll examine the almost contact metric ones. As remarked by Pedersen, Poon and Swann ([241]), compact locally conformal Kähler manifolds with parallel Lee form describe compact Hermitian Einstein–Weyl spaces.

By definition, a *Hermitian–Weyl space* is an even dimensional manifold M with a Weyl structure $([h], D)$ such that there exists a D-parallel almost complex structure J, compatible with h. Such a space, denoted by (M, J, h), is said *closed* (*exact*) if the 1-form ω such that $Dh = \omega \otimes h$ is closed (exact).

The almost complex structure J of a Hermitian–Weyl space (M, J, h) is integrable, since it is parallel with respect to the symmetric connection D; (5.25) implies that the fundamental form Ω satisfies $d\Omega = \omega \wedge \Omega$. Thus, if $\dim M = 2n \geq 6$, then ω is closed and coincides with the Lee form of (M, J, h). Combining with Proposition 3.30, in the $2n$-dimensional case, $n \geq 3$, the class of Hermitian–Weyl spaces is just the class of the locally conformal Kähler manifolds and any space in this class is closed. This is not true in the 4-dimensional case; in fact the 4-dimensional l.c.K. manifolds set up the class of closed Hermitian–Weyl spaces.

Now, we consider a compact l.c.K. manifold (M, J, h). A result of I. Vaisman (Theorem 1.1 in [311]) states that the Lee form of M is exact if and only if M carries a global Kähler metric. Thus, the compact, closed and non-exact Hermitian–Weyl spaces are just the compact l.c.K. manifolds which do not admit any Kähler metric. Furthermore, a theorem of P. Gauduchon states that the compact l.c.K. manifold (M, J, h) admits a metric g in $[h]$, uniquely determined up to homotheties, whose corresponding Lee form is coclosed, and hence harmonic; this metric is called *distinguished* ([106; 107]). So, there is no restriction in fixing a $2n$-dimensional compact l.c.K. manifold (M, J, g), with g distinguished. Since the Lee form is closed and coclosed, the Ricci tensor ρ^D of D is related to the Ricci tensor ρ of g by:

$$\rho^D = \rho + (n-1)\left(\nabla\omega + \frac{1}{2}\omega \otimes \omega\right) - \frac{1}{2}(n-1)||\omega||^2 g\,. \qquad (5.29)$$

In particular, ρ^D is symmetric. Therefore, (g, ω) determines a Hermitian–Weyl structure provided that there exists a differential function $\tilde{\Lambda}$ on M such that:

$$\rho + (n-1)\left(\nabla\omega + \frac{1}{2}\omega \otimes \omega\right) = \frac{1}{2}((n-1)||\omega||^2 + \tilde{\Lambda})g\,. \qquad (5.30)$$

The function $\tilde{\Lambda}$ depends on the scalar curvature τ of (M, g) according to the formula $n\tilde{\Lambda} = \tau - \frac{1}{2}(2n-1)(n-1)||\omega||^2$; and the corresponding Einstein–Weyl function is

$$\Lambda = \frac{\tau}{2n} + \frac{n-1}{4n}||\omega||^2\,.$$

Moreover, since ω is closed and M is compact, $\tilde{\Lambda}$ never vanishes or $\tilde{\Lambda} \equiv 0$ (Lemma 2 in [108]).

Theorem 5.10 *Let (M, J, g) be a compact connected l.c.K. manifold, with g distinguished. If g determines an Einstein–Weyl structure, then one of the cases can occur:*

a) *(M, J, g) is a Kähler–Einstein manifold;*
b) *the Weyl connection is Ricci-flat and (M, J, g) is a generalized Hopf manifold with positive constant scalar curvature and Ricci tensor $\rho = \frac{1}{2}(n-1)(\|\omega\|^2 g - \omega \otimes \omega)$.*

Proof. Since g is a distinguished metric, the Lee form ω is harmonic. If ω is exact, then $\omega = df$, where f is a harmonic function and the compactness of M implies its constancy, that is ω vanishes. This corresponds to the case a); in fact (M, J, g) is Kähler and Einstein, since the Levi-Civita and the Weyl connections coincide. When ω is not exact, the function $\tilde{\Lambda}$ vanishes and one proves that $\|\omega\|$ is constant (see Lemmas 2, 2' in [108]). Then, via (5.30) one has $\rho(B, B) = 0$, B being the Lee vector field. So, $\int_M \rho(B, B) dv_g = 0$ and the Bochner theorem implies $\nabla B = 0$. Thus, ω is ∇-parallel and never vanishes; from (5.30) and (5.29) one has:

$$\rho = \frac{1}{2}(n-1)(\|\omega\|^2 g - \omega \otimes \omega), \ \ \rho^D = 0, \ \ \tau = \frac{1}{2}(2n-1)(n-1)\|\omega\|^2 ,$$

and the case b).

The above theorem is useful for the construction of *k-generalized Hopf manifolds*. In fact, we can consider a compact Hermitian–Weyl manifold (M, J, g), with g distinguished. Assume that the Weyl connection D is flat and the Lee form ω is not exact. Thus, (M, J, g) is a generalized Hopf manifold and, using the relation between the curvatures of D and the Riemannian connection of g, ([307]), we have:

$$\begin{aligned} R(X,Y,Z,W) = & \tfrac{1}{4}\|\omega\|^2(g(X,Z)g(Y,W) - g(Y,Z)g(X,W)) \\ & -\tfrac{1}{4}(g(X,Z)\omega(Y)\omega(W) - g(Y,Z)\omega(X)\omega(W) \\ & -g(X,W)\omega(Y)\omega(Z) + g(Y,W)\omega(X)\omega(Z)) \,. \end{aligned}$$

As in Proposition 3.33, we put $c = \|\omega\|/2$ and consider the c-Sasakian structure (φ, ξ, η, g) on a leaf S of the foliation defined by $\omega = 0$. Fixed $x \in S$ and an orthonormal basis $\{X, \varphi X\}$ of a φ-holomorphic 2-plane in $T_x S$, we have $g(JB, X) = 0$, $\varphi X = JX$ and the above formula implies $R(X, \varphi X, X, \varphi X) = \|\omega\|^2/4$. Thus (M, J, g) is a k-g.H.m., $k = \|\omega\|^2/4$.

Now, we report some results, originally due to I. Vaisman and then improved by K. Tsukada, concerning submersions from a compact g.H.m. onto a suitable Kähler manifold. As remarked by Vaisman ([310]), with any connected compact g.H.m. (M, J, g) is naturally associated the so-called *vertical foliation* $\mathcal{L}$, spanned by the Lee and the anti-Lee vector fields B, JB. Since the metric g on M is determined up to homotheties, we can assume $\|\omega\| = 1$. It is well known that B is analytic, *i.e.* $L_B J = 0$. This implies that $\mathcal{L}$ is a complex analytic foliation, with complex 1-dimensional leaves, whose holomorphic tangent spaces are generated by $B - \mathrm{i}JB$. Thus, the leaves are complex parallelizable and M admits local complex coordinates $\{z^\alpha\}_{1\le\alpha\le n}$ so that the equations $dz^k = 0$, $k \in \{1, \dots, n-1\}$, locally define $\mathcal{L}$ and $B - \mathrm{i}JB = \lambda\frac{\partial}{\partial z^n}$, $\lambda = \lambda(z^1, \dots, z^n)$ being a local analytic function without zero's ([310]). Since moreover $\nabla_{JB}JB = 0$, $\nabla_B B = 0$, $\nabla_{JB}B = 0$, $\nabla_B JB = 0$, the leaves of $\mathcal{L}$ are totally geodesic and locally flat submanifolds of M. Let $\mathcal{L}^\perp$ denote the orthogonal distribution of $\mathcal{L}$, which is defined by: $\omega = 0$, $\theta = 0$, where $\theta = -\omega \circ J$ is the anti-Lee form. Note that B, JB are Killing vector fields, so in particular $(L_X g)(Y, Z) = 0$, for any $X \in \mathcal{L}$, $Y, Z \in \mathcal{L}^\perp$. This means that the metric g can be locally expressed, in the above considered complex coordinates, as

$$ds^2 = 2\sum_{h,k=1}^{n-1} g_{h\overline{k}} dz^h \otimes d\overline{z}^k + (\omega + \mathrm{i}\theta) \otimes (\omega - \mathrm{i}\theta) ,$$

and the $g_{h\overline{k}}$ only depend on $\{z^1, \dots, z^{n-1}, \overline{z}^1, \dots, \overline{z}^{n-1}\}$, *i.e.* g is a bundle-like metric with respect to $\mathcal{L}$. Moreover, the Kähler form Ω of (M, J, g) is locally given by:

$$\Omega = -\mathrm{i}\sum_{h,k=1}^{n-1} g_{h\overline{k}} dz^h \otimes d\overline{z}^k - \omega \wedge \theta .$$

Since in a generalized Hopf manifold, $\Omega = d\theta - \omega \wedge \theta$, one locally obtains

$$d\theta = -\mathrm{i}\sum_{h,k=1}^{n-1} g_{h\overline{k}} dz^h \otimes d\overline{z}^k .$$

So, $d\theta$ is just the transversal component of Ω and Ω is transversally closed. Now, assume that $\mathcal{L}$ is a regular foliation. According to Vaisman ([310; 311]), (M, J, g) is called *strongly regular* and the leaves space $N = M/\mathcal{L}$ is a complex manifold. Then N inherits from M a Kähler structure and the canonical projection $\mu : M \to N$ acts as the projection of a principal complex torus bundle. In fact, in this case, the leaves of $\mathcal{L}$ are embedded as totally geodesic, complex parallelizable manifolds, so they need to be complex 1-dimensional tori and, applying Theorem 4.19 (see also [35]), M

is a principal $T^1_{\mathbf{C}}$-bundle over N. Moreover, the bundle-like character of g means that $h = g - \omega \otimes \omega - \theta \otimes \theta$ projects to a metric g' on N; the complex structure J of M projects to a complex structure J' compatible with g' and $\mu : (M, J, g) \to (N, J', g')$ is an l.c.K. submersion with totally geodesic fibres. Hence, applying Proposition 3.38, one gets that (N, J', g') is Kähler. We remark (see Proposition 3.36) that the invariant A of μ acts as

$$A_X Y = -\frac{1}{2}\Omega(X, Y)JB, \qquad X, Y \in \mathcal{L}^\perp .$$

This is useful to compute the Ricci tensor ρ' of (N, J', g') when the Weyl connection D of (M, J, g) is Ricci-flat, that is (M, J, g) falls in class b) of Theorem 5.10. In fact, in this case, considering the horizontal lifts X, Y of given vector fields X', Y' on N, via (5.29) and (5.30), we get:

$$\rho'(X', Y') \circ \mu = \rho(X, Y) + 2 \sum_{i=1}^{2(n-1)} g(A_X X_i, A_Y X_i) = \frac{n}{2} g'(X', Y') \circ \mu .$$

Therefore, (N, J', g') is a Kähler–Einstein manifold with scalar curvature $\tau' = n(n-1)||\omega||^2$. This proves the following result [310; 241].

Theorem 5.11 *Let (M, J, g) be a compact, connected, strongly regular generalized Hopf manifold with* $\dim M \geq 6$. *Then, the natural projection $\mu : M \to N = M/\mathcal{L}$ acts as the projection of a principal $T^1_{\mathbf{C}}$-bundle and N inherits from M a Kähler structure making μ a Vaisman submersion. In particular, if the Weyl connection of M is Ricci flat, then N is a Kähler–Einstein manifold with positive scalar curvature.*

Remark 5.3 With the previous notation, the complex 1-form $\psi = \theta - \mathrm{i}\omega$ can be regarded as a principal connection on the $T^1_{\mathbf{C}}$-bundle $\mu : M \to N$. The vector fields B, JB are the fundamental vector fields on M corresponding to the complex numbers $-\mathrm{i}, 1$, respectively (**C** is considered as the Lie algebra of $T^1_{\mathbf{C}}$). Moreover, one has:

i) $\psi \circ J = \mathrm{i}\psi$,
ii) the curvature form Ψ of ψ is given by $\Psi = d\theta = \mu^*\Omega'$, Ω' being the Kähler form of N.

In [301], K. Tsukada improves the above theorem. In fact, he constructs a structure of g.H.m. on the total space of a $T^1_{\mathbf{C}}$-principal bundle over a compact Kähler manifold equipped with a connection form satisfying i) and ii). For a detailed proof, see Theorem 4.1 in [301].

Remark 5.4 We recall that a compact homogeneous generalized Hopf manifold is strongly regular ([311]). Thus, Proposition 3.41 and Theorem 3.8 can be regarded as an improvement of Theorem 5.11.

In [219; 220], Narita points out the interrelation between almost contact, Einstein–Weyl structures and Riemannian submersions.

Let $\pi : (M, g) \to (B, g')$ be a Riemannian submersion with totally geodesic 1-dimensional fibres, denote by V a unit vertical vector field and by η its dual 1-form. In this case, since the O'Neill tensor T vanishes, one has $\nabla_V V = v(\nabla_V V) = g(\nabla_V V, V)V = 0$. Any Einstein–Weyl structure (g, ω) on M with ω vertical is determined by the function f such that $\omega = f\eta$. Such a structure is called *standard* according to [242; 219]. The corresponding tensor field $\mathcal{D}\omega$, defined in Sec. 5.5, vanishes on the pairs of horizontal vector fields. In fact, given $X, Y \in \mathcal{X}^h(M)$, we have:

$$\begin{aligned}\mathcal{D}\omega(X,Y) &= (\nabla_X\omega)Y + (\nabla_Y\omega)X + \omega(X)\omega(Y) = -\omega(\nabla_X Y + \nabla_Y X)\\ &= fg(V, A_X Y + A_Y X) = 0\ .\end{aligned}$$

This implies $\rho(X, Y) = \Lambda g(X, Y)$, where Λ is the Einstein–Weyl function and, combining with (1.36), the Ricci tensor ρ' of B is given by:

$$\rho'(X', Y') \circ \pi = \Lambda g'(X', Y') \circ \pi + 2g(A_X V, A_Y V)\ , \tag{5.31}$$

X, Y denoting the horizontal lifts of X', Y', respectively.

This motivates the investigation of the submersions like the one just considered whose base space is an Einstein manifold. In fact, in [219] the following result is proved.

Theorem 5.12 *Let $\pi : (M, g) \to (B, g')$ be a non-trivial Riemannian submersion with 1-dimensional, totally geodesic fibres over the Einstein manifold (B, g'), $\dim B \geq 3$. Assuming that M is compact, connected, and (g, ω) defines a standard Einstein–Weyl structure with constant Einstein–Weyl function, then M admits a Sasakian structure $(\varphi, \xi, \overline{\eta}, kg)$, $k \in \mathbf{R}^*_+$, and B has a Kähler structure (J, kg') which makes π a contact-complex Riemannian submersion.*

Proof. Let E' be the Einstein constant of B, so $\rho' = E'g'$. From (5.31) we get

$$g(A_X V, A_Y V) = \frac{1}{2}(E' \circ \pi - \Lambda)g(X, Y),\ X, Y \in \mathcal{X}^h(M)\ .$$

In particular, this implies $\|A\|^2 = \frac{1}{2}(E' \circ \pi - \Lambda) \dim B$, so the real number $k = \frac{1}{2}(E' \circ \pi - \Lambda)$ is positive and on the horizontal distribution one has:

$$g(A_X V, A_Y V) = kg(X, Y) . \tag{5.32}$$

Now, we prove that, for any basic vector field X, $A_X V$ is basic too, and

$$A_{A_X V} V = -kX . \tag{5.33}$$

Considering X basic, we show that for any basic vector field Y, the function $g(A_X V, Y)$ is constant on any fibre; this allows to state that $A_X V$ is projectable, hence basic. In fact, via (1.26)(g) and Lemma 1.4, one has:

$$\begin{aligned} 0 &= g((\nabla_V A)(X, Y), V) \\ &= g(\nabla_V (A_X Y), V) - g(A_{\nabla_V X} Y, V) - g(A_X(\nabla_V Y), V) \\ &= V(g(A_X Y, V)) - g(A_{A_X V} Y, V) - g(A_X(A_Y V), V) \\ &= -V(g(A_X V, Y)) , \end{aligned}$$

hence $g(A_X V, Y)$ is constant on any fibre. Moreover, combining with (5.32), we have:

$$g(\nabla_V (A_X V), Y) = -g(A_X V, \nabla_V Y) = -g(A_X V, A_Y V) = -kg(X, Y) ,$$

and:

$$\begin{aligned} 0 &= g((\nabla_V A)(X, V), Y) \\ &= g(\nabla_V (A_X V), Y) - g(A_{\nabla_V X} V, Y) - g(A_X(\nabla_V V), Y) \\ &= -kg(X, Y) - g(A_{A_X V} V, Y) , \end{aligned}$$

and then (5.33). Considering now the tensor field φ on M such that $\varphi F = \frac{1}{\sqrt{k}} A_F V$, for any $F \in \mathcal{X}(M)$, one proves that (φ, V, η, g) is an almost contact metric structure. In fact, the relations $\varphi V = 0$, $\eta \circ \varphi = 0$ are obvious; via (5.33), for any basic vector field X, one has $\varphi^2 X = \frac{1}{k} A_{A_X V} V = -X$, hence $\varphi^2 = -I + \eta \otimes V$. The compatibility of g with the given structure follows from (5.32). Putting:

$$\overline{g} = kg \ , \ \ \xi = -\frac{1}{\sqrt{k}} V \ , \ \ \overline{\eta} = -\sqrt{k}\eta \ ,$$

$(\varphi, \xi, \overline{\eta}, \overline{g})$ defines an ACM-structure on M having again ∇ as Levi-Civita connection. We are going to prove that φ induces a complex structure J on B such that (B, J, kg') is a Kähler manifold, $\pi_* \circ \varphi = J \circ \pi_*$, $\pi^*(kg') = kg - \overline{\eta} \otimes \overline{\eta}$ and $(M, \varphi, \xi, \overline{\eta}, kg)$ is Sasakian. In fact, given a vector field X' on B with horizontal lift X, since $A_X \xi = \frac{1}{\sqrt{k}} A_X V$ is basic, then $JX' = \pi_*(A_X \xi) = \pi_*(\varphi X)$ is well defined. Obviously, one has $\pi_* \circ \varphi = J \circ \pi_*$ and this also implies that J is an almost complex structure. Moreover, since obviously $\overline{g} = kg$ is compatible with $(\varphi, \xi, \overline{\eta})$ and $\pi^*(kg') = kg - \overline{\eta} \otimes \overline{\eta}$, (B, J, kg') is an almost Hermitian manifold whose Kähler form Ω acts as:

$$\begin{aligned}\Omega(X', Y') \circ \pi &= kg(X, A_Y \xi) = \sqrt{k} g(A_X Y, V) = \tfrac{\sqrt{k}}{2} g(v[X, Y], V) \\ &= -\sqrt{k} d\eta(X, Y) \ ,\end{aligned}$$

X, Y denoting the horizontal lifts of $X', Y' \in \mathcal{X}(B)$. Hence Ω is closed and B is an almost Kähler manifold. A well-known theorem of Sekigawa states that any compact almost Kähler manifold with non-negative scalar curvature is Kähler. Therefore, it is enough to prove that B has positive scalar curvature with respect to the metric g'; in fact the scalar curvature of (B, kg') is given by $\frac{1}{k}\tau'$, τ' scalar curvature of g'. To this aim, firstly we prove the formula:

$$\tau' \circ \pi = (n+2)||A||^2 + \frac{1}{4} n(n-1)||\omega||^2 + 2\delta\omega \ , \tag{5.34}$$

where $n = \dim B$. In fact, from (1.36) and (5.26) one obtains:

$$\begin{aligned}||A||^2 &= \textstyle\sum_{i=1}^n g(A_{X_i} V, A_{X_i} V) = \rho(V, V)\Lambda - \frac{1}{4}(n-1)\mathcal{D}\omega(V, V) \\ &= \Lambda - \tfrac{1}{4}(n-1)\{2(\nabla_V \omega)V + \omega(V)^2\} \ ,\end{aligned}$$

where $\{X_i\}_{1 \leq i \leq n}$ is a local orthonormal horizontal frame. Considering $\{X_1, \dots, X_n, V\}$ as a local orthonormal frame on M, we have:

$$\delta\omega = -f \sum_{i=1}^n g(V, \nabla_{X_i} X_i) + (\nabla_V \omega)V = (\nabla_V \omega)V \ ,$$

and

$$||\omega||^2 = \omega(V)^2 \ ,$$

so that

$$\Lambda = \frac{1}{4}(n-1)(2\delta\omega + ||\omega||^2) + ||A||^2 \ .$$

Since moreover $\tau' \circ \pi = nE' \circ \pi = 2||A||^2 + n\Lambda$, we obtain (5.34). Assuming now $\tau' \leq 0$ and applying the divergence theorem, from (5.34) one has

$$(n+2)||A||^2 + \frac{1}{4} n(n-1)||\omega||^2 = 0 \ ,$$

and then $A = 0$ which contradicts the non-triviality of π. Therefore, $\tau' > 0$ and (B, J, g') is a Kähler manifold. This also implies that $(\varphi, \xi, \overline{\eta}, \overline{g} = kg)$

is a Sasakian structure; in fact, this condition is equivalent to:

$$(\nabla_\xi \varphi)\xi = 0,\ (\nabla_\xi \varphi)X = 0,\ (\nabla_X \varphi)\xi = -X,\ (\nabla_X \varphi)Y = \overline{g}(X,Y)\xi\ ,$$

for any $X, Y \in \mathcal{X}^h(M)$. The first relations follow from (5.32) and $\nabla_\xi \xi = 0$. Finally, considering basic vector fields X, Y, π-related to X', Y', the horizontal component of $(\nabla_X \varphi)Y$ vanishes, since it is π-related to $(\nabla'_{X'} J')Y'$; applying (5.32) we have:

$$(\nabla_X \varphi)Y = v(\nabla_X \varphi Y) = \frac{1}{\sqrt{k}} A_X(A_Y V) = -\frac{1}{\sqrt{k}} g(A_X V, A_Y V) = \overline{g}(X,Y)\xi\ ,$$

and hence the statement.

Corollary 5.3 *Let $\pi : (M, g) \to (B, g')$ be a Riemannian submersion with 1-dimensional, totally geodesic fibres over an Einstein manifold. If M is even dimensional, compact, connected and admits a standard Einstein–Weyl structure (g, ω) with constant Einstein–Weyl function, then π is trivial and M is, locally, a product manifold.*

Example 5.12 One considers the Hopf manifold $\mathbf{H^n} = (\mathbf{C^n})^*/\Delta$, where Δ is the group generated by the transformation $z \in (\mathbf{C^n})^* \mapsto e^2 z \in (\mathbf{C^n})^*$. The canonical complex structure on $\mathbf{C^n}$ induces a complex structure J on $\mathbf{H^n}$ and $(\mathbf{H^n}, J, g)$ is a generalized Hopf manifold. In terms of the usual coordinates on $\mathbf{C^n}$, the metric g and the Lee form ω are expressed as:

$$g = \frac{\sum_{j=1}^n dz^j \otimes d\overline{z}^j}{\sum_{j=1}^n z^j \overline{z}^j}\ , \qquad \omega = -\frac{1}{\sum_{j=1}^n z^j \overline{z}^j} \sum_{j=1}^n (z^j d\overline{z}^j + \overline{z}^j dz^j)\ ,$$

so (g, ω) defines an Einstein–Weyl structure with flat Weyl connection and Einstein–Weyl function $\Lambda = n - 1$. Since, for any $m \in \mathbf{Z}$ and $z \in (\mathbf{C^n})^*$, one has:

$$\frac{\exp(m) z}{\sqrt{\sum_{j=1}^n \exp(2m) z^j \overline{z}^j}} = \frac{z}{\sqrt{\sum_{j=1}^n z^j \overline{z}^j}}\ ,$$

it is meaningful to consider the map $p : \mathbf{H^n} \to \mathbf{S^{2n-1}}$ such that

$$p([z]) = \frac{z}{\sqrt{\sum_{j=1}^n z^j \overline{z}^j}}\ .$$

Then, p is a Riemannian submersion from $(\mathbf{H^n}, g)$ onto the standard unit sphere $\mathbf{S^{2n-1}}$, its fibres are isometric to the circle $\mathbf{S^1}(\frac{1}{\pi})$ and the Lee form

ω is vertical. Indeed, p is a trivial submersion, since the map

$$F : \mathbf{H^n} \to \mathbf{S^1}\left(\frac{1}{\pi}\right) \times \mathbf{S^{2n-1}} \, ,$$

defined by

$$F([z]) = \left(\frac{1}{\pi}\exp\left(-\mathrm{i}\frac{\pi}{2}\sum_{j=1}^{n} z^j \overline{z}^j\right), p([z])\right) \, ,$$

realizes an isometry between $(\mathbf{H^n}, g)$ and $\mathbf{S^1}(\frac{1}{\pi}) \times \mathbf{S^{2n-1}}$ with the product of the standard metrics ([306]).

Now, we prove a preliminary result due to Narita ([220]), concerning the existence of Einstein–Weyl structures on almost contact manifolds which are the total space of a submersion.

Proposition 5.5 *Let $(M^{2n+1}, \varphi, \xi, \eta, g)$ be an almost contact metric manifold such that $\nabla_X \xi = -\varphi X$. Assume that (M^{2n+1}, g) has Ricci tensor given by $\rho = \beta g + \gamma \eta \otimes \eta$, with β, γ constant and $\gamma \leq 0$. Then $(g, a\eta)$, with $a^2 = -4\gamma/(2n-1)$, determines an Einstein–Weyl structure with Einstein–Weyl function β.*

Proof. Putting $\omega = a\eta$, where $a^2 = -4\gamma/(2n-1)$, we prove that

$$\rho + \frac{1}{4}(2n-1)\mathcal{D}\omega = \beta g \, .$$

Firstly, observe that $\nabla_\xi \xi = 0$, so that:

$$\rho(\xi,\xi) + \frac{1}{4}(2n-1)\mathcal{D}\omega(\xi,\xi) = \beta + \gamma + \frac{1}{4}(2n-1)a^2 = \beta \, .$$

Moreover, given $X, Y \in \mathcal{X}^h(M^{2n+1})$, one has:

$$\begin{aligned} \mathcal{D}\omega(X,Y) &= a^2\{g(\nabla_X \xi, Y) + g(\nabla_Y \xi, X)\} \\ &= -a^2\{g(\varphi X, Y) + g(\varphi Y, X)\} = 0 \, , \end{aligned}$$

hence

$$\rho(X,Y) + \frac{1}{4}(2n-1)\mathcal{D}\omega(X,Y) = \beta g(X,Y) \, .$$

Finally, given $X \in \mathcal{X}^h(M)$:

$$\mathcal{D}\omega(X,\xi) = a^2\{g(\nabla_X \xi, \xi) + g(\nabla_\xi \xi, X)\} = 0 \, ,$$

so $\rho(X,\xi) + \frac{1}{4}(2n-1)\mathcal{D}\omega(X,\xi) = 0$, and then the statement.

Remark 5.5 Any Sasakian manifold M^{2n+1} with constant holomorphic φ-sectional curvature $k \geq 1$ satisfies the hypotheses of the previous proposition. In fact, (see [350]), its Ricci tensor is given by

$$\rho = \frac{1}{2}((n+1)k + 3n - 1)g - \frac{1}{2}(n-1)(k-1)\eta \otimes \eta .$$

Moreover, the Sasaki condition implies $\nabla_X \xi = -\varphi X$, hence $(g, a\eta)$, with $a^2 = 2(n+1)(k-1)/(2n+1)$, determines an Einstein–Weyl structure.

Theorem 5.13 *Let $\pi : (M^{2n+1}, g) \to (B^{2n}, g')$, $n \geq 2$, be a Riemannian submersion with totally geodesic fibres over an Einstein manifold of scalar curvature $\tau' \geq 4n(n+1)$. If M^{2n+1} admits a Sasakian structure (φ, ξ, η, g), with ξ vertical, then M^{2n+1} has a standard Einstein–Weyl structure with constant Einstein–Weyl function.*

Proof. Assume that $(M^{2n+1}, \varphi, \xi, \eta, g)$ is a Sasakian manifold with ξ vertical. We prove that its Ricci tensor is given by:

$$\rho = \left(\frac{\tau'}{2n} - 2\right) g + \left(2n + 2 - \frac{\tau'}{2n}\right) \eta \otimes \eta , \tag{5.35}$$

so obtaining the statement via Proposition 5.5. Since M is Sasakian, the invariant A of π acts as:

$$A_X \xi = h(\nabla_X \xi) = -\varphi X , \ A_X Y = -g(A_X \xi, Y)\xi = g(\varphi X, Y)\xi .$$

These formulas imply $(\nabla_X A)(X, \xi) = 0$, $X \in \mathcal{X}^h(M)$, and, considering a local orthonormal horizontal frame $\{X_i, \varphi X_i\}_{1 \leq i \leq n}$ one has:

$$\sum_{i=1}^{n} \{g(A_{X_i}\xi, A_{X_i}\xi) + g(A_{\varphi X_i}\xi, A_{\varphi X_i}\xi)\} = 2n .$$

From (1.36), since g' is an Einstein metric, we have:

$$\rho(\xi, \xi) = 2n , \ \rho(X, Y) = \left(\frac{\tau'}{2n} - 2\right) g(X, Y) , \ \rho(X, \xi) = 0 ,$$

for any $X, Y \in \mathcal{X}^h(M)$, and then (5.35).

Example 5.13 Let M^{2n+1} be a compact, regular, contact manifold. A theorem of Boothby and Wang ([33]) states that M^{2n+1} is the total space of a principal circle bundle over a symplectic manifold B^{2n} with fundamental form Ω such that $[\Omega] \in H^2(B^{2n}, \mathbf{Z})$. Denote by (J, g') the almost Kähler structure such that $g'(X, JY) = \Omega(X, Y)$, by $\pi : M^{2n+1} \to B^{2n}$ the projection of the bundle, which is called a Boothby–Wang fibration, and by (φ, ξ, η, g) the ACM-structure on M^{2n+1} defined in Example 4.2. In this case, $(M^{2n+1}, \varphi, \xi, \eta, g)$ is a K-contact manifold, so $\nabla_X \xi = -\varphi X$, $X \in \mathcal{X}(M^{2n+1})$. Assuming that B^{2n} is an Einstein manifold with scalar curvature $\tau' \geq 4n(n+1)$, the Sekigawa theorem implies that (B^{2n}, J, g') is a Kähler manifold and, equivalently, $(M^{2n+1}, \varphi, \xi, \eta, g)$ is Sasakian ([33]). Via Theorem 5.13, $(g, a\eta)$, where $a^2 = 2(\tau' - 4n(n+1))/(n(2n-1))$, defines a standard Einstein–Weyl structure on M^{2n+1}.

Example 5.14 Let G be a compact, simply connected, semisimple Lie group, K a connected closed subgroup of G and assume that the Lie algebra of G splits as $\mathfrak{g} = \mathfrak{k} \oplus \mathfrak{p} \oplus \mathfrak{m}$, where $\mathfrak{k}$ is the Lie algebra of K and $\mathfrak{p}$, $\mathfrak{m}$ are subspaces of $\mathfrak{g}$, $\dim \mathfrak{p} = 1$, such that

$$[\mathfrak{k}, \mathfrak{p}] \subseteq \mathfrak{p}\ ,\ [\mathfrak{k}, \mathfrak{m}] \subseteq \mathfrak{m}\ ,\ [\mathfrak{p}, \mathfrak{p}] \subseteq \mathfrak{k}\ ,\ [\mathfrak{m}, \mathfrak{m}] \subseteq \mathfrak{k} \oplus \mathfrak{m}\ ,\ [\mathfrak{p}, \mathfrak{m}] \subseteq \mathfrak{m}\ .$$

Let $H \subseteq K$ be the connected subgroup of G with Lie algebra $\mathfrak{h} = \mathfrak{k} \oplus \mathfrak{p}$ and consider the Riemannian submersion $\pi : (G/K, g) \to (G/H, g')$, where g, g' are the left-invariant metrics determined by $-B_{|\mathfrak{p}\oplus\mathfrak{m}}$,and $-B_{|\mathfrak{m}}$, respectively, B denoting the Killing–Cartan form of $\mathfrak{g}$. Assume the existence of a positive function λ on G/K such that $[[X, W], W] = 4\lambda B(W, W)X$, for any X in $\mathfrak{m}$ and W in $\mathfrak{p}$. This condition implies the relations:

$$R(X, W, X, W) = \lambda g(X, X) g(W, W)\ ,$$

$$A_{A_X W} W = \tfrac{1}{4}[[X, W], W]_{\mathfrak{m}} = -\lambda g(W, W) X\ ,$$

$$g(A_X W, A_Y W) = \lambda g(W, W) g(X, Y)\ ,$$

X, Y being horizontal, W vertical. In [216], Narita proves that the function λ is constant, G/K admits a locally symmetric Sasakian structure $(\varphi, \xi, \eta, \lambda g)$, with ξ vertical, and G/H is a Hermitian symmetric space. Putting $\dim G/K = 2n + 1$, we prove that, if $\lambda < 1/(4n + 4)$, then $(\lambda g, a\eta)$ is an Einstein–Weyl structure on G/K, where

$$a^2 = 2\frac{1 - 4(n+1)\lambda}{(2n-1)\lambda}\ .$$

According to Proposition 5.5, it is enough to prove that the Ricci tensor ρ of λg, which coincides with the Ricci tensor of g, is given by:

$$\rho = \left(\frac{1}{2\lambda} - 2\right) \lambda g + \frac{1}{2\lambda}(4(n+1)\lambda - 1)\eta \otimes \eta\ . \tag{5.36}$$

Firstly, observe that π has totally geodesic fibres and G/K is a Sasakian manifold; hence, on the horizontal distribution, we have:

$$A_X \xi = -\varphi X ,$$

$$\overline{g}((\nabla_X A)(Y, Z), \xi) = \overline{g}((\nabla_X \varphi)Y, Z) + \overline{g}(\varphi X, A_Y Z) = 0 ,$$

and then, applying (1.36), $\rho(X, \xi) = 0$. Moreover,

$$\rho(\xi, \xi) = \sum_{i=1}^{2n} R(X_i, \xi, X_i, \xi) = 2n\lambda g(\xi, \xi) = 2n .$$

Finally, the Ricci tensor of the Hermitian symmetric space G/H is $\rho' = \frac{1}{2}g'$ and on the horizontal distribution we have:

$$\rho(X, Y) = \left(\frac{1}{2} - 2\lambda\right) g(X, Y) .$$

These relations give (5.36).

Chapter 6

Riemannian Submersions and Submanifolds

Some topics involving the theories of submersions and submanifolds are presented in this chapter.

In Sec. 6.1, CR-submersions are considered. More precisely, let $\tilde{M}$ be an almost Hermitian manifold with complex structure J and M a CR-submanifold of $\tilde{M}$ (Bejancu, [17]). We consider a submersion $\pi : M \to B$ onto an almost Hermitian manifold such that J interchanges $\ker \pi_*$ with the normal bundle of M in $\tilde{M}$ and π_* is a complex isometry from the orthogonal complement of $\ker \pi_*$ in TM onto TB. We state a theorem of S. Kobayashi on CR-submersions from a CR-submanifold of a Kähler manifold, also relating the second fundamental form of M in $\tilde{M}$ and the holomorphic sectional curvatures of the 2-planes spanned by (X, JX) and $(\pi_* X, \pi_* JX)$, X denoting a horizontal vector. Then, we state similar results, due to F. Narita ([217]), under the hypothesis that $\tilde{M}$ is a locally conformal Kähler manifold. More general results obtained by other authors are also presented.

The second section is devoted to show how the theory of Riemannian submersions allows to state links between some submanifolds of Sasakian manifolds and submanifolds of Kähler manifolds (these results are due to K. Yano, M. Kon and others). If $i : M' \to M$ is a horizontal isometric immersion, then its second fundamental form is closely related to the second fundamental form of its projection $F = \pi \circ i$. For instance, F is totally geodesic, minimal or totally umbilical, if and only if i has the corresponding property. Applying this theory to the Boothby–Wang fibration of a Sasakian manifold, we obtain a correspondence between the anti-invariant submanifolds of Sasakian and Kählerian manifolds (Reckziegel, [251]). Some examples are also given.

Section 6.3 deals with Riemannian submersions with isometric reflections with respect to the fibres. The concept of reflection with respect to a submanifold is introduced by B. Y. Chen and L. Vanhecke ([61]). Under some geometric conditions, we consider Riemannian submersions with fibres of dimension 1 and prove that the reflections with respect to the fibres are isometries if and only if M admits a locally φ-symmetric Sasakian structure.

6.1 Submersions of CR-submanifolds

In this section we consider Riemannian submersions whose total space is a CR-submanifold of a (nearly) Kähler or locally conformal Kähler manifold or a semi-invariant submanifold of a Sasakian manifold.

We begin with the following remark.

Let $(\tilde{M}, g)$ be a Riemannian manifold and M a Riemannian submanifold which is also the total space of a Riemannian submersion π over a Riemannian manifold (N, g'). Then, the Gauss equation:

$$\begin{aligned}\tilde{R}(X,Y,Z,E) = R(X,Y,Z,E) &+ g(\alpha(Y,Z),\alpha(X,E)) \\ &- g(\alpha(Y,E),\alpha(X,Z))\ ,\end{aligned}$$

combined with (1.28) gives:

$$\begin{aligned}\tilde{R}(X,Y,Z,E) = & R^*(X,Y,Z,E) - 2g(A_X Y, A_Z E) \\ & + g(A_Y Z, A_X E) - g(A_X Z, A_Y E) \\ & + g(\alpha(Y,Z),\alpha(X,E)) - g(\alpha(Y,E),\alpha(X,Z))\ ,\end{aligned} \tag{6.1}$$

where A is the integrability tensor of π, α denotes the second fundamental form of M in $\tilde{M}$ and X, Y, Z, E are π-horizontal vector fields.

Now, let $(\tilde{M}, J, g)$ be an almost Hermitian manifold and M a real submanifold of $\tilde{M}$. We consider the bundle $\mathcal{H} = TM \cap J(TM)$, which is J-invariant *i.e.* $J_p(\mathcal{H}_p) = \mathcal{H}_p$, for any $p \in M$. We recall that M is called a *CR-submanifold* of $\tilde{M}$ if $\mathcal{H}$ is a subbundle of $(T\tilde{M})_{|M}$ and its orthogonal complement in $T(M)$ is a totally real subbundle ([17; 60; 350]).

Definition 6.1 Let M be a CR-submanifold of $\tilde{M}$ and (N, J', g') an almost Hermitian manifold. A *CR-submersion* $\pi : M \to N$ is a Riemannian submersion such that:

a) the vertical distribution $\mathcal{V} = \ker \pi_*$ is the orthogonal complement of $\mathcal{H}$ in TM;

b) J interchanges $\mathcal{V}$ and $TM^{\perp}$;
c) for any $p \in M$, $\pi_{*p} : \mathcal{H}_p \to T_{\pi(p)}N$ is a complex isometry *i.e.* $\pi_* \circ J = J' \circ \pi_*$.

The following result, due to Kobayashi ([171]), represents the analogue of the symplectic reduction theorem of Marsden–Weinstein.

Theorem 6.1 *Let $\tilde{M}$ be a Kähler manifold and $\pi : M \to N$ a CR-submersion. Then, N is Kähler. Moreover, for any unit horizontal vector X, one has:*

$$\tilde{H}(X) = H'(\pi_* X) - 4\|\alpha(X, X)\|^2 ,$$

where $\tilde{H}$ and H' are the holomorphic sectional curvatures of $\tilde{M}$ and N respectively, and α denotes the second fundamental form of M in $\tilde{M}$.

Proof. Let X, Y be basic vector field π-related to X', Y' on N. Then, using the Gauss formula, the Kähler condition for $\tilde{M}$ gives:

$$h(\nabla_X JY) + v(\nabla_X JY) + \alpha(X, JY) = J(h(\nabla_X Y) + v(\nabla_X Y) + \alpha(X, Y)) ,$$

and, separating the horizontal, vertical and normal components, we get:

$$\begin{aligned} h(\nabla_X JY) - J(h(\nabla_X Y)) = 0 ; \\ v(\nabla_X JY) - J(\alpha(X, Y)) = 0 ; \\ \alpha(X, JY) - J(v(\nabla_X Y)) = 0 . \end{aligned}$$

The first relation projects on $\nabla'_{X'} J'Y' - J'(\nabla'_{X'} Y') = 0$, implying that N is Kähler. From the last two equations, by the definition of the tensor A, we obtain $A_X JY = J(\alpha(X, Y))$ and $\alpha(X, JY) = J(A_X Y)$. Then $\alpha(X, JX) = 0$ and by polarization we get:

$$\alpha(JX, JY) = \alpha(X, Y) . \tag{6.2}$$

Now, the symmetry of α implies:

$$A_X JY = J(\alpha(X, Y)) = A_Y JX , \tag{6.3}$$

and then

$$A_{JX} JY = A_X Y , \tag{6.4}$$

which, as well as (6.2), obviously holds for horizontal vector fields. Finally, given a unit horizontal vector X, (6.1) implies:

$$\begin{aligned} \tilde{H}(X) = R^*(X, JX, X, JX) - 3\|A_X JX\|^2 + \|\alpha(X, JX)\|^2 \\ -g(\alpha(JX, JX), \alpha(X, X)) . \end{aligned}$$

Then, using (6.2) and (6.3), we get the link between the holomorphic sectional curvatures.

Now, we assume that $\tilde{M}$ is a locally conformal Kähler, non-Kähler manifold, M a CR-submanifold of $\tilde{M}$ and $\pi : M \to N$ a CR-submersion, (N, J', g') being an almost Hermitian manifold. As proved by Narita ([217]), the given structure on N specializes according to the behavior of the Lee vector field $\tilde{B}$ of $\tilde{M}$.

Lemma 6.1 *Let $\tilde{M}$ be an l.c. Kähler manifold and $\pi : M \to N$ a CR-submersion. Then the vertical component of $\tilde{B}$ vanishes. Moreover, for any unit horizontal vector X, one has:*

$$\tilde{H}(X) = H'(\pi_* X) - 3\|A_X JX\|^2 - \|\alpha(X,X)\|^2 ,$$

where $\tilde{H}$ and H' are the holomorphic sectional curvatures of $\tilde{M}$ and N respectively, and α denotes the second fundamental form of M in $\tilde{M}$.

Proof. Obviously, the bundle $T\tilde{M}$ restricted to M orthogonally splits as the Whitney sum:

$$TM \oplus TM^{\perp} = \mathcal{V} \oplus \mathcal{H} \oplus TM^{\perp} ,$$

and along M we have the decomposition $\tilde{B} = v(\tilde{B}) + h(\tilde{B}) + n(\tilde{B})$. Using the definition of the invariant A, the Gauss formula for M becomes:

$$\tilde{\nabla}_X Y = h(\nabla_X Y) + A_X Y + \alpha(X,Y) , \tag{6.5}$$

for any X, Y horizontal vector fields. Then, from Corollary 3.11, we have:

$$\begin{aligned} A_X JY &= v(J(\tilde{\nabla}_X Y)) + \tfrac{1}{2} g(X,Y) v(J\tilde{B}) - \tfrac{1}{2} g(X,JY) v(\tilde{B}) \\ &= J(\alpha(X,Y)) + \tfrac{1}{2} g(X,Y) J(n(\tilde{B})) - \tfrac{1}{2} g(X,JY) v(\tilde{B}) . \end{aligned} \tag{6.6}$$

Putting $X = JY$, we get

$$J(\alpha(JY,Y)) = \frac{1}{2} g(JY,JY) v(\tilde{B}) ,$$

and replacing JY with Y,

$$-J(\alpha(Y,JY)) = \frac{1}{2} g(Y,Y) v(\tilde{B}) .$$

Then, the symmetry of α implies $v(\tilde{B}) = 0$ and $\alpha(Y,JY) = 0$. It follows that the Lee form $\tilde{\omega}$ vanishes on $\mathcal{V}$. By polarization, we get $\alpha(JX,Y) = -\alpha(X,JY)$ and then $\alpha(JX,JY) = \alpha(X,Y)$. Now, (6.6) easily implies $A_{JX} JY = A_X Y$ and a direct computation gives the statement on the holomorphic sectional curvatures ([217]).

Theorem 6.2 *Let $\tilde{M}$ be an l.c. Kähler manifold and $\pi : M \to N$ a CR-submersion. If $h(\tilde{B})$ is basic and $\dim N \geq 4$, then N is a locally conformal Kähler manifold. Moreover, if $\tilde{M}$ is a generalized Hopf manifold and $\tilde{B}$ is basic, then N is a g.H.m., also.*

Proof. We put $B' = \pi_*(h(\tilde{B}))$ and denote by ω' its dual 1-form with respect to g'. Then, denoting by ω the restriction to M of the Lee form $\tilde{\omega}$ of $\tilde{M}$, for any $X' \in \mathcal{X}(N)$, with horizontal lift X, we have:

$$(\pi^*\omega')(X) = \omega'(X') \circ \pi = g'(X', B') \circ \pi = g(X, h(\tilde{B})) = \tilde{\omega}(X) = \omega(X) .$$

Thus $\pi^*\omega' = \omega$ on $\mathcal{H}$ and, since d commutes with π_* and π is a Riemannian submersion, ω' is a closed 1-form. Then, considering the Weyl connection on N given by:

$$D_{X'}Y' = \nabla'_{X'}Y' - \frac{1}{2}(\omega'(X')Y' + \omega'(Y')X' - g'(X', Y')B') ,$$

a direct computation gives $DJ' = 0$ and N is locally conformal Kähler. Finally, assuming that $\tilde{M}$ is a g.H.m. and $\tilde{B}$ is basic, we have $\tilde{\nabla}_X\tilde{B} = 0$ for any $X \in \mathcal{H}$, and the Gauss formula gives $\nabla_X\tilde{B} = 0$. Then, since $h(\nabla_XY)$ is basic π-related to $\nabla'_{X'}Y'$, for any $X', Y' \in \mathcal{X}(N)$ with horizontal lifts X, Y, we get:

$$\begin{aligned} g'(\nabla'_{X'}B', Y') \circ \pi &= (X'(g'(B', Y')) - g'(B', \nabla'_{X'}Y')) \circ \pi \\ &= X(g(\tilde{B}, Y)) - g(\tilde{B}, \nabla_XY) \\ &= g(\nabla_X\tilde{B}, Y) = 0 . \end{aligned}$$

Thus, $\nabla'B' = 0$ and N is a generalized Hopf manifold.

We refer to [217] for some examples. Here, we present the following one.

Example 6.1 Consider a Euclidean space $E^{2n-1}(-3)$ with the standard Sasakian structure making it a Sasakian space form with constant φ-sectional curvature -3.

Denoting by $\mathbf{S^1}(r_i)$ a circle of radius r_i, it is well known that the Pythagorean product $\tilde{E} = E^{2(n-p)-1}(-3) \times \mathbf{S^1}(r_1) \times \cdots \times \mathbf{S^1}(r_p)$, $p \geq 2$, is a pseudo-umbilical generic submanifold of $E^{2n-1}(-3)$ ([349]). Then, $\mathbf{S^1} \times E^{2n-1}(-3)$ is a generalized Hopf manifold with Lee form ω given by the length element of $\mathbf{S^1}$ and having $\mathbf{S^1} \times \tilde{E}$ as a CR-submanifold tangent to the Lee vector field. Moreover, the projection:

$$\pi : \mathbf{S^1} \times \tilde{E} \to \mathbf{S^1} \times E^{2(n-p)-1}(-3)$$

is a CR-submersion, so $\mathbf{S^1} \times E^{2(n-p)-1}(-3)$ is also a g.H.m.

Theorem 6.3 *Let $\tilde{M}$ be an l.c. Kähler manifold and $\pi : M \to N$ a CR-submersion. If $\tilde{B}$ is a section of $T(M)^{\perp}$, then N is a Kähler manifold.*

Proof. Obviously, $\tilde{\omega}$ vanishes on $\mathcal{X}(M)$, so, using Corollary 3.11, we have:

$$\tilde{\nabla}_X JY = J(\tilde{\nabla}_X Y) - \frac{1}{2}g(X, JY)\tilde{B} + \frac{1}{2}g(X,Y)J\tilde{B} \,,$$

for any X, Y horizontal vector fields. Taking the horizontal component, we get $h(\tilde{\nabla}_X JY) = h(J(\tilde{\nabla}_X Y))$, and the Gauss formula implies $h(\nabla_X JY) = J(h(\nabla_X Y))$, since $\mathcal{H}$ is J-invariant. Finally, since $h(\nabla_X Y)$ is basic π-related to $\nabla'_{X'}Y'$ if X and Y are basic π-related to X' and Y', we have:

$$\nabla'_{X'} J'Y' = J'(\nabla'_{X'}Y') \,,$$

and N is Kähler.

In [271], one can find analogous results for submersions from a CR-submanifold of a non-Kähler, nearly Kähler manifold $\tilde{M}$, of dimension $n > 6$.

Theorem 6.4 *Let $\tilde{M}$ be a nearly Kähler manifold and $\pi : M \to N$ a CR-submersion over an almost Hermitian manifold N. Then, N is nearly Kähler. Moreover, for any unit horizontal vector X, one has:*

$$\tilde{H}(X) = H'(\pi_* X) - 4\|\alpha(X,X)\|^2 \,,$$

where $\tilde{H}$ and H' are the holomorphic sectional curvatures of $\tilde{M}$ and N, respectively, and α denotes the second fundamental form of M in $\tilde{M}$.

Proof. Let X, Y be basic vector field π-related to X', Y' on N. Then, using the Gauss formula, the nearly Kähler condition for $\tilde{M}$ gives:

$$h(\nabla_X JX) + v(\nabla_X JX) + \alpha(X, JX) = J(h(\nabla_X X) + v(\nabla_X X) + \alpha(X,X)) \,,$$

and, separating the horizontal, vertical and normal components, we get:

$$\begin{aligned} h(\nabla_X JX) - J(h(\nabla_X X)) &= 0 \,; \\ v(\nabla_X JX) - J(\alpha(X,X)) &= 0 \,; \\ \alpha(X, JX) - J(v(\nabla_X X)) &= 0 \,. \end{aligned}$$

The first relation projects on $\nabla'_{X'}J'X' - J'(\nabla'_{X'}X') = 0$, implying that N is nearly Kähler. From the last two equations, by the definition of the tensor

A, we obtain $A_X JX = J(\alpha(X,X))$ and $\alpha(X, JX) = 0$. By polarization we get:

$$\alpha(JX, JY) = \alpha(X, Y) , \tag{6.7}$$

$$A_X JY + A_Y JX = 2J(\alpha(X,Y)) . \tag{6.8}$$

Now, imposing the nearly Kähler condition to the basic vector field X and to any vertical vector field U, and computing the horizontal component, we get:

$$h(\nabla_X JU) - J(A_X U) + h(\nabla_U JX) - J(\nabla_U X) = 0 ,$$

which reduces to

$$h(\nabla_X JU) = A_{JX} U - 2J(A_X U) , \tag{6.9}$$

since $[U, X]$ and $[U, JX]$ are vertical vector fields. Taking the scalar product of (6.8) with U, we have

$$J(A_X U) + A_{JX} U = -2h(\nabla_X JU) ,$$

and, comparing with (6.9), $A_{JX} U = J(A_X U)$. The scalar product of this equation with JY allows to state:

$$A_{JX} JY = A_X Y , \tag{6.10}$$

which, as well as (6.7), obviously holds for horizontal vector fields. Finally, using (6.1), (6.7) and (6.10), a direct computation gives the link between the holomorphic sectional curvatures.

An extension to the context of submersions from contact CR-submanifolds of Sasakian manifolds can be found in [237].

6.2 Links Between Submanifolds of Sasakian and Kähler Manifolds

As remarked in Sec. 2.1, horizontal lifts of curves on the base space of a Riemannian submersion exist, at least locally, and the completeness of the metric on the total space implies the global existence of such lifts.

In [250] H. Reckziegel treats a more general question, *i.e.* the existence, with respect to a Riemannian submersion $\pi : (M, g) \to (B, g')$, of lifts of submanifolds of B. We prove that the local existence of such lifts corresponds to a particular property of a family of operators defined by means of the invariant A of π; necessary conditions for the global existence are also given.

Interesting applications of these results are obtained in [251], considering π as a contact-complex Riemannian submersion from a Sasakian manifold to a Kähler one, with vertical distribution spanned by the Reeb vector field ξ. This is the subject of the second part of this section.

Definition 6.2 Given a Riemannian submersion $\pi : M \to B$, an isometric immersion $i : M' \to M$ is *horizontal* if for any $x \in M'$, $i_*(T_x M')$ is a subspace of the horizontal space $\mathcal{H}_{i(x)}$.

Lemma 6.2 *Let $\pi : M \to B$ be a Riemannian submersion and consider a horizontal isometric immersion $i : M' \to M$. Then, one has:*

i) *the integrability tensor A of π vanishes on the vector fields tangent along i;*
ii) *$i' = \pi \circ i : M' \to B$ is an isometric immersion;*
iii) *the second fundamental form α of i takes values on the horizontal distribution of π; it is related to the second fundamental form α' of i' by:*

$$\pi_*(\alpha(X,Y)) = \alpha'(\pi_* X, \pi_* Y) ,$$

for any X, Y vectors tangent to i;
iv) *any section σ of the normal bundle of i projects , via π, to a section $\pi_*\sigma$ of the normal bundle of i' and the Weingarten operators a_σ, $a'_{\pi_*\sigma}$ are related by:*

$$\pi_*(a_\sigma X) = a'_{\pi_*\sigma}(\pi_* X) ,$$

for any X vector tangent to i;
v) *if σ is a horizontal section of the normal bundle of i, for any vector X tangent to i, one has:*

$$v(\nabla^{\perp}_X \sigma) = A_X \sigma , \quad \pi_*(\nabla^{\perp}_X \sigma) = {\nabla'}^{\perp}_{\pi_* X} \pi_* \sigma .$$

Proof. The proof is given in details in [250]. We only observe that, in the hypothesis of the statement, from the Gauss formula, for any X, Y tangent along i, we have $A_X Y = v(\nabla_X Y) = v(\alpha(X,Y))$. Since the first term is skew-symmetric and the last one symmetric, one obtains $A_X Y = 0$ and $\alpha(X,Y)$ is horizontal.

Remark 6.1 In the hypotheses of the previous Lemma, from iii) and iv) we get that i is totally geodesic, minimal or, possibly, totally umbilical, if and only if $i' = \pi \circ i$ has the corresponding property.

Now, we consider a Riemannian submersion $\pi : (M, g) \to (B, g')$. For any $p \in M$, $v \in \mathcal{V}_p$, let ψ^v be the endomorphism of $T_{\pi(p)}B$ such that

$$g'_{\pi(p)}(\psi^v(\pi_* X), \pi_* Y) = -g_p(A_X v, Y) , \tag{6.11}$$

for any $X, Y \in \mathcal{H}_p$. From (1.25), given $p \in M$, $X \in \mathcal{H}_p$, $V \in \mathcal{X}^v(M)$, we get:

$$\psi^{V_p}(\pi_* X) = -\pi_*(\nabla_X V) . \tag{6.12}$$

Putting, for any $x \in B$,

$$\mathcal{I}_x = \{\psi^v \mid v \in \mathcal{V}_p \,, p \in \pi^{-1}(x)\} ,$$

the family $\mathcal{I} = (\mathcal{I}_x)_{x \in B}$ is called the *horizontal integrability structure* of π. If the fibres of π are connected and locally homogeneous, for any $x \in B$ the map:

$$p \in \pi^{-1}(x) \mapsto \{\psi^v \mid v \in \mathcal{V}_p\}$$

is locally constant and $\mathcal{I}$ is a family of linear spaces ([250], Theorem 3).

In [250], Reckziegel proves the following theorems.

Theorem 6.5 *Let $\pi : M \to B$ be a Riemannian submersion and assume that $i' : M' \to B$ is an isometric immersion. The following conditions are equivalent:*

a) *for any $x' \in M'$, $\psi \in \mathcal{I}_{i'(x')}$, the spaces $i'_*(T_{x'}M')$ and $\psi(i'_*(T_{x'}M'))$ are mutually orthogonal;*
b) *for any $(x', q) \in M' \times M$, with $i'(x') = \pi(q)$, there exist a neighborhood U of x' in M' and a horizontal isometric immersion $i : U \to M$ such that $i(x') = q$ and $\pi \circ i = i'_{|U}$.*

Proof. Assuming b), we consider $x' \in M'$, $q \in \pi^{-1}(i'(x'))$, $v \in \mathcal{V}_q$, X', Y' tangent vectors to M' at x' and prove that $\psi^v(i'_* X')$, $i'_* Y'$ are orthogonal. Indeed, denote by $i : U \to M$ a local horizontal lift of i' such that $i(x') = q$ and by X, Y the horizontal lifts of X', Y' at q. Since X, Y are tangent to i, we have $A_X Y = 0$ and, via (6.11), $g'_{x'}(\psi^v(X'), Y') = -g_q(A_X v, Y) = 0$. Vice versa, assuming a), the set:

$$\tilde{M} = \{(x', q) \in M' \times M \mid i'(x') = \pi(q)\} ,$$

is a regular submanifold of $M' \times M$ and we consider the canonical maps $\tilde{\pi} : \tilde{M} \to M'$, $\tilde{f} : \tilde{M} \to M$ such that $\pi \circ \tilde{f} = i' \circ \tilde{\pi}$. Note that, for any $(x', q) \in \tilde{M}$, the restriction of $\tilde{f}$ to $\tilde{\pi}^{-1}(x')$ realizes a diffeomorphism onto $\pi^{-1}(i'(x'))$, so $(\tilde{f}_*)_{(x',q)}$ is an isomorphism from the $\tilde{\pi}$-vertical space $\tilde{\mathcal{V}}_{(x',q)}$ onto the π-vertical space $\mathcal{V}_{x'}$. This allows to define a Riemannian metric $\tilde{g}$ on $\tilde{M}$ which makes $\tilde{f}$ an isometric immersion and $\tilde{\pi}$ a Riemannian submersion with horizontal distribution $\tilde{\mathcal{H}}$, where at any $(x', q) \in \tilde{M}$, $\tilde{\mathcal{H}}_{(x',q)} = \{Z \in T_{(x',q)}\tilde{M} \mid \tilde{f}_* Z \in \mathcal{H}_q\}$. Then, via the Gauss formula and (1.19), one proves that the invariants $\tilde{A}, A$ of $\tilde{\pi}, \pi$ respectively, pointwise are related by $\tilde{f}_*(\tilde{A}_X Y) = A_{\tilde{f}_* X} \tilde{f}_* Y$, for any sections X, Y of $\tilde{M}$. Hence, for any vertical vector field V, we have:

$$g(\tilde{f}_*(\tilde{A}_X Y), V) = g'(\psi^v(i'(\tilde{\pi}_* X)), i'(\tilde{\pi}_* Y)) = 0 ,$$

so implying the integrability of the distribution $\tilde{\mathcal{H}}$. Fixed $x = (x', q) \in \tilde{M}$, we can consider an integral manifold $\hat{M}$ of $\tilde{\mathcal{H}}$, through x, such that the restriction τ of $\tilde{\pi}$ to $\hat{M}$ realizes an isometry between $\hat{M}$ and a neighborhood U of x' in M'. Then $i = \tilde{f} \circ \tau^{-1}$ is a local horizontal lift of i', with $i(x') = q$.

Now, we recall that the horizontal distribution of a Riemannian submersion $\pi : M \to B$ is Ehresmann complete if any curve $\gamma : [a, b] \to B$ can be horizontally lifted to a curve on M, starting at any point of $\pi^{-1}(\gamma(a))$ (see Sec. 2.1)

Theorem 6.6 *Let $\pi : M \to B$ be a Riemannian submersion with Ehresmann complete horizontal distribution. Then, for every connected Riemannian submanifold $i' : M' \to B$ satisfying condition* a) *in Theorem* 6.5 *and for every $(x', q) \in M' \times M$ with $i'(x') = \pi(q)$, there exist a Riemannian manifold $\hat{M}$, a covering map $\tau : \hat{M} \to M'$, a point $p \in \hat{M}$ with $\tau(p) = x'$ and a horizontal isometric immersion $j : \hat{M} \to M$ such that $j(p) = q$ and $\pi \circ j = i' \circ \tau$.*

Proof. Fixed the isometric immersion i' satisfying the hypotheses, we proceed as in Theorem 6.5, step a)$\Rightarrow$ b), considering the Riemannian manifold $\tilde{M}$ and the canonical projections $\tilde{\pi}$ and $\tilde{f}$. Given $(x', q) \in \tilde{M}$, we denote by τ and j the restrictions of $\tilde{\pi}$ and $\tilde{f}$ to $\hat{M}$, respectively, $\hat{M}$ being the maximal integral manifold of the horizontal (integrable) distribution of $\tilde{\pi}$ through (x', q). Then τ is a local isometry, j is a horizontal isometric immersion and $\pi \circ j = i' \circ \tau$, $\tau(x', q) = x'$, $j(x', q) = q$. To complete the proof, one only needs to prove that τ acts as a covering map; for this, it is enough to state the existence of a τ-lift of any curve $\gamma : [0, 1] \to M'$, starting at every $x \in \tau^{-1}(\gamma(0))$. In fact, given $\gamma : [0, 1] \to M'$, $x = (x'', q'') \in \tau^{-1}(\gamma(0))$, we consider the curve $i' \circ \gamma : [0,1] \to B$, and $q'' \in \pi^{-1}(i'(\gamma(0)))$. Since the horizontal distribution of π is Ehresmann complete, there exists the horizontal lift $\tilde{\gamma} : [0,1] \to M$, of $i' \circ \gamma$ such that $\tilde{\gamma}(0) = q''$. For any $t \in [0,1]$, $\tilde{\gamma}(t) \in \pi^{-1}(i'(\gamma(t)))$ and the map $c = (\gamma, \tilde{\gamma})$ is a curve on $\tilde{M}$ with $c(0) = (x'', q'') \in \hat{M}$, $\dot{c}(t) \in \tilde{\mathcal{H}}_{c(t)}$, $t \in [0,1]$. Thus, c takes values on $\tilde{M}$ and $\tau \circ c = \gamma$.

Remark 6.2 Examples of Riemannian submersions with Ehresmann complete horizontal distribution are the submersions $\pi : M \to B$ such that there exists a connected Lie group acting on M by isometries whose orbits coincide with the fibres of π. For instance, this is realized by the Hopf submersion $\pi : \mathbf{S^{2n+1}} \to \mathbf{P_n(C)}$ and by the generalized Hopf submersion $\pi : \mathbf{S^{4n+3}} \to \mathbf{P_n(Q)}$, considering the action of $U(n+1)$ on $\mathbf{S^{2n+1}}$ and of $Sp(n+1)$ on $\mathbf{S^{4n+3}}$, respectively (see Examples 5.3, 5.4).

Now, we consider a submersion $\pi : M \to B$ whose total space has a Sasakian structure (φ, ξ, η, g) and ξ spans the vertical distribution. Then, B inherits from M a Kähler structure (J, g') which makes π a contact-complex Riemannian submersion. In fact, as in [206], for any $X', Y' \in \mathcal{X}(B)$ with horizontal lifts X, Y, we put:

$$g'(X', Y') \circ \pi = g(X, Y) \ , \quad JX' = \pi_*(\varphi X) \ .$$

It is easy to prove that (J, g') is a Kähler structure and $\pi_* \circ \varphi = J \circ \pi_*$. Following [251] we call *canonical fibration* of a Sasakian manifold a contact-complex Riemannian submersion $\pi : M \to B$, where M is a Sasakian manifold whose Reeb vector field ξ spans the vertical distribution and B is Kähler.

Proposition 6.1 *Let (B, J, g') be a Kähler manifold. If the Kähler form of B is exact, then $M = B \times \mathbf{R}$ admits a Sasakian structure (φ, ξ, η, g) which makes the first projection $\pi : M \to B$ a canonical fibration.*

Proof. Let ω be a 1-form on B such that $d\omega = \Omega$ is the Kähler form of B and consider the first projection $\pi : M \to B$. Putting:

$$\eta = \pi^* \omega + dt \ , \quad g = \pi^* g' + \eta \otimes \eta \ , \quad \xi = \frac{\partial}{\partial t} \ ,$$

t being the coordinate on $\mathbf{R}$, g turns out to be a Riemannian metric on M which makes π a Riemannian submersion and ξ is the vector field dual of η with respect to g. Moreover, ξ is a Killing vector field which spans the vertical distribution of π and, denoting by ∇ the Levi-Civita connection on M, we put $\varphi = -\nabla\xi$. Then, (φ, ξ, η, g) is an almost contact metric structure, $\pi_* \circ \varphi = J \circ \pi_*$ and, since B is Kähler, we get:

$$(\nabla_X \varphi)Y = g(X, Y)\xi - \eta(Y)X \, ,$$

so M is a Sasakian manifold.

Now, we recall that an isometric immersion $i : M' \to M$ into an almost contact metric manifold $(M, \varphi, \xi, \eta, g)$ is *anti-invariant* if for any $x \in M'$, $i_*(T_x M')$ and $\varphi(i_*(T_x M'))$ are mutually orthogonal. The concept of *totally real*, or *anti-invariant* submanifolds of an almost Hermitian manifold (M, J, g) is introduced analogously, simply replacing the tensor φ with the almost complex structure J.

Proposition 6.2 *Let $\pi : M \to B$ be a canonical fibration of a Sasakian manifold. Then, we have:*

1) *an isometric immersion $i : M' \to M$ is horizontal if and only if ξ is a section of the normal bundle $TM'^{\perp}$ and i is anti-invariant;*
2) *an isometric immersion $i' : M' \to B$ has local horizontal lifts if and only if i' is totally real.*

Proof. Since ξ spans the vertical distribution of π, M' is a horizontal submanifold of M if and only if ξ is a section of the normal bundle $TM'^{\perp}$ and this means that M' is anti-invariant ([348] Proposition 7.5, Chapter 2). Thus, we obtain 1). Now, we fix an isometric immersion $i' : M' \to B$. Given $p \in M$, $x \in M'$ with $\pi(p) = i'(x)$, the operator ψ^{ξ_p} defined in (6.11), via (6.12) acts as:

$$\psi^{\xi_p}(\pi_* X_p) = -\pi_*(\nabla_{X_p}\xi) = \pi_*(\varphi X_p) = J_{i'(x)}(\pi_* X_p) \, ,$$

hence:

$$\mathcal{I}_{i'(x)} = \{\psi^v \mid v \in \mathcal{V}_p \, , \pi(p) = i'(x)\} = \{\lambda J_{i'(x)} \mid \lambda \in \mathbf{R}\}.$$

The equivalence in 2) follows from Theorem 6.5.

Remark 6.3 Let $\pi : M \to B$ be a canonical fibration of a Sasakian manifold and, in addition, suppose that ξ is a complete vector field. Since ξ is Killing and spans the vertical distribution, the fibres of π are just the orbits of the 1-parameter group of isometries generated by ξ. Applying Theorem 6.6 and the previous proposition, any totally real submanifold $i' : M' \to B$ admits global horizontal lifts. In fact, for any $(x', q) \in M' \times M$, with $i'(x') = \pi(q)$, there exist a Riemannian covering map $\tau : \hat{M} \to M'$, a point $\hat{p} \in \hat{M}$ with $\tau(\hat{p}) = x'$ and an anti-invariant submanifold $j : \hat{M} \to M$, orthogonal to ξ, such that $j(\hat{p}) = q$ and $\pi \circ j = i' \circ \tau$.

Remark 6.4 Another consequence of Proposition 6.2 is that $i' = \pi \circ i$ is a totally real isometric immersion into B, provided that $i : M' \to M$ is a horizontal isometric immersion; i' is also called the *totally real image* of i by π. By Lemma 6.2, the link between the second fundamental forms α, α' of i, i' implies that the mean curvature vector field of i is the π-horizontal lift of the mean curvature vector field of i'. Moreover, given a vector field X tangent to i, π-related to X', which is tangent to i', for any section σ of the normal bundle of i, we have:

$$\eta(\nabla^{\perp}_{X}\sigma) = X(\eta(\sigma)) + g(\varphi X, \sigma) , \tag{6.13}$$

$$\pi_*(\nabla^{\perp}_{X}\sigma) = \nabla'^{\perp}_{X}(\pi_*\sigma) - \eta(\sigma)J(X') . \tag{6.14}$$

In fact, since ξ is normal to i and M is Sasakian, we have:

$$\eta(\nabla^{\perp}_{X}\sigma) = X(g(\sigma,\xi)) - g(\sigma, \nabla^{\perp}_{X}\xi) = X(\eta(\sigma)) + g(\varphi X, \sigma) ,$$

and then (6.13). Equation (6.14) holds for horizontal vector fields normal to i by v) in Lemma 6.2. Finally, for any differentiable map λ on M, we have:

$$\pi_*(\nabla^{\perp}_{X}(\lambda\xi)) = -\lambda\pi_*(\varphi X) = -\eta(\lambda\xi)J(\pi_*X) ,$$

which gives (6.14) for vertical vector fields.

The following results are consequence of the above formulas.

Proposition 6.3 *Let $\pi : M \to B$ be a canonical fibration of a Sasakian manifold, $i : M' \to M$ a horizontal isometric immersion and i' its totally real image by π. The second fundamental forms α, α' of i, i' pointwise satisfy:*

$$\eta((\nabla_X\alpha)(Y,Z)) = g'(JX', \alpha'(Y',Z')) \circ \pi ,$$
$$\pi_*((\nabla_X\alpha)(Y,Z)) = (\nabla'_{X'}\alpha')(Y',Z') ,$$

for any vector fields X, Y, Z tangent along i and π-related to X', Y', Z'. In particular, α' is parallel if and only if

$$(\nabla_X\alpha)(Y,Z) = g(\varphi X, \alpha(Y,Z))\xi ,$$

for any X, Y, Z tangent along i.

Proof. We recall that the covariant derivative of α acts as:

$$(\nabla_X\alpha)(Y,Z) = \nabla^{\perp}_X(\alpha(Y,Z)) - \alpha(\overline{\nabla}_X Y, Z) - \alpha(Y, \overline{\nabla}_X Z) ,$$

$\overline{\nabla}$ denoting the Levi-Civita connection of M'. Since α takes values in the horizontal distribution, via (6.13) we have:

$$\begin{aligned}\eta((\nabla_X\alpha)(Y,Z)) &= X(\eta(\alpha(Y,Z))) + g(\varphi X, \alpha(Y,Z)) \\ &= g'(JX', \alpha'(Y',Z')) \circ \pi .\end{aligned}$$

Now, we observe that for any X, Y tangent to i, $\overline{\nabla}_X Y$ is the π-horizontal lift of $\overline{\nabla}'_{X'}Y'$, where $\overline{\nabla}'$ is the Levi-Civita connection of i'; so, via (6.14), we obtain:

$$\begin{aligned}\pi_*((\nabla_X\alpha)(Y,Z)) &= \nabla'^{\perp}_{X'}(\alpha'(Y',Z')) - \alpha'(\overline{\nabla}'_{X'}Y', Z') - \alpha'(Y', \overline{\nabla}'_{X'}Z') \\ &= (\nabla_{X'}\alpha')(Y',Z') .\end{aligned}$$

Proposition 6.4 *Let $\pi : M \to B$ be a canonical fibration of a Sasakian manifold, $i : M' \to M$ a horizontal isometric immersion and i' its totally real image by π. Then, i' has parallel mean curvature vector field if and only if the mean curvature vector field H of i satisfies:*

$$\nabla^{\perp}_X H = g(\varphi X, H)\xi ,$$

for any X tangent along i.

We end this section showing the interrelation between the f-structures on the normal bundle of $i : M' \to M$ and $i' = \pi \circ i$, i being a horizontal isometric immersion into the Sasakian manifold M and π a canonical fibration. Firstly, we recall that an f-structure on the normal bundle $\nu(M')$ of an anti-invariant submanifold $i : M' \to M$ is defined as follows. For any $x \in M'$, we consider the orthogonal decomposition

$$T_x M'^{\perp} = \varphi(T_x M') \oplus N_x(M') , \tag{6.15}$$

where $N_x(M')$ is the orthogonal complement of $\varphi(T_x M')$ in $T_x M'^{\perp}$. Then, for any section σ of the normal bundle of i, one puts $\varphi\sigma = \tilde{P}\sigma + \tilde{f}\sigma$, where $\tilde{P}\sigma$, $\tilde{f}\sigma$ respectively denote the tangential and the normal components of $\varphi\sigma$ with respect to the splitting given in (6.15). Then $\tilde{f}$, which has constant rank and satisfies $\tilde{f}^3 + \tilde{f} = 0$, determines an f-structure on the normal bundle of i, provided that it does not vanish ([348]). The covariant derivative $\nabla^{\perp}_X \tilde{f}$ is given by:

$$(\nabla^{\perp}_X \tilde{f})\sigma = -\alpha(X, \tilde{P}\sigma) - \varphi(a_\sigma X) ,$$

for any X tangent along i, σ normal to i, a_σ being the Weingarten operator. Analogously, one can define an f-structure on the normal bundle of a totally real submanifold $i' : M' \to B$ of a Kähler manifold (B, J, g'), considering at any $x \in M'$, the orthogonal splitting

$$T_x M'^{\perp} = J(T_x M') \oplus N_x(M') ,$$

and putting for any section σ of the normal bundle of i', $J\sigma = P\sigma + f\sigma$, where $P\sigma$, $f\sigma$ respectively denote the tangent and the normal components of $J\sigma$. Again f satisfies $f^3 + f = 0$, so f defines an f-structure if it does not vanish.

From Lemma 6.2 one easily obtains:

Proposition 6.5 *Let $\pi : M \to B$ be a canonical fibration of a Sasakian manifold, $i : M' \to M$ a horizontal isometric immersion and $i' = \pi \circ i$ its totally real image. Then, the f-structures $\tilde{f}$ and f on the normal bundles of i and i' are related by $\pi_* \circ \tilde{f} = f \circ \pi_*$. Moreover, $\tilde{f}$ is parallel if and only if f is parallel.*

6.3 Riemannian Submersions and Isometric Reflections with Respect to the Fibres

Reflections with respect to submanifolds have been introduced by B. Y. Chen and L. Vanhecke in [61]. We present some results concerning reflections with respect to the fibres in a Riemannian submersion.

Definition 6.3 Let $\tilde{M}$ be a Riemannian manifold and M a relatively compact, connected submanifold. Let $p \in M$ and $\gamma :]-\varepsilon, \varepsilon[\to \tilde{M}$ a geodesic such that $\gamma(]-\varepsilon, \varepsilon[) \subset M$, $\gamma(0) = p$, $\dot{\gamma}(0) = X \in T_p M^{\perp}$, $\|X\| = 1$. The map ψ_M acting as $\exp(tX) \mapsto \exp(-tX)$ is a local diffeomorphism of a local tubular neighborhood of M in $\tilde{M}$ and is called *the local reflection with respect to M*.

The following result is due to Chen and Vanhecke ([61]).

Theorem 6.7 *Let $\tilde{M}$ be a Riemannian manifold and M a relatively compact, connected submanifold. The local reflection with respect to M is a local isometry if and only if:*

a) *M is totally geodesic;*

b) $(\nabla^{2k}_{X\cdots X} R)(X, Y, X, V) = 0\,;$

$$(\nabla^{2k+1}_{X\cdots X}R)(X,Y,X,Z)=0\,;$$

$$(\nabla^{2k+1}_{X\cdots X}R)(X,V,X,W)=0\,;$$

where X, Y, Z are vector fields normal to M, V, W are tangent to M, $k \in \mathbf{N}$ and R denotes the Riemannian curvature of $\tilde{M}$.

Definition 6.4 A Sasakian manifold $(M^{2n+1}, \varphi, \xi, \eta, g)$ is called *locally φ-symmetric* if $\varphi^2((\nabla_X R)(Y,Z,H)) = 0$ for any vector fields X, Y, Z, H orthogonal to ξ.

We report the following result ([37; 282]).

Proposition 6.6 *Let M be a Sasakian manifold. Then, M is locally φ-symmetric if and only if the local φ-geodesic symmetries are isometries.*

Results in the case of Riemannian submersions with 1-dimensional fibres are obtained by Narita ([216]).

Theorem 6.8 *Let (M,g) be a connected, orientable $(2n+1)$-dimensional Riemannian manifold and $\pi : M \to N$ a Riemannian submersion having 1-dimensional fibres. Assume that there exists a positive function σ on M such that*

$$A_{A_X W}W = -\sigma g(W,W)X\ ,$$

for any horizontal vector field X and a vertical vector field W. Then, M admits a locally φ-symmetric Sasakian structure if and only if the reflections with respect to the fibres are isometries.

Proof. We show that, if the reflections with respect to the fibres are isometries, then M admits a Sasakian structure. Such a structure turns out to be locally φ-symmetric by the previous proposition since in this case the reflections with respect to the fibres are just the local φ-geodesic symmetries. The hypothesis on the reflections implies that the fibres are totally geodesic submanifolds of M. Let ξ be a unit vertical vector field such that $\nabla_\xi \xi = 0$ and define

$$\varphi(E) = -\frac{1}{\sqrt{\sigma}}A_E\xi\ , \tag{6.16}$$

for any $E \in \mathcal{X}(M)$. Obviously, we get $\varphi(\xi) = 0$ and $\eta(\varphi(E)) = 0$, η being the 1-form dual to ξ. Moreover, using the properties of the tensor A, we have:

$$\varphi^2(E) = \varphi^2(h(E)) = \frac{1}{\sigma} A_{A_{h(E)}\xi}\xi = -g(\xi,\xi)h(E) = -h(E) = -E + \eta(E)\xi \ ,$$

$$\begin{aligned} g(\varphi E, \varphi F) &= -\tfrac{1}{\sigma} g(\xi, A_E(A_F \xi)) = \tfrac{1}{\sigma} g(\xi, A_{A_{h(F)}\xi} h(E)) = g(h(E), h(F)) \\ &= g(E,F) - g(v(E), v(F)) = g(E,F) - \eta(E)\eta(F) \ . \end{aligned}$$

Hence, M admits an almost contact metric structure. Now, since $T = 0$, (1.26, g) implies

$$g((\nabla_\xi A)_X Y, \xi) = 0 \ . \tag{6.17}$$

From (1.29) and (6.16) it follows that

$$R(X,\xi,Y,\xi) = g(A_X \xi, A_Y \xi) = \sigma g(X,Y) \ . \tag{6.18}$$

In particular, putting $X = Y$, $\|X\| = 1$, we obtain that σ describes the mixed sectional curvatures. Now, we prove that σ is constant. Namely, since the reflections are isometries, we have $R(X,Y,Z,\xi) = 0$ and $(\nabla_X R)(X,\xi,X,\xi) = 0$ ([61]), then, from (6.18), for any horizontal X, one obtains

$$0 = (\nabla_X R)(X,\xi,X,\xi) = X(\sigma) g(X,X) \ ,$$

so $X(\sigma) = 0$. On the other hand, by the hypothesis, we have $A_{A_X\xi}\xi = -\sigma X$ and assuming $\|X\| = 1$, using (6.17) and the skew-symmetry of A, we get:

$$\begin{aligned} \xi(\sigma) &= \xi(g(A_X \xi, A_X \xi)) = 2g(A(\nabla_\xi X, \xi), A_X \xi) \\ &= -2g(A_{A_X\xi}\xi, \nabla_\xi X) = 2\sigma g(X, \nabla_\xi X) = 0 \ . \end{aligned}$$

Thus σ is constant. Putting:

$$\tilde{g} = \sigma g \ , \quad \tilde{\xi} = \frac{1}{\sqrt{\sigma}}\xi \ , \quad \tilde{\eta} = \sqrt{\sigma}\,\eta \ ,$$

we obtain an almost contact metric structure $(\varphi, \tilde{\xi}, \tilde{\eta}, \tilde{g})$ on M, with $\nabla = \tilde{\nabla}$ and $A = \tilde{A}$. By a long computation, ([216]), one has

$$(\tilde{\nabla}_E \varphi) F = \tilde{g}(E,F)\tilde{\xi} - \tilde{\eta}(F) E \ ,$$

and $(M, \varphi, \tilde{\xi}, \tilde{\eta}, \tilde{g})$ is a Sasakian manifold.

Theorem 6.9 *Let M be a connected orientable 3-dimensional manifold and $\pi : M \to N$ a Riemannian submersion with 1-dimensional fibres, such that there exist a horizontal vector field X and a vertical vector field U satisfying $A_X U \neq 0$ everywhere. Then, the reflections with respect to the* *fibres are isometries if and only if M admits a locally φ-symmetric Sasakian structure.*

Proof. As before, it is enough to prove the necessary condition. Without loss of generality, we can assume that $||U|| = 1 = ||X||$. It is easy to verify that $\{U, X, A_X U\}$ is a global orthogonal basis of TM. Furthermore, $A_{A_X U} U$ is orthogonal to U and to $A_X U$, so

$$A_{A_X U} U = g(A_{A_X U} U, X) X = -g(A_X U, A_X U) X = -\sigma_X X \ ,$$

where, by (1.32), σ_X is a positive function describing the sectional curvatures of the 2-planes spanned by U and X. Now, we prove that $A_{A_Y U} U = -\sigma Y$, for any horizontal vector field Y, and σ does not depend on Y. Namely, for $Y = A_X U$ we have:

$$A_{A_{A_X U} U} U = -\sigma_X A_X U = -\sigma_X Y$$

and $\sigma_{A_X U} = \sigma_X$. Finally, for $Y = \alpha X + \beta A_X U$, $\alpha \neq 0$, $\beta \neq 0$, we get:

$$A_{A_Y U} U = -\sigma_X(\alpha X + \beta A_X U) = -\sigma_X Y$$

and $\sigma_Y = \sigma_X$. Hence, we apply Theorem 6.8.

Corollary 6.1 *Let G be a 3-dimensional semisimple, compact, connected Lie group with a bi-invariant metric. Let K be a 1-dimensional closed subgroup of G and suppose that the Riemannian submersion $\pi : G \to G/K$ satisfies the conditions $R(X, Y, Z, U) = 0$ and $A_X U \neq 0$ where X, Y, Z are horizontal vector fields and U is a vertical one. Then, G admits a locally φ-symmetric Sasakian structure.*

Proof. Since G is symmetric and $R(X, Y, Z, U) = 0$ holds, then, as proved in [61], the reflections with respect to the fibres are isometries and we can apply Theorem 6.9.

Example 6.2 Consider the Riemannian submersion

$$\pi : SU(2) \to SU(2)/S(U(1) \times U(1)).$$

The Lie algebra $\mathfrak{su}(2)$ is the direct sum $\mathfrak{g}_1 \oplus \mathfrak{g}_2$, where

$$\mathcal{H} = \mathfrak{g}_1 = \left\{ \begin{pmatrix} 0 & -\overline{z} \\ z & 0 \end{pmatrix} ; z \in \mathbf{C} \right\} ,$$

$$\mathcal{V} = \mathfrak{g}_2 = \left\{ \begin{pmatrix} \alpha & 0 \\ 0 & -\alpha \end{pmatrix} ; \alpha + \overline{\alpha} = 0 \, , \alpha \in \mathbf{C} \right\} .$$

Then, for any X, Y, Z horizontal and U vertical, we have $R(X, Y, Z, U) = 0$ and $A_X U \neq 0$; hence, by the previous Corollary, $SU(2)$ admits a locally φ-symmetric Sasakian structure.

Example 6.3 Consider the Heisenberg group

$$H = \left\{ \begin{pmatrix} 1 & x & y \\ 0 & 1 & z \\ 0 & 0 & 1 \end{pmatrix} ; x, y, z \in \mathbf{R} \right\}$$

identified with $\mathbf{R}^3$ with the multiplication:

$$(x, y, z)(x', y', z') = (x + x', y + y' + xz', z + z')$$

and the left-invariant metric

$$g = dx^2 + dz^2 + (dy - xdz)^2 .$$

Note that H is not semisimple, the metric is not bi-invariant, so we cannot use Corollary 6.1. The vector fields $X_1 = \frac{\partial}{\partial x}$, $X_2 = \frac{\partial}{\partial z} + x\frac{\partial}{\partial y}$, $U = \frac{\partial}{\partial y}$ are left invariant, orthonormal and verify $[X_1, X_2] = U$, $[X_1, U] = [X_2, U] = 0$. Let K be the 1-dimensional closed subgroup obtained putting $x = z = 0$ and consider the Riemannian submersion $\pi : H \to H/K$ having totally geodesic fibres. Then, by direct computation, ([216]), one has $(\nabla_{X_1} R)(X_1, U, X_1, X_2) = \frac{1}{2}$ and H is not a locally symmetric space. Furthermore, in [216] Narita proved that conditions a) and b) in Theorem 6.7 are satisfied, hence the reflections with respect to the fibres are isometries. Finally, since $A_{X_1} U = h(\nabla_{X_1} U) = h(-\frac{1}{2}X_2) = -\frac{1}{2}X_2 \neq 0$ and $A_{X_2} U = h(\nabla_{X_2} U) = h(\frac{1}{2}X_1) = \frac{1}{2}X_1 \neq 0$, we can apply Theorem 6.9 and H admits a locally φ-symmetric Sasakian structure.

We end with a classification theorem obtained combining Theorem 6.9, the classification of naturally reductive homogeneous 3-dimensional spaces ([295]) and the fact that simply connected, complete, locally φ-symmetric Sasakian spaces are naturally reductive homogeneous spaces ([37]).

Theorem 6.10 *Let M be a 3-dimensional orientable, connected, simply connected, complete Riemannian manifold and $\pi : M \to N$ a Riemannian submersion, with 1-dimensional fibres and assume that $A_X U \neq 0$ for a horizontal vector field X and a vertical vector field U. Then, the reflections with respect to the fibres are isometries if and only if M is isometric to one of the following spaces:*

a) *the unit sphere $\mathbf{S}^3$ in $\mathbf{R}^4$;*

b) $SU(2)$;
c) *the Heisenberg group* H ;
d) *the universal covering space of* $SL(2, \mathbf{R})$.

Chapter 7

Semi-Riemannian Submersions

Semi-Riemannian manifolds play a relevant role in many areas of Mathematics and Physics. A detailed exposition of the subject, together with its applications, is given by J.K. Beem, P.E. Ehrlich ([16]) and by B. O'Neill ([228]). Recent developments involving semi-Riemannian maps are explained in details by E. Garcia-Rio, D. Kupeli ([105]) and by A. Bejancu, K. Duggal ([18]).

The first two sections of this chapter deal with definitions, examples and basic properties of semi-Riemannian manifolds and semi-Riemannian submersions.

In Sec. 7.3 we discuss Lorentzian submersions, *i.e.* submersions from a Lorentz manifold M over a Riemannian manifold B. If M has negative sectional curvature and $\pi : M \to B$ is a semi-Riemannian submersion, then the fibres are 1-dimensional and totally geodesic. We prove that if M is an anti-de Sitter space, then B is holomorphically isometric to the complex hyperbolic space $\mathbf{CH^n}$ ([189]).

The following section presents recent results on the classification of semi-Riemannian submersions from real or complex pseudo-hyperbolic spaces ([12]). We classify the semi-Riemannian submersions with totally geodesic fibres, from a pseudo-hyperbolic space of index $s \geq 1$ onto a Riemannian manifold. These results are linked to the ones stated in Chapter 2, regarding the Riemannian case.

Section 7.5 is devoted to the study of semi-Riemannian submersions with totally umbilical fibres. We present the basic formulas and the curvature equations in the case $s \neq 0$. Some integral formulas are obtained, involving sectional curvatures, scalar curvatures and mixed scalar curvatures. These formulas give some obstructions to the existence of semi-Riemannian submersions. For instance, there are no semi-Riemannian submersions $\pi : M \to B$ with totally umbilical fibres, when M is a semi-Riemannian manifold of index $\dim M - \dim B \geq 2$ with constant positive sectional curvature and B is a compact, orientable Riemannian manifold ([11]).

Finally, recent results on semi-Riemannian submersions with minimal fibres are explained.

7.1 Semi-Riemannian Manifolds

We begin with some algebraic preliminaries on non-degenerate bilinear forms on an m-dimensional real vector space V.

Let $g : V \times V \to \mathbf{R}$ be a symmetric bilinear form. We say that g is *non-degenerate* if $g(u, v) = 0$ for each $v \in V$ implies $u = 0$, otherwise g is called *degenerate*. A non-degenerate symmetric bilinear form on V is called a *semi-Euclidean metric* on V. It may induce either a non-degenerate or a degenerate symmetric bilinear form on a subspace W of V; then W is said to be a *non-degenerate* or a *degenerate* subspace, respectively.

We say that g is *positive* (*negative*) definite provided that $u \neq 0$ implies $g(u, u) > 0 \quad (< 0)$. If g is non-degenerate, there exists an ordered basis $(e_1, e_2, \ldots, e_m)$ of V such that:

$$\begin{aligned} g(e_i, e_i) &= -1\,, \quad 1 \leq i \leq s\,, \\ g(e_i, e_i) &= 1\,, \quad s+1 \leq i \leq m\,, \\ g(e_i, e_j) &= 0\,, \quad i \neq j\,, \end{aligned}$$

where s is uniquely determined and $(s, m - s)$ is the signature of g. Obviously, in the case $s = 0$ or $s = m$, the first or the second condition has to be dropped. The integer s is called the *index* of g on V and it is the largest dimension of a subspace $W \subset V$ on which the induced metric is negative definite.

A *semi-Riemannian metric* g on an m-dimensional manifold M is a symmetric tensor field of type $(0, 2)$ on M such that for any $p \in M$ the tensor g_p is a non-degenerate symmetric bilinear form on T_pM of constant index. We call (M, g) a *semi-Riemannian manifold*. Frequently, we denote by M_s^m an m-dimensional semi-Riemannian manifold of index s. In the particular case $m \geq 2$ and $s = 1$, we call (M, g) a *Lorentzian manifold*. Obviously, if $s = 0$, (M, g) is a Riemannian manifold.

A tangent vector v at a point $p \in M$ is said to be:
spacelike, if $g_p(v, v) > 0$ or $v = 0$,

timelike, if $g_p(v, v) < 0$,
null (or lightlike), if $g_p(v, v) = 0$ and $v \neq 0$.
The set of null vectors of T_pM is called the *null cone* at $p \in M$.

With any semi-Riemannian manifold is associated the so-called Levi-Civita connection, denoted by ∇; it is the unique symmetric connection preserving the semi-Riemannian metric. The curvature R of ∇ is defined as in the Riemannian case and satisfies the same formal properties, including the Bianchi identities. On the other hand, the sectional curvatures can be considered for non-degenerate 2-planes, only. Namely, given X, Y in a non-degenerate 2-plane $\alpha \subset T_p(M)$, $p \in M$, it is easy to see that $\{X, Y\}$ is a basis of α if and only if $\Delta = g_p(X, X)g_p(Y, Y) - g_p(X, Y)^2 \neq 0$. In this case,

$$K(\alpha) = \frac{1}{\Delta} R_p(X, Y, X, Y)$$

is the sectional curvature at p with respect to α (it is independent of the choice of the basis of α). It is well known that all non-degenerate planes have sectional curvature $c \in \mathbf{R}$ if and only if

$$R(X, Y, Z) = c\{g(Y, Z)X - g(X, Z)Y\} .$$

Now, we describe some basic examples of semi-Riemannian manifolds.

Example 7.1 **The flat model space.** Let g_s^m be the symmetric bilinear form on $\mathbf{R^m}$ given by

$$g_s^m(x, y) = -\sum_{i=1}^{s} x^i y^i + \sum_{i=s+1}^{m} x^i y^i ,$$

for any $x = (x^1, x^2, \dots, x^m), y = (y^1, y^2, \dots, y^m) \in \mathbf{R^m}$. Then, $(\mathbf{R^m_s}, g_s^m)$ is called the m-dimensional *semi-Euclidean* space of index s. It is of constant sectional curvature $c = 0$.

One can obtain many examples through the following general construction. Let $\tau : M \to M'$ be an immersion and g' a semi-Riemannian metric on M'. If for any $p \in M$, $\tau_*(T_pM)$ is a non-degenerate subspace of $T_{\tau(p)}M'$, then $g = \tau^* g'$ is a semi-Riemannian metric on M.

Example 7.2 **The pseudosphere** of radius $r > 0$. Putting:

$$\mathbf{S^m_s}(r) = \{x \in \mathbf{R^{m+1}_s} \mid g_s^{m+1}(x, x) = r^2\} ,$$

one obtains an m-dimensional semi-Riemannian manifold of index s and of constant curvature $c = 1/r^2$. It is easy to see that $\mathbf{S^m_s}(r)$ is diffeomorphic to $\mathbf{R^s} \times \mathbf{S^{m-s}}(r)$. We remark that $\mathbf{S^m_s}(r)$ is simply connected for $s \neq m-1$ and $s \neq m$; $\mathbf{S^m_{m-1}}(r)$ is connected with infinite cyclic fundamental group and $\mathbf{S^m_m}(r)$ has two simply connected components ([340]). In the theory of general relativity, $\mathbf{S^4_1}(r)$ is called the *de Sitter space-time*.

Example 7.3 **The pseudo-hyperbolic space** of radius $r > 0$. Putting:

$$\mathbf{H^m_s}(r) = \{x \in \mathbf{R^{m+1}_{s+1}} \mid g^{m+1}_{s+1}(x,x) = -r^2\} ,$$

one obtains an m-dimensional semi-Riemannian manifold of index s and of constant curvature $c = -1/r^2$. We remark that $\mathbf{H^m_s}(r)$ is simply connected for $2 \leq s \leq m$, whereas $\mathbf{H^m_1}(r)$ is not simply connected. $\mathbf{H^{2n+1}_1}(r)$ is called the *anti-de Sitter space*.

In the following we take $r = 1$, omitting the corresponding symbol.

Remark 7.1 Let g be a symmetric tensor field of type $(0,2)$ on an m-dimensional manifold M. Then, the set of points where g is non-degenerate of index s is an open set of M ([105]).

Remark 7.2 In contrast to the Riemannian case, there are topological obstructions to the existence of a Lorentz metric on a manifold M. Such a metric exists if either M is non-compact, or M is compact and has Euler number $\chi(M) = 0$ ([228]).

7.2 Semi-Riemannian Submersions. Examples.

Given a C^∞-submersion $\pi : M \to B$, with (M, g) semi-Riemannian manifold, at any point $p \in M$, one can consider the vertical space which is a non-degenerate subspace of $T_p(M)$ and then admits a g_p-orthogonal complement, called the *horizontal* space at p.

Definition 7.1 A C^∞-submersion $\pi : M \to B$ between two semi-Riemannian manifolds (M, g) and (B, g') is called a *semi-Riemannian submersion* if it satisfies the conditions:

a) the fibres are semi-Riemannian submanifolds of M ,
b) for any $p \in M$, π_{*p} preserves the length of the horizontal vectors.

Remark 7.3 This definition agrees with the definition of Riemannian submersion, since the condition a) is automatically satisfied if g and g' are Riemannian metrics. Thus, one obtains the vertical and horizontal distributions, $\mathcal{V}, \mathcal{H}$, which allow to introduce the fundamental tensors T

and A, ([226]), as well as in the Riemannian case. Thus, the properties of T, A stated in Sec. 1.3 are still valid and the curvature formulas (1.27), (1.28), (1.29) hold, too. It is easy to see that the restriction of g to the horizontal distribution $\mathcal{H}$ is a semi-Riemannian metric.

Remark 7.4 In [105] Garcia-Rio and Kupeli defined a new class of maps between two semi-Riemannian manifolds, called the *semi-Riemannian maps*, which may admit degenerate fibres, so including the semi-Riemannian submersions.

Many explicit examples of semi-Riemannian submersions can be given following a standard construction ([228] and [26] for the Riemannian case). Let G be a Lie group and K, H two compact Lie subgroups of G such that $K \subset H$. Let $\pi : G/K \to G/H$ be the bundle, with fibre H/K, associated with the H-principal bundle $\rho : G \to G/H$. Let $\mathfrak{g}$ be the Lie algebra of G and $\mathfrak{k} \subset \mathfrak{h}$ the corresponding Lie algebras of K and H. We choose an $Ad(H)$-invariant complement $\mathfrak{m}$ to $\mathfrak{h}$ in $\mathfrak{g}$, and an $Ad(K)$-invariant complement $\mathfrak{p}$ to $\mathfrak{k}$ in $\mathfrak{h}$. An $Ad(H)$-invariant non-degenerate bilinear symmetric form on $\mathfrak{m}$ defines a G-invariant semi-Riemann metric g' on G/H and an $Ad(K)$-invariant non-degenerate bilinear symmetric form on $\mathfrak{p}$ defines an H-invariant semi-Riemannian metric $\hat{g}$ on H/K. The orthogonal direct sum of such non-degenerate bilinear symmetric forms on $\mathfrak{p} \oplus \mathfrak{m}$ defines a G-invariant semi-Riemannian metric g on G/K.

Extending Theorem 5.2 to the semi-Riemannian case, one obtains:

Theorem 7.1 *The map $\pi : (G/K, g) \to (G/H, g')$ is a semi-Riemannian submersion with totally geodesic fibres.*

As an application of this result, we get the following examples.

Example 7.4 Let $G = SU(1,n), H = S(U(1)U(n)), K = SU(n)$. We have the semi-Riemannian submersion

$$\pi : \mathbf{H_1^{2n+1}} = SU(1,n)/SU(n) \to \mathbf{CH^n} = SU(1,n)/S(U(1)U(n)) \ ,$$

from the anti-de Sitter space onto the disk D^n in $\mathbf{C^n}$ with the Bergman metric, also called the *complex hyperbolic space* ([172] Vol. II), ([189]).

Example 7.5 Let $G = Sp(1,n), H = Sp(1)Sp(n), K = Sp(n)$. We get the semi-Riemannian submersion

$$\pi : \mathbf{H_3^{4n+3}} = Sp(1,n)/Sp(n) \to \mathbf{QH^n} = Sp(1,n)/Sp(1)Sp(n) \ ,$$

$\mathbf{QH^n}$ denoting the quaternionic hyperbolic space with constant quaternionic sectional curvature -4 ([340]).

Example 7.6 Let $G = Spin(1,8), H = Spin(8), K = Spin(7)$. We have the semi-Riemannian submersion

$$\pi : \mathbf{H_7^{15}} = Spin(1,8)/Spin(7) \to \mathbf{H^8}(-4) = Spin(1,8)/Spin(8)\,.$$

Example 7.7 Let $G = Sp(1,n), H = Sp(1)Sp(n), K = U(1)Sp(n)$. We obtain the semi-Riemannian submersion

$$\pi : \mathbf{CH_1^{2n+1}} = Sp(1,n)/U(1)Sp(n) \to \mathbf{QH^n} \doteq Sp(1,n)/Sp(1)Sp(n)\,.$$

We end this section with an example due to Kwon and Suh ([183]).

Example 7.8 For any $r, n \in \mathbf{N}$, $1 \leq r \leq n$, one can define a semi-Riemannian submersion Θ with totally geodesic fibres from the (real) pseudo-hyperbolic space $\mathbf{H_{2r-1}^{2n+1}}$ onto the complex pseudo-hyperbolic space $\mathbf{CH_{r-1}^n}$. We put $\mathbf{C_r^{n+1}} = (\mathbf{C^{n+1}}, F)$, where F is the semi-Hermitian form on $\mathbf{C^{n+1}}$ defined by:

$$F(z,w) = -\sum_{i=1}^{r} z^i \overline{w}^i + \sum_{j=r+1}^{n+1} z^j \overline{w}^j\,,$$

for any $z = (z^1, \dots, z^{n+1})$, $w = (w^1, \dots, w^{n+1}) \in \mathbf{C^{n+1}}$. With respect to the semi-Riemannian metric $g' = \mathrm{Re}(F)$ of index $2r$, $\mathbf{C_r^{n+1}}$ is a $(2n+1)$-dimensional semi-Riemannian manifold isometric to the space $\mathbf{R_{2r}^{2n+2}}$ and the real hypersurface $\{z \in \mathbf{C_r^{n+1}} \mid F(z,z) = -1\}$, with the metric g induced by g', is identified with the unit pseudo-hyperbolic space $\mathbf{H_{2r-1}^{2n+1}}$. Through the parallel displacement in $\mathbf{C_r^{n+1}}$, the tangent space $T_z(\mathbf{C_r^{n+1}})$ at any point z can be identified with $\{w \in \mathbf{C_r^{n+1}} \mid \mathrm{Re}(F)(z,w) = 0\}$, so that $T_z(\mathbf{C_r^{n+1}}) = < \mathrm{i}z > \oplus T'_z$, where T'_z denotes the orthogonal complement of $< \mathrm{i}z >$. Then $\mathbf{H_{2r-1}^{2n+1}}$ can be considered as the total space of a principal $\mathbf{S^1}$-bundle over a complex pseudo-hyperbolic space $\mathbf{CH_{r-1}^n}$ with projection Θ. Moreover, there exists an $\mathbf{S^1}$-invariant connection on $\mathbf{H_{2r-1}^{2n+1}}$ whose horizontal space at any point z is just T'_z; since the map $w \in T'_z \mapsto \mathrm{i}w \in T'_z$ is compatible with the $\mathbf{S^1}$-action, a complex structure J on $\mathbf{CH_{r-1}^n}$ is induced via Θ. Furthermore, $\mathbf{CH_{r-1}^n}$ inherits from g a semi-Riemannian metric g_0 of index $2(r-1)$ making Θ a semi-Riemannian submersion with totally geodesic fibres isometric to $\mathbf{H_1^1}$. It is easy to prove that $(\mathbf{CH_{r-1}^n}, J, g_0)$ has constant holomorphic sectional curvature -4. In particular, if $r = 1$,

$(\mathbf{CH^n}, J, g_0)$ is the complex hyperbolic space with the Bergman metric and Θ is just the submersion of Example 7.4.

7.3 Lorentzian Submersions

A *Lorentzian submersion* is a semi-Riemannian submersion whose total space is a Lorentz manifold.

The following results are obtained by A. Magid ([189]).

Theorem 7.2 *Let $\pi : (M_1^{m+k}, g) \to (B^m, g')$ be a Lorentzian submersion with totally geodesic fibres. If M_1^{m+k} has negative sectional curvatures and B is Riemannian, then $k = 1$.*

Moreover, if M_1^{m+k} has constant sectional curvature -1, then $m = 2n$ for some $n > 0$, M_1^{2n+1} admits an almost contact indefinite metric structure, and B^{2n} is a Kähler manifold of constant sectional curvature -4.

Proof. From the hypotheses, any horizontal space $\mathcal{H}_p$, $p \in M_1^{m+k}$ is m-dimensional with positive definite induced g. Hence, $g(X, X) > 0$ for any horizontal vector $X \neq 0$. Let $K(\alpha)$ be the sectional curvature of the 2-plane α spanned by $X \in \mathcal{H}_p$, and $V \in \mathcal{V}_p$. Since $T = 0$, by (1.32), we have:

$$\frac{g(A_X V, A_X V)}{g(X, X)g(V, V)} = K(\alpha) < 0 \,. \tag{7.1}$$

Thus, $A_X V$ being horizontal, we have $g(V, V) < 0$, *i.e.* any vertical vector is timelike. Since M_1^{m+k} is a Lorentzian manifold, then the fibres are 1-dimensional, so that $k = 1$. Now, we suppose that the total space has constant sectional curvature -1.

The fibres of the projection π define a foliation by timelike geodesics. Let ξ be the vertical vector field such that $g(\xi, \xi) = -1$. For any p in M_1^{2n+1}, we define the operator $\varphi_p : \mathcal{H}_p \to \mathcal{H}_p$, by $\varphi_p X = A_X \xi_p$. By Lemma 1.1, φ_p is a skew-symmetric operator. Since $g(\xi, \xi) = -1$, (7.1) implies $g_p(\varphi_p X, \varphi_p X) = g_p(X, X)$, for any $X \in \mathcal{H}_p$, *i.e.* φ_p is an isometry and then a complex structure on $\mathcal{H}_p$. Hence, m is even, say $m = 2n$. Now, we consider a $(1,1)$-tensor field φ on M_1^{2n+1} putting $\varphi(E) = A_E \xi$. Obviously, $\varphi(V) = 0$ for any vertical vector V. Let η be the 1-form dual to ξ, so that $\eta(\xi) = -1$. It is not difficult to prove that (φ, ξ, η, g) is an almost contact indefinite metric structure on M_1^{2n+1}. Finally, to end the proof, we claim that:

Step I φ projects in an almost complex structure J on B^{2n} .
Step II J is integrable.
Step III B^{2n} is Kähler.

Step I. We prove that if X is a basic vector field on M_1^{2n+1}, then φX, *i.e.* $A_X\xi$, is a basic vector field, too. This means that for any basic vector field Y, $g(\varphi X, Y)$ is a constant function on any fibre $\pi^{-1}(x), x \in B^{2n}$. Namely, we have:

$$\xi(g(\varphi X, Y)) = g(\nabla_\xi \varphi X, Y) + g(\varphi X, \nabla_\xi Y) , \tag{7.2}$$

and $R(\xi, X)\xi = -X$, since M_1^{2n+1} has constant curvature -1. We know that $\nabla_\xi \xi = 0$ and $\nabla_{[X,\xi]}\xi = 0$ so that $\nabla_\xi \nabla_X \xi = -X$. Moreover, $\nabla_\xi \varphi X = \nabla_\xi (A_X \xi) = \nabla_\xi (\nabla_X \xi) = -X$. Then (7.2) becomes:

$$\xi(g(\varphi X, Y)) = -g(X, Y) + g(\varphi X, \varphi Y) = 0 ,$$

since $[\xi, Y]$ is vertical and φ is an isometry on $\mathcal{H}_p$. Consequently, φX is a basic vector field, and putting $J(X') = \pi_*(\varphi X)$, with X π-related to X', one obtains an almost complex structure on B^{2n}.
Step II. Firstly, we prove that, if X, Y are horizontal, then

$$h(\nabla_X \varphi Y) = \varphi \nabla_X Y . \tag{7.3}$$

Namely, from (1.28), since $g(R(Y, Z)\xi, X) = 0$, with X, Y, Z horizontal vector fields, we obtain:

$$0 = g(R(Y, Z)\xi, X) = g((\nabla_X A)_Y Z, \xi) ,$$

which expands to:

$$g(\nabla_X (A_Y Z), \xi) - g(A_{\nabla_X Y} Z, \xi) - g(A_Y (\nabla_X Z), \xi) = 0 .$$

On the other hand, we have

$$A_Y Z = -g(A_Y Z, \xi)\xi = g(A_Y \xi, Z)\xi = g(\varphi Y, Z)\xi ,$$

and by a long but simple computation, one obtains (7.3). Now, to prove that the almost complex structure J on B^{2n} is integrable, observe that if X, Y are basic vector fields π-related to X', Y', then the basic vector field corresponding to $N_J(X', Y')$ is given by:

$$S(X, Y) = h[\varphi X, \varphi Y] - h[X, Y] - \varphi[X, \varphi Y] - \varphi[\varphi X, Y] ,$$

since φ vanishes on the vertical distribution. Therefore, it is enough to prove that $S(X,Y)$ vanishes or, equivalently,

$$h(\nabla_{\varphi X}\varphi Y) - h(\nabla_{\varphi Y}\varphi X) - h(\nabla_X Y) + h(\nabla_Y X)$$
$$= \varphi(\nabla_X \varphi Y) - \varphi(\nabla_{\varphi Y} X) + \varphi(\nabla_{\varphi X} Y) - \varphi(\nabla_Y \varphi X)\,.$$

With a straightforward computation this equality follows from (7.3).
Step III. Since the metric g' on B^{2n} verifies $g'(X',Y')\circ\pi = g(X,Y)$, for any two basic vector fields X, Y on M_1^{2n+1}, π-related to X' and Y', one easily sees that g' is a Hermitian metric. Thus, in order to show that B^{2n} is Kähler, we only prove that: $\nabla_{X'} JY' = J\nabla'_{X'}Y'$. This is equivalent to (7.3) for any basic vector fields X, Y, π-related to X' and Y'. Let $K(\alpha)$ be the sectional curvature of the φ-holomorphic horizontal 2-plane α spanned by the unit vectors $X, \varphi X$ at a point $p \in M_1^{2n+1}$, and let $K'(\alpha')$ be the sectional curvature of the holomorphic 2-plane α' spanned by (X', JX') at $\pi(p)$. Then, by (1.31) we have:

$$K'(\alpha') = K(\alpha) + 3g(A_X\varphi X, A_X\varphi X)\,.$$

But

$$A_X\varphi X = A_X A_X\xi = -g(A_X A_X\xi,\xi)\xi = g(A_X\xi, A_X\xi)\xi = g(\varphi X,\varphi X)\xi = \xi\,.$$

Thus, since $g(\xi,\xi) = -1$, we obtain $K'(\alpha') = -4$ and B^{2n} is locally holomorphically isometric to $\mathbf{CH^n}$.

Theorem 7.3 *Let $\pi : \mathbf{H_1^{2n+1}} \to B^{2n}$ be a Lorentzian submersion with totally geodesic fibres, from the anti-de Sitter space onto a Riemannian manifold. Then, B^{2n} is a Kähler manifold holomorphically isometric to* $\mathbf{CH^n}$.

Proof. We apply the above theorem. Moreover, since the total space $\mathbf{H_1^{2n+1}}$ is diffeomorphic to $\mathbf{S^1} \times \mathbf{R^{2n}}$ ([228]), using the exact sequence of homotopy groups induced by the fibration π, one has $\pi_1(B^{2n}) = 0$, *i.e.* B^{2n} is simply connected. Hence, B^{2n} is holomorphically isometric to $\mathbf{CH^n}$.

In analogy with the Hopf fibration, one can prove that $\mathbf{H_1^{2n+1}}$ is an indefinite Sasakian manifold ([281]).

7.4 Submersions from Pseudo-hyperbolic Spaces

Proposition 7.1 *Let $\pi : (M_s^{n+s}, g) \to (B^n, g')$ be a semi-Riemannian submersion from an $(n+s)$-dimensional semi-Riemannian manifold of index*

$s \geq 1$ onto an n-dimensional Riemannian manifold. We assume that M_s^{n+s} is geodesically complete and simply connected. Then B^n is complete and simply connected. Moreover, if B^n has non-positive curvatures, then the fibres are simply connected.

Proof. Since M_s^{n+s} is geodesically complete, the base space B^n is complete. Let $\tilde{g}$ be the Riemannian metric on M_s^{n+s} defined by

$$\tilde{g}(E,F) = g(hE,hF) - g(vE,vF) .$$

Since $\tilde{g}$ is a horizontally complete Riemannian metric and B^n is a complete Riemannian manifold, then $\mathcal{H}$ is an Ehresmann connection for π (Theorem 1 in [353]) and, by a theorem of Ehresmann (Theorem 9.40, in [26]), π is a locally trivial fibration. Then, using the exact homotopy sequence of π we get $\pi_1(B^n) = 0$, since M_s^{n+s} is simply connected. Finally, if B^n has non-positive sectional curvatures, applying the Cartan–Hadamard theorem, B^n is diffeomorphic to $\mathbf{R^n}$ and $\pi_2(B^n) = 0$. Using again the exact homotopy sequence we obtain that any fibre is simply connected.

We consider the m-dimensional (real) pseudo-hyperbolic space $\mathbf{H_s^m}$ of index $s \geq 2$, since the case $s = 1$ falls in the above section. We recall that $\mathbf{H_s^m}$ has constant sectional curvature -1 and the curvature tensor is given by

$$R(X,Y,X,Y) = -g(X,X)g(Y,Y) + g(X,Y)^2 ,$$

where g is the metric tensor given in Example 7.3. The following proposition is a semi-Riemannian version of Proposition 2.6.

Proposition 7.2 *Let $\pi : \mathbf{H_s^m} \to B^n$ be a semi-Riemannian submersion with totally geodesic fibres from a pseudo-hyperbolic space of index $s \geq 2$ onto an n-dimensional Riemannian manifold. Then $m = n+s$, the induced metrics on the fibres are negative definite and B^n has negative sectional curvatures. Moreover, B^n is a non-compact Riemannian symmetric space of rank one and the fibres are diffeomorphic to $\mathbf{S^s}$ with $s \in \{3,7\}$.*

Proof. The definition of semi-Riemannian submersion implies $g(X,X) > 0$ for any horizontal vector $X \neq 0$. Using (1.29), we get $g(V,V) < 0$ for any vertical vector $V \neq 0$, so that the index of g is $s = m - n$. Let $\{X,Y\}$ be an orthonormal basis of a horizontal 2-plane α and α' the plane spanned by $\{\pi_* X, \pi_* Y\}$. From (1.31), we have $K'(\alpha') = 3g(A_X Y, A_X Y) - 1$. On the other hand, $g(A_X Y, A_X Y) \leq 0$ since $A_X Y$ is a vertical vector, so that $K'(\alpha') < 0$. As in Theorem 2.4, we get that B^n is locally symmetric and Proposition 7.1 implies that it is a non-compact Riemannian symmetric space of rank one. Now, we prove that any fibre is diffeomorphic to the standard s-sphere $\mathbf{S^s}$, $s \geq 2$. Obviously, any fibre is geodesically complete as a totally geodesic submanifold of a geodesically complete manifold and it is simply connected by Proposition 7.1. Since it has dimension s, index s and constant sectional curvature -1, then it is isometric to $\mathbf{H^s_s}$ ([228]). Hence, any fibre is diffeomorphic to $\mathbf{S^s}, s \geq 2$. As for the Riemannian case (Theorem 2.6, step B), we can prove that the tangent bundle of any fibre is trivial (a detailed proof can be found in Lemma 2.5 in [12]). Hence, a well-known result of Adams implies that $s \in \{3, 7\}$.

The classification of the Riemannian symmetric spaces of rank one and non-compact type allows to state that the base B^n is isometric to one of the following spaces:

1) $\mathbf{H^n}(c)$, the real hyperbolic space with constant sectional curvature c .
2) $\mathbf{CH^k}(c)$, the complex hyperbolic space with constant holomorphic sectional curvature c .
3) $\mathbf{QH^k}(c)$, the quaternionic hyperbolic space with constant quaternionic sectional curvature c .
4) $\mathbf{CayH^2}(c)$, the Cayley hyperbolic plane with constant Cayley sectional curvature c .

The following result relates the dimension of the fibres to the geometry of the base space.

Proposition 7.3 *Let* $\pi : \mathbf{H^{n+s}_s} \to (B^n, g')$ *be a semi-Riemannian submersion with totally geodesic fibres and* $s \geq 2$. *Then we have:*

a) *if* $s = 3$, *then* $n = 4k$ *and* B^n *is isometric to* $\mathbf{QH^k}(-4)$ *;*
b) *if* $s = 7$, *then one the following cases occurs:*
   i) $n = 8$ *and* B^n *is isometric to* $\mathbf{H^8}(-4)$*;*
   ii) $n = 16$ *and* B^n *is isometric to* $\mathbf{CayH^2}(-4)$.

Proof. We know that the metrics induced on the fibres are negative definite. Let Y and Z be two linear independent horizontal vectors at any point p and α' the 2-plane spanned by $\pi_* Y$ and $\pi_* Z$. By (1.31), we obtain $K'(\alpha') \leq -1$. Using again (1.31) and applying the Schwartz inequality to the positive definite metric induced on $\mathcal{H}$ one has $K'(\alpha') \geq -4$, so that $K' \in [-4,-1]$. Via the curvature equations of a semi-Riemannian submersion, one can prove that, if B^n has constant sectional curvature c, then $c = -4$. Now, for any $p \in \mathbf{H}^{\mathbf{n+s}}_{\mathbf{s}}$ and $X \in \mathcal{H}_p$, we consider the map $A^*_X : \mathcal{H}_p \to \mathcal{V}_p$, given by $A^*_X(Y) = A_X Y$ for any horizontal vector Y. Then $Y \in \ker A^*_X$ if and only if, at $x = \pi(p)$, we have:

$$R'(\pi_*X, \pi_*Y, \pi_*X, \pi_*Y) = -g'(\pi_*X, \pi_*X)g'(\pi_*Y, \pi_*Y) + g'(\pi_*X, \pi_*Y)^2 .$$

For any $X' \in T_xB^n$, we denote by $\mathcal{L}_{X'}$ the space:

$$\{Y' \in T_xB^n \mid R'(X', Y', X', Y') = -g'(X', X')g'(Y', Y') + g'(X', Y')^2\} .$$

With this notation, $\pi_*(\ker A^*_X) = \mathcal{L}_{\pi_*X}$ and since π_* maps isometrically $\mathcal{H}_p$ onto T_xB^n, we have: $\dim \mathcal{H}_p - \dim \mathcal{V}_p = \dim \ker A^*_X = \dim \mathcal{L}_{\pi_*X}$. We compute $\dim \mathcal{L}_{X'}$ from the geometry of B^n, according to the following possibilities for the base space.

Case 1. $B^n = \mathbf{H}^{\mathbf{n}}(-4)$. Since

$$R'(X', Y', X', Y') = -4(g'(X', X')g'(Y', Y') - g'(X', Y')^2) ,$$

one has $\mathcal{L}_{X'} = \{\lambda X' \mid \lambda \in \mathbf{R}\}$, $\dim \mathcal{L}_{X'} = 1$ and $\dim \mathcal{H} = \dim \mathcal{V} + 1$. If $s = 3$, then B^4 is isometric to $\mathbf{H}^{\mathbf{4}}(-4)$, which falls in the case a), since $\mathbf{H}^{\mathbf{4}}(-4)$ is isometric to $\mathbf{QH}^{\mathbf{1}}(-4)$. If $s = 7$, then $\dim \mathcal{H} = 8$ and this is the case i) in b).

Case 2. $B^n = \mathbf{CH}^{\mathbf{k}}(-4)$. Let I_0 be the natural complex structure on $\mathbf{CH}^{\mathbf{k}}(-4)$. The curvature tensor is given by:

$$R'(X', Y', X', Y') = -g'(X', X')g'(Y', Y') + g'(X', Y')^2 - 3g'(I_0X', Y')^2 ,$$

and we get $\mathcal{L}_{X'} = \{I_0X'\}^{\perp}$. So, $\dim \mathcal{L}_{X'} = 2k - 1 = \dim \mathcal{H} - 1$, which implies $s = 1$ and this case cannot occur.

Case 3. $B^n = \mathbf{QH}^{\mathbf{k}}(-4)$. Let I_0, J_0, K_0 be the local almost complex structures which give rise to the quaternionic structure on $\mathbf{QH}^{\mathbf{k}}(-4)$. The curvature tensor is given by:

$$R'(X', Y', X', Y') = -g'(X', X')g'(Y', Y') + g'(X', Y')^2$$

$$-3g'(I_0X', Y')^2 - 3g'(J_0X', Y')^2 - 3g'(K_0X', Y')^2 .$$

So, $Y' \in \mathcal{L}_{X'}$ if and only if $g'(I_oX', Y') = g'(J_0X', Y') = g'(K_0X', Y') = 0$. Therefore $\mathcal{L}_{X'} = \{I_0X', J_0X', K_0X'\}^{\perp}$ and $\dim \mathcal{L}_{X'} = 4k - 3 = \dim \mathcal{H} - 3$. We get $s = \dim \mathcal{V} = 3$ and then the case a).

Case 4. $B^n = \mathbf{CayH}^{\mathbf{2}}(-4)$. Let I_0, J_0, K_0, M_0, M_0I_0, M_0J_0, M_0K_0 be the local almost complex structures which give rise to the Cayley structure on $\mathbf{CayH^2}(-4)$. The curvature tensor is given by:

$$R'(X',Y',X',Y') = -g'(X',X')g'(Y',Y') + g'(X',Y')^2 - 3g'(I_0X',Y')^2$$
$$-3g'(J_0X',Y')^2 - 3g'(K_0X',Y')^2 - 3g'(M_0I_0X',Y')^2$$
$$-3g'(M_0J_0X',Y')^2 - 3g'(M_0K_0X',Y')^2 \; .$$

We get $\mathcal{L}_{X'} = \{I_0X', J_0X', K_0X', M_0X', M_0I_0X', M_0J_0X', M_0K_0X'\}^{\perp}$. So, $\dim \mathcal{L}_{X'} = \dim \mathcal{H} - 7$, $s = \dim \mathcal{V} = 7$ and we obtain ii) in b).

To state the classification theorem, we need the following lemma.

Lemma 7.1 *Let $\pi : \mathbf{H_3^{4k+3}} \to \mathbf{QH^k}$ be a semi-Riemannian submersion with totally geodesic fibres. Given $p \in \mathbf{H_3^{4k+3}}$, let $\mathcal{U} : \mathcal{V}_p \to End(\mathcal{H}_p)$ be the representation defined by $\mathcal{U}(v)(X) = A_X v$. Then, there exists an orthonormal basis of $(\mathcal{V}_p, -g_p)$ such that $A^{v_1}A^{v_2}A^{v_3} = I_{\mathcal{H}_p}$.*

Proof. We denote $\mathcal{U}(v)$ by A^v. It is trivial to see that A^v is skew-symmetric and $g_p(A^vX, A^vX) = -g_p(X,X)g_p(v,v)$. This implies $g_p(A^vA^vX, X) = g_p(X,X)g_p(v,v)$ and, by polarization, we get $g_p(A^vA^vX, Y) = g_p(X,Y)g_p(v,v)$ for any $Y \in \mathcal{H}_p$. So $A^vA^vX = g_p(v,v)X$. Again by polarization we get

$$A^vA^w + A^wA^v = 2g_p(v,w)I_{\mathcal{H}_p} = -2\tilde{g}_p(v,w)I_{\mathcal{H}_p} \; , \qquad (7.4)$$

where $\tilde{g}$ is the Riemannian metric introduced in the proof of Proposition 7.1. This condition allows to extend $\mathcal{U}$ to a representation of the Clifford algebra $Cl(\mathcal{V}_p, \tilde{g}_p)$ of $\mathcal{V}_p$. We denote again by $\mathcal{U}$ such an extension. Since $\dim \mathcal{V}_p = 3$, $Cl(\mathcal{V}_p, \tilde{g}_p)$ has at most two types of irreducible representations. We notice that $\mathcal{H}_p$ is a $Cl(\mathcal{V}_p, \tilde{g}_p)$-module which splits in simple modules of dimension 4. The next step is to show that any two such simple modules in the decomposition of $\mathcal{H}_p$ are equivalent. Let $\{v_1, v_2, v_3\}$ be an orthonormal basis of $(\mathcal{V}_p, \tilde{g}_p)$. Since the affiliation of a simple $Cl(\mathcal{V}_p, \tilde{g}_p)$-module to one of the two possible types is decided by the action of $v_1v_2v_3$, it is sufficient to check that $A^{v_1}A^{v_2}A^{v_3} = I_{\mathcal{H}_p}$. Considering the function $X \to g_p(A^{v_1}A^{v_2}A^{v_3}X, X)$ defined on the unit sphere in $\mathcal{H}_p$, we have:

$$g_p(A^{v_1}A^{v_2}A^{v_3}X, X) = -g_p(A^{v_2}A^{v_3}X, A^{v_1}X) = g_p(A_XA_{A_Xv_3}v_2, v_1) \; ,$$

and a straightforward computation shows that $A_XA_{A_Xv_3}v_2$ is orthogonal to v_2 and v_3. Hence $A_XA_{A_Xv_3}v_2$ is a multiple of v_1. By polarization, from the relation $A_X A_X v = g_p(X,X)v$, we get:

$$A_X A_Y + A_Y A_X = 2g(X,Y)I_{\mathcal{V}_p} \tag{7.5}$$

for any horizontal vectors X and Y. In particular, we have:

$$A_X A_{A_X v_3} v_2 = -A_{A_X v_3} A_X v_2 + 2g(X, A_X v_3)v_2 = -A_{A_X v_3} A_X v_2 \ .$$

Now, let S be the linear subspace of $\mathcal{H}_p$ spanned by the orthonormal vectors $\{X, A_X v_1, A_X v_2, A_X v_3\}$. Using (7.5), considering for any $i \in \{1,2,3\}$ the 2-plane α_i spanned by $\{\pi_* X, \pi_* A_X v_i\}$ we get $K'(\alpha_i) = -4$. By the geometry of $\mathbf{QH^n}$, there exists a unique totally geodesic hyperbolic line $\mathbf{QH^1}$ passing through $\pi(p)$ such that $T_{\pi(p)}\mathbf{QH^1} = \pi_* S$. Notice that for any 2-plane α in $T_{\pi(p)}\mathbf{QH^1}$, $K'(\alpha) = -4$. In particular, we have $K'(\alpha') = -4$ for the 2-plane spanned by $\{\pi_* A_X v_2, \pi_* A_X v_3\}$. Hence $g(A_{A_X v_3} A_X v_2, A_{A_X v_3} A_X v_2) = -1$ and then $A_X A_{A_X v_3} v_2 = \pm v_1$. Thus, $g(A^{v_1} A^{v_2} A^{v_4} X, X) = \pm 1$ for any unit vector X. Since the function $X \to g(A^{v_1} A^{v_2} A^{v_3} X, X)$ defined on the unit sphere in $\mathcal{H}_p$ is continuous, we get either

$$\text{(i)} \quad g(A^{v_1} A^{v_2} A^{v_3} X, X) = 1 \ ,$$

or

$$\text{(ii)} \quad g(A^{v_1} A^{v_2} A^{v_3} X, X) = -1 \ .$$

We may assume the case (i). Namely, in the case (ii), we replace the orthonormal basis $\{v_1, v_2, v_3\}$ of $(\mathcal{V}_p, \tilde{g}_p)$ with the orthonormal basis $\{v_1, v_2, -v_3\}$. Since $A^{v_1} A^{v_2} A^{v_3}$ is an isometry, we have $g(A^{v_1} A^{v_2} A^{v_3} X, A^{v_1} A^{v_2} A^{v_3} X) = 1$ for any unit horizontal vector X. So, the Schwartz inequality in $(\mathcal{H}_p, g_p)$

$$g(A^{v_1} A^{v_2} A^{v_3} X, X)^2 \leq g(A^{v_1} A^{v_2} A^{v_3} X, A^{v_1} A^{v_2} A^{v_3} X) g(X,X)$$

becomes an equality. It follows that $A^{v_1} A^{v_2} A^{v_3} X = \lambda X$ for some λ and, since $A^{v_1} A^{v_2} A^{v_3}$ is an isometry, we obtain $\lambda = 1$. Obviously, $A^{v_1} A^{v_2} A^{v_3} X = X$ for any $X \in \mathcal{H}_p$.

The previous results help in proving the following classification theorem ([12]).

Theorem 7.4 (Baditoiu, Ianus). *Let $\pi : \mathbf{H^m_s} \to B$ be a semi-Riemannian submersion with totally geodesic fibres from a pseudo-hyperbolic space onto a Riemannian manifold. Then π is equivalent to one of the following canonical semi-Riemannian submersions, given in Examples* 7.4, 7.5, 7.6*:*

(a) $\pi : \mathbf{H^{2k+1}_1} \to \mathbf{CH^k}$,

(b) $\pi : \mathbf{H_3^{4k+3}} \to \mathbf{QH^k}$,

(c) $\pi : \mathbf{H_7^{15}} \to \mathbf{H^8}(-4)$.

Proof. The index of the pseudo-hyperbolic space cannot be $s = 0$, otherwise we get $0 \leq g(A_X V, A_X V) = -g(X,X)g(V,V) < 0$ for any X horizontal and V vertical. As remarked by Magid ([189]), any semi-Riemannian submersion with totally geodesic fibres from a pseudo-hyperbolic space of index 1 onto a Riemannian manifold is equivalent to the canonical semi-Riemannian submersion $\pi : \mathbf{H_1^{2k+1}} \to \mathbf{CH^k}$. Now, assuming $s \geq 2$, by Proposition 7.3 we have the following possibilities for π:

(1) $\pi : \mathbf{H_3^{4k+3}} \to \mathbf{QH^k}$,

(2) $\pi : \mathbf{H_7^{15}} \to \mathbf{H^8}(-4)$,

(3) $\pi : \mathbf{H_7^{23}} \to \mathbf{CayH^2}$.

We prove that any two semi-Riemannian submersions in one of the categories (1) or (2) are equivalent, while there are no semi-Riemannian submersions with totally geodesic fibres in the category (3). Firstly, we discuss the case (1). Let $\pi' : \mathbf{H_3^{4k+3}} \to \mathbf{QH^k}$ be another semi-Riemannian submersion with totally geodesic fibres. Fixed $p \in \mathbf{H_3^{4k+3}}$, let $\{v_1, v_2, v_3\}$ be an orthonormal basis of $\mathcal{V}_p$ such that $v_1v_2v_3$ acts on $\mathcal{H}_p$ as the identity. For an arbitrarily chosen point $q \in \mathbf{H_3^{4k+3}}$ with $\pi'(q) = \pi(p)$, we consider the horizontal and vertical subspaces $\mathcal{H}'_q$ and $\mathcal{V}'_q$. By Lemma 7.1 there exists an orthonormal basis $\{v'_1, v'_2, v'_3\}$ in $\mathcal{V}'_q$ such that $v'_1v'_2v'_3$ acts on $\mathcal{H}'_q$ as the identity. Let $L_1 : \mathcal{V}'_q \to \mathcal{V}_p$ be the isometry given by $L_1(v'_i) = v_i$ for any $i \in \{1,2,3\}$ and $Cl(L_1) : Cl(\mathcal{V}'_q) \to Cl(\mathcal{V}_p)$ the extension of L_1 to the Clifford algebras. Therefore the composition $\mathcal{U} \circ Cl(L_1) : Cl(\mathcal{V}'_q) \to End(\mathcal{H}_p)$ makes $\mathcal{H}_p$ a $Cl(\mathcal{V}'_q)$-module of dimension $4k$. Let $\mathcal{H}_p = H_1 \oplus \cdots \oplus H_k$ and $\mathcal{H}'_q = H'_1 \oplus \cdots \oplus H'_k$ be the decompositions of $\mathcal{H}_p$ and $\mathcal{H}'_q$ in simple modules. For any $i \in \{1,2,3\}$ there exists an isomorphism $f_i : H'_i \to H_i$ such that

$$f_i \circ A'^{v'_j} = A^{v_j} \circ f_i \ .$$

As in Theorem 2.7, after a rescaling by a constant, any f_i is an isometry preserving the O'Neill's integrability tensor. The direct sum of all these isometries gives an isometry $L_2 : \mathcal{H}'_q \to \mathcal{H}_p$. Therefore

$$L = L_1 \oplus L_2 : T_q\mathbf{H}_3^{4k+3} \to T_p\mathbf{H}_3^{4k+3}$$

is an isometry which maps $\mathcal{H}'_q$ onto $\mathcal{H}_p$ and preserves the O' Neill's integrability tensor. Since $\mathbf{H}_3^{4k+3}$ is a simply connected complex symmetric space, there exists an isometry $f : \mathbf{H}_3^{4k+3} \to \mathbf{H}_3^{4k+3}$ such that $f(q) = p$ and $f_{*q} = L$ ([340]). Therefore, with the same argument as in Theorem 2.2 in [86] we get that π and π' are equivalent. Now, we prove that any two semi-Riemannian submersions with totally geodesic fibres, $\pi, \pi' : \mathbf{H}_7^{15} \to \mathbf{H}^8(-4)$, are equivalent. The proof, analogous to the case (1), is easier. Chosen $p, q \in \mathbf{H}_7^{15}$ with $\pi'(q) = \pi(p)$, let $\mathcal{H}_p, \mathcal{V}_p$ be the horizontal and vertical subspaces in $T_p\mathbf{H}_7^{15}$ for π and $\mathcal{H}'_q, \mathcal{V}'_q$ the horizontal and vertical subspaces in $T_q\mathbf{H}_7^{15}$ for π'. As in Lemma 7.1 we can consider an orthonormal basis $\{v_1, \dots, v_7\}$ of $(\mathcal{V}_p, \tilde{g}_p)$ and an orthonormal basis $\{v'_1, \dots, v'_7\}$ of $(\mathcal{V}'_q, \tilde{g}_q)$ such that $A^{v_1}A^{v_2}\cdots A^{v_7} = I_{\mathcal{H}_p}$ and $A^{v'_1}A^{v'_2}\cdots A^{v'_7} = I_{\mathcal{H}'_q}$. Since $\dim \mathcal{V}_p = 7$, the irreducible modules are 8-dimensional and then $\dim \mathcal{H}_p = 8$ implies that $\mathcal{H}_p$ is simple. Thus, $\mathcal{H}'_q$ and $\mathcal{H}_p$ are equivalent modules. Analogously to the case (1), we can construct an isometry $f : \mathbf{H}_7^{15} \to \mathbf{H}_7^{15}$ such that $f(q) = p$ and $f_{*q} = L$ and we conclude that π and π' are equivalent. The non-existence of semi-Riemannian submersions $\pi : \mathbf{H}_7^{23} \to \mathbf{CayH^2}$, with totally geodesic fibres, can be proved with a technique similar to the one used by Ranjan (Proposition 5.1 in [246]).

In 1978 R. Escobales classified Riemannian submersions from complex projective spaces under the assumption that the fibres are connected, complex, totally geodesic submanifolds (Sec. 2.3 and [87]). Using Theorem 7.4, we obtain a classification of semi-Riemannian submersions from a complex pseudo-hyperbolic space $\mathbf{CH_s^m}$, with a metric of index s and constant holomorphic sectional curvature $c = -4$, onto a Riemannian manifold under the same conditions on the fibres.

Proposition 7.4 *Let $\pi : \mathbf{CH_s^m} \to B^n$ be a semi-Riemannian submersion with complex, connected, totally geodesic fibres. Then $2m = n + 2s$, the induced metrics on the fibres are negative definite and the fibres are diffeomorphic to $\mathbf{P_s(C)}$.*

Proof. We denote by J the natural almost complex structure on $\mathbf{CH_s^m}$. Obviously, the fibres have constant holomorphic sectional curvature -4 and, since they are complex submanifolds, we get:

$$g(A_XU, A_XU) = -g(U,U)g(X,X) + 3g(X,JU)^2 = -g(U,U)g(X,X) .$$

We obtain that $g(U,U) \leq 0$ for any vertical vector U, the induced metrics on the fibres are negative definite and $n = 2m - 2s$. Furthermore, for horizontal vectors X, Y, we have:

$$\begin{aligned} R'(\pi_*X, \pi_*Y, \pi_*X, \pi_*Y) &= R(X,Y,X,Y) + 3g(A_XY, A_XY) \\ &= -(g(X,X)g(Y,Y) - g(X,Y)^2) \\ &\quad -3g(X,JY)^2 + 3g(A_XY, A_XY) \leq 0 \,, \end{aligned}$$

so B^n has non-positive curvatures. Applying Proposition 7.1, the fibres are simply connected and, since they are complete, complex manifolds with constant holomorphic curvature -4, they are isometric to $\mathbf{CH^s_s}$ and then diffeomorphic to $\mathbf{P_s(C)}$.

Theorem 7.5 *Let $\pi : \mathbf{CH^m_s} \to B$ be a semi-Riemannian submersion with connected, complex, totally geodesic fibres from a complex pseudo-hyperbolic space onto a Riemannian manifold. Then, up to equivalence, π is the canonical semi-Riemannian submersion $\pi : \mathbf{CH^{2n+1}_1} \to \mathbf{QH^n}$ in Example 7.7.*

Proof. Let $\Theta : \mathbf{H^{2m+1}_{2s+1}} \to \mathbf{CH^m_s}$ be the canonical semi-Riemannian submersion with totally geodesic fibres described in Example 7.8. As in Proposition 2.11, we can prove that $\tilde{\pi} = \pi \circ \Theta$ is a semi-Riemannian submersion with totally geodesic fibres. Since the dimension of the fibres of $\tilde{\pi}$ is $2s + 1 \geq 3$, we get, by Theorem 7.4, the following possible cases:

i) $m = 2k + 1$, $2s + 1 = 3$ and B is isometric to $\mathbf{QH^k}$,
ii) $m = 7, 2s + 1 = 7$ and B is isometric to $\mathbf{H^8(-4)}$.

Firstly, we prove the equivalence of any two semi-Riemannian submersions $\pi, \pi' : \mathbf{CH^{2k+1}_1} \to \mathbf{QH^k}$ with connected, complex, totally geodesic fibres. As in Proposition 7.4 we have $g(A_XU, A_XU) = -g(U,U)g(X,X)$. Given a point $p \in \mathbf{CH^{2k+1}_1}$, like in the proof of Lemma 7.1, via (7.4) we have:

$$A^vA^w + A^wA^v = -2\tilde{g}(v,w)I_{\mathcal{H}_p} \,,$$

and the extension of $\mathcal{U} : \mathcal{V}_p \to End(\mathcal{H}_p)$ to the Clifford algebra $Cl(\mathcal{V}_p, \tilde{g}_p)$ makes $\mathcal{H}_p$ a $Cl(\mathcal{V}_p, \tilde{g}_p)$-module which splits in k irreducible modules of real dimension 4. By the classification of the irreducible representations in the case $\dim \mathcal{V}_p = 2$, we have that any two such irreducible modules are equivalent. As in Theorem 7.4, considering a point $q \in \mathbf{CH^{2k+1}_1}$ such that $\pi'(q) = \pi(p)$, we may construct an isometry

$$L = L_1 \oplus L_2 : T_q\mathbf{CH^{2k+1}_1} \to T_p\mathbf{CH^{2k+1}_1} \,,$$

with $f(q) = p$ and $f_{*q} = L$, and we get that π and π' are equivalent. Finally, the case ii) cannot occur. In fact, in [12] Baditoiu and Ianus, following the technique of Ranjan ([246]), proved the following result.

Proposition 7.5 *There are no semi-Riemannian submersions*

$$\pi : \mathbf{CH}_3^7 \to \mathbf{H}^8(-4)$$

with connected, complex, totally geodesic fibres.

Recently, in [9] G. Baditoiu extended the previous results, considering semi-Riemannian submersions whose base space is a strictly semi-Riemannian manifold.

7.5 Submersions with Totally Umbilical Fibres

Let $\pi : (M, g) \to (B, g')$ be a semi-Riemannian submersion from an m-dimensional connected semi-Riemannian manifold of index s onto an n-dimensional connected semi-Riemannian manifold of index s' $(s' \leq s)$. Since the O'Neill's tensor T acts on the fibres as the second fundamental form, the fibres are totally umbilical if $T_U V = g(U, V)H$, for any vertical vector fields U, V on M, H being the mean curvature vector field. Let $\{e_1, \dots, e_m\}$ be a local orthonormal frame on M such that $e_1, \dots, e_r$ are vertical and $e_{r+1}, \dots, e_m$ are basic, $r \geq 1$ denoting the dimension of the fibres. Vertical vectors are indexed by latin indices and horizontal ones by greek indices. Firstly, we have $rH = \sum_{i=1}^{r} \varepsilon_i T_{e_i} e_i$. We also have the following notation:

$$g(A, A) = \sum_{\alpha,\beta} \varepsilon_\alpha \varepsilon_\beta g(A_{e_\alpha} e_\beta, A_{e_\alpha} e_\beta) = \sum_{i,\alpha} \varepsilon_i \varepsilon_\alpha g(A_{e_\alpha} e_i, A_{e_\alpha} e_i)\ , \quad (7.6)$$

$$g(T, T) = \sum_{i,j} \varepsilon_i \varepsilon_j g(T_{e_i} e_j, T_{e_i} e_j) = \sum_{i,\alpha} \varepsilon_i \varepsilon_\alpha g(T_{e_i} e_\alpha, T_{e_i} e_\alpha)\ , \quad (7.7)$$

where $\varepsilon_a = g(e_a, e_a)$, for any $a \in \{1, \dots, m\}$.

Using the curvature equations for a semi-Riemannian submersion one has:

Lemma 7.2 *Let $\pi : (M, g) \to (B, g')$ be a semi-Riemannian submersion with totally umbilical fibres. Then, for any vertical vector fields U, V and horizontal vector fields X, Y one has:*

a) $R(U,V,U,V) = \hat{R}(U,V,U,V) + (g(U,V)^2 - g(U,U)g(V,V))g(H,H)$;
b) $R(X,U,X,U) = g(U,U)(g(\nabla_X H, X) - g(X,H)^2) + g(A_X U, A_X U)$;
c) $R(X,Y,X,Y) = R^*(X,Y,X,Y) - 3g(A_X Y, A_X Y)$.

Example 7.9 We describe an important class of semi-Riemannian submersions which occur in general relativity ([228]). Suppose that (B,g') and $(F,\hat{g})$ are semi-Riemannian manifolds and let f be a never vanishing smooth function on B and $\hat{g}_{x'} = f^2(x')\hat{g}$ a smooth family of Riemannian metrics on F, indexed by B. Then, the resulting semi-Riemannian manifold $(B \times F, g' + f^2\hat{g})$ is the *warped product* of (B,g') and $(F,\hat{g})$ by f^2, denoted by $M = B \times_{f^2} F$. Putting $g = g' + f^2\hat{g}$, the projection $\pi : (M,g) \to (B,g')$ onto the first factor is a semi-Riemannian submersion having integrable horizontal distribution and $(F, f^2(x')\hat{g})$ as fibre at $x' \in B$. On the other hand, T acts on the fibres as $T_U V = -g(U,V)\frac{1}{f}\mathrm{grad} f$, the mean curvature vector field H is given by $H = -\frac{1}{f}\mathrm{grad} f$, so that the fibres are totally umbilical (they are not totally geodesic if f is a non-constant function) and H is a basic vector field.

The following result is an immediate extension to the semi-Riemannian case of Proposition 9.104 in [26].

Proposition 7.6 *Let $\pi : (M,g) \to (B,g')$ be a semi-Riemannian submersion with totally umbilical fibres. If the horizontal distribution $\mathcal{H}$ is integrable and the mean curvature vector field H is basic, then M is, locally, a warped product.*

Furthermore, applying the O'Neill formulas and in particular Lemma 7.2, we obtain the following generalization of Proposition 9.106 in [26] to the semi-Riemannian submersions.

Proposition 7.7 *The Ricci tensor ρ of the warped product $M = B \times_{f^2} F$ satisfies:*

i) $\rho(U,V) = \hat{\rho}(U,V) - ((\frac{1}{f}\Delta' f + \frac{(r-1)}{f^2}\|df\|^2) \circ \pi) g(U,V)$;
ii) $\rho(U,X) = 0$;
iii) $\rho(X,Y) = (\rho'(X',Y') - \frac{r}{f}(\nabla' df)(X',Y')) \circ \pi$,

for any vertical U,V and any basic X,Y, π-related to X', Y', where $\hat{\rho}$ is the Ricci tensor on $(F,\hat{g})$ and ρ' the Ricci tensor of (B,g').

Now, we report a result essentially due to Escobales and Parker ([91]).

Proposition 7.8 *Let $\pi : (M,g) \to (B,g')$ be a semi-Riemannian submersion with totally umbilical fibres and X,Y basic vector fields. Then*

$A_X Y$ is a Killing vector field along the fibres if and only if $g(\nabla_X H, Y) = g(\nabla_Y H, X)$.

Proof. Let U, V be vertical vector fields. Since the fibres are totally umbilical, via Lemma 1.4 and (1.26)g, we have:

$$\begin{aligned} g(\nabla_U(A_X Y), V) + g(\nabla_V(A_X Y), U) &= g((\nabla_U A)_X Y, V) + g((\nabla_V A)_X Y, U) \\ &= g(U, V)(g(\nabla_Y H, X) - g(\nabla_X H, Y)), \end{aligned}$$

and the statement follows.

Corollary 7.1 *Let $\pi : (M, g) \to (B, g')$ be a semi-Riemannian submersion with totally umbilical fibres, such that the mean curvature vector field H is basic, π-related to H'. Then, for any X, Y basic vector fields, $A_X Y$ is a Killing vector field along the fibres if and only if the 1-form on B associated with H' is closed.*

Proof. Let ω be the 1-form associated with H'. Since:

$$2d\omega(X', Y') = g'(\nabla'_{X'} H', Y') - g'(\nabla'_{Y'} H', X'), \quad X', Y' \in \mathcal{X}(B),$$

the statement follows from Proposition 7.8.

We recall that the mixed scalar curvature τ^{HV} of a semi-Riemannian manifold M_s^m is the function on M_s^m such that at any point p:

$$\tau^{HV}(p) = \sum_{i,\alpha} \varepsilon_i \varepsilon_\alpha R_p(e_\alpha, e_i, e_\alpha, e_i). \tag{7.8}$$

Proposition 7.9 *Let $\pi : (M, g) \to (B, g')$ be a semi-Riemannian submersion with totally umbilical fibres. Then:*

$$\tau^{HV} = r\mathrm{div}(H) + r(r-1)g(H, H) + g(A, A).$$

Proof. The statement follows from the following generalization to the semi-Riemannian case, of a Ranjan's formula ([249])

$$\tau^{HV} = r\mathrm{div}(H) + r^2 g(H, H) + g(A, A) - g(T, T),$$

since $g(T, T) = rg(H, H)$.

As a consequence of Proposition 7.9, we have:

Proposition 7.10 *If $\pi : (M, g) \to (B, g')$ is a semi-Riemannian submersion with totally umbilical fibres and M is compact and orientable, then:*

$$\int_M \tau^{HV} dv_g = r(r-1) \int_M g(H, H) dv_g + \int_M g(A, A) dv_g.$$

Corollary 7.2 *If $\pi : M \to B$ is a Riemannian submersion with totally umbilical fibres and M is a compact and orientable manifold, then*

$$\int_M \tau^{HV} dv_g \geq 0 \, .$$

By Proposition 7.10, we have the following splitting theorem, which is a generalization of Proposition 3.1 of R. Escobales ([86]) and is proved in [11].

Theorem 7.6 *Let $\pi : (M, g) \to (B, g')$ be a Riemannian submersion with totally umbilical fibres, where M is a compact, orientable manifold with non-positive mixed sectional curvatures. Then, one has:*

a) *$R(X, U, Y, V) = 0$ for any X, Y horizontal vector fields and for any U, V vertical vector fields;*
b) *the horizontal distribution is integrable;*
c) *the fibres are totally geodesic.*

Now, we suppose that $\mathcal{F}$ is a bundle-like foliation on a Riemannian manifold (M, g) ([288]) and define the *normal curvature* 1*-form* H^b setting $H^b(E) = g(H, E)$ for any vector field E on M, where H is the mean curvature vector field of $\mathcal{F}$. We denote by $\mathcal{V}$ the distribution defined by the foliation $\mathcal{F}$ and put $\mathcal{H} = \mathcal{V}^\perp$. We say that H^b is a *basic* 1*-form* if $H^b(X) = 0$ and $L_X(H^b) = 0$ for any $X \in \mathcal{V}^\perp$ ([91]). For any $W \in \mathcal{V}$ let $\rho^{\mathcal{V}}(W, W) = \sum_{k=1}^{r} R(W, e_k, W, e_k)$, where $(e_1, \ldots, e_r)$ is a local orthonormal frame of $\mathcal{V}$, $r \geq 1$ denoting the dimension of the leaves of $\mathcal{F}$.

We report now some results of Escobales and Parker ([91]).

Theorem 7.7 *Let (M, g) be a connected Riemannian manifold with an umbilical bundle-like foliation $\mathcal{F}$ having codimension $q \leq m - 1$, where $m = \dim M$. Assume that:*

a) *either M is compact or all the leaves of $\mathcal{F}$ are compact;*
b) *the normal curvature* 1*-form H^b is closed or, equivalently, under the above compactness hypotheses, H is a basic vector field;*
c) *$\rho^{\mathcal{V}}(W, W) \leq (1 + q - n)||H||^2$ at each point of M, for any vector $W \in \mathcal{V}$ and $\rho^{\mathcal{V}}(W, W) < (1 + q - n)||H||^2$ at at least one point of each leaf of $\mathcal{F}$.*

Then, the orthogonal distribution $\mathcal{H}$ is integrable with totally geodesic integral manifolds and, locally, M is a warped product.

Corollary 7.3 *Let (M, g) be a connected m-dimensional Riemannian manifold with an umbilical bundle-like foliation $\mathcal{F}$ whose leaves are surfaces. Assume that:*

a) *either* (a.1): *M is compact, or* (a.2): *all the leaves of $\mathcal{F}$ are compact;*
b) *$K(V, W) \leq -||H||^2$ on each leaf of $\mathcal{F}$, at each point, and $K(V, W) < -||H||^2$ at at least one point on each leaf of $\mathcal{F}$.*

Then, $\mathcal{H}$ is integrable with totally geodesic integral manifolds and, locally, M is an umbilical product. If, additionally, H^b is basic, then, locally, M is a warped product. Finally, assuming that M is complete and a.2) *and* b) *hold, then the universal covering space of M is diffeomorphic to a product of the form $\mathbf{R}^2 \times B$ and, if H^b is basic, this universal covering is in fact isometric to a warped product.*

7.6 Semi-Riemannian Submersions with Minimal Fibres

We define the *second fundamental form* α_π of a map $\pi : (M, g) \to (B, g')$ between two semi-Riemannian manifolds by:

$$\alpha_\pi(X, Y) = \tilde{\nabla}_X \pi_* Y - \pi_* \nabla_X Y \ , \tag{7.9}$$

for any local vector fields X, Y on M, where ∇ is the Levi-Civita connection of M and $\tilde{\nabla}$ is the pullback of the connection ∇' of B to the induced vector bundle $\pi^{-1}(TB)$.

The *tension field* $\tau(\pi)$ of π is defined as the trace of α_π, *i.e.*:

$$\tau(\pi)_p = \sum_{a=1}^{m} \varepsilon_a \alpha_\pi(e_a, e_a) \ , \tag{7.10}$$

where $\{e_a\}_{1 \leq a \leq m}$ is a local orthonormal frame around a point $p \in M$.

We say that π is a *harmonic map* if and only if $\tau(\pi)$ vanishes at each point $p \in M$. For the theory of the harmonic maps we refer to ([80; 81; 304]).

Now, we suppose that $\pi : (M, g) \to (B, g')$ is a semi-Riemannian submersion and let $\{e_1, \ldots, e_m\}$ be a local orthonormal frame on M such that $\{e_i\}$, $1 \leq i \leq r$, are vertical and $\{e_\alpha\}$, $r + 1 \leq \alpha \leq m$ are basic. Then, we have:

$$\tilde{\nabla}_{e_\alpha} \pi_* e_\alpha = \pi_* \nabla_{e_\alpha} e_\alpha \ , \quad \alpha \in \{r + 1, \ldots, m\} \ , \tag{7.11}$$

$$\tilde{\nabla}_{e_i}\pi_* e_i = 0\,, \quad i \in \{1, \ldots, r\}\,, \tag{7.12}$$

since the e_i are vertical. Consequently, we obtain

$$\tau(\pi)_p = -\pi_* \sum_{i=1}^{r} \varepsilon_i \nabla_{e_i} e_i = -r\pi_*(H_p)\,, \tag{7.13}$$

where $H_p = \frac{1}{r}\sum_{i=1}^{r} \varepsilon_i T_{e_i} e_i = \frac{1}{r}\sum_{i=1}^{r} \varepsilon_i h(\nabla_{e_i} e_i)$ is the mean curvature vector at a point $p \in M$.

This gives the following well-known theorem.

Theorem 7.8 *A semi-Riemannian submersion $\pi : (M, g) \to (B, g')$ is a harmonic map if and only if each fibre is a minimal submanifold.*

Remark 7.5 Any Riemannian submersion $\pi : (M, g) \to (B, g')$ is a harmonic morphism, *i.e.* it pulls back any harmonic function on (B, g') to a harmonic function on (M, g). In other words, π preserves Laplacian's equation. For this subject we refer to the book of P. Baird and J.C. Wood ([14]).

Remark 7.6 The restriction of α_π to $\mathcal{X}^h \times \mathcal{X}^v$ or $\mathcal{X}^v \times \mathcal{X}^h$ vanishes if and only if the horizontal distribution of the semi-Riemannian submersion π is integrable.

Now, we report some results of J. H. Kwon and Y. J. Suh ([183]) about semi-Riemannian submersions with minimal fibres. The following proposition has to be compared with Proposition 2.3.

Proposition 7.11 *Let $\pi : (M, g) \to (B, g')$ be a semi-Riemannian submersion with minimal fibres and total space an $(n + s)$-dimensional semi-Riemannian manifold of index s. We suppose that the sectional curvatures of all non-degenerate plane sections of (M, g) are non-negative (non-positive) and B is an n-dimensional manifold with a positive (negative) definite metric. Then, the fibres of π are totally geodesic and the horizontal distribution is integrable.*

Proposition 7.12 *Let (M, g) be an $(n + s)$-dimensional semi-Riemannian manifold of index s and constant sectional curvature $c > 0$ $(c < 0)$. If B is an n-dimensional manifold with positive (negative) definite metric g', then semi-Riemannian submersions $\pi : (M, g) \to (B, g')$ with minimal fibres do not exist.*

Finally, in [183], geometric properties of the horizontal distribution of a semi-Riemannian submersion with minimal fibres in terms of Ricci curvatures are stated.

Chapter 8

Applications of Riemannian Submersions in Physics

8.1 Gauge Fields, Instantons and Riemannian Submersions

Gauge theory first appeared in the early attempt by H. Weyl to unify the general relativity and the theory of electromagnetism. Later, in 1954, C. N. Yang and R. L. Mills ([345]) proposed a variational approach as classical model to describe strong interactions (responsible of the cohesion of the nucleus). The Lagrangian of the interaction should involve a potential with values in the Lie algebra of $SU(2)$, and should be invariant under the group of the local internal symmetries called *gauge transformations*. In physics' theories there are four types of interactions: strong, weak, electromagnetic and gravitational, with different groups of symmetries. The theory of gauge fields was developed by physicists to explain and unify these four fundamental forces of the nature.

A modern formulation of Yang–Mills theory requires the use of principal fibre bundles. We refer to [172] for the basic tools on such bundles. In Physics, the base manifold M of a principal G-bundle (P, M, G, π) is usually a space-time manifold, but more generally M may be any m-dimensional (semi)-Riemannian manifold, and G a compact Lie group.

The gauge group of the principal G-bundle P, denoted by $\mathcal{G}(P)$, is defined as the group of all *inner automorphisms* of P *i.e.* the automorphisms ψ of P satisfying $\psi(pa) = \psi(p)a$, for $p \in P$ and $a \in G$.

A connection on P is a 1-form ω on P with values in the Lie algebra $\mathfrak{g}$ of G, such that:

i) $R_a^*\omega = Ad(a^{-1})\omega$, for any $a \in G$;
ii) $\omega(V) = \overline{V}$, at each point of P and for any $\overline{V} \in \mathfrak{g}$,

where R denotes the right action of G and the vertical vector field V generates the one-parameter group of diffeomorphisms of P given by the action of the subgroup $\mathrm{Exp}(t\overline{V})$ of G. It follows that $\mathcal{H} = \ker \omega$ is a G-equivariant distribution on P and for any $p \in P$, one has $T_pP = \mathcal{V}_p \oplus \mathcal{H}_p$, where $\mathcal{V}_p$ is the tangent space to the fibre $\pi^{-1}(\pi(p))$ through p. The curvature 2-form Ω on P is $\mathfrak{g}$-valued and defined by:

$$\Omega = d\omega + \frac{1}{2}[\omega, \omega] . \tag{8.1}$$

For a given principal G-bundle P, with a standard construction ([172]), one considers a G-vector bundle $E = P \times_\rho \mathbf{R}^\mathbf{r}$ associated with P by a faithful orthogonal representation $\rho : G \to O(r)$. The gauge group $\mathcal{G}(P)$ is identified with the group $\mathcal{G}(E)$ of all the automorphisms of E inducing the identity map of M by the projection $\pi : E \to M$ ([49]). Now, one can associate with a connection 1-form on P a unique G-connection ∇ on E ([172; 49]). The corresponding curvature tensor R^∇ is the 2-form in $\Gamma(\Lambda^2 T^* \otimes \mathfrak{g})$ defined by:

$$R^\nabla(X, Y) = [\nabla_X, \nabla_Y] - \nabla_{[X,Y]} , \tag{8.2}$$

for any vector fields X, Y on M.

On the space $\mathcal{C}(E)$ of the G-connections on E, we define the Yang–Mills functional $\mathcal{YM} : \mathcal{C}(E) \to \mathbf{R}_+$, putting:

$$\mathcal{YM}(\nabla) = \frac{1}{2}\int_M ||R^\nabla||^2 dv_g , \tag{8.3}$$

where the norm is defined in terms of a Riemannian metric g on M and of a fixed $Ad(G)$-invariant inner product on $\mathfrak{g}$. Relation (8.3) gives a measure of the total curvature of ∇. The Yang–Mills functional is invariant under the gauge group $\mathcal{G}(E)$ as well as the set of its critical points ([159]).

A connection ∇ is called a *Yang–Mills connection* if it is a critical point of $\mathcal{YM}$, *i.e.* if for any smooth family $\{\nabla^t\}, -\varepsilon < t < \varepsilon$, with $\nabla^0 = \nabla$, one has:

$$\left(\frac{d}{dt}\mathcal{YM}(\nabla^t)\right)_{|t=0} = 0 . \tag{8.4}$$

The space of the Yang–Mills connections on a given (metric) vector bundle E is infinite dimensional, unless empty. It is known that the Euler–Lagrange equation for the Yang–Mills functional (8.3) is given by the so-called *Yang–Mills equation*:

$$\delta^\nabla R^\nabla = 0 , \tag{8.5}$$

where δ^{∇} is the codifferential operator with respect to ∇. By the Bianchi identity we have $d^{\nabla}R^{\nabla} = 0$, so that (8.5) is equivalent to:

$$\triangle^{\nabla}R^{\nabla} = 0 , \tag{8.6}$$

where $\triangle^{\nabla}$ is the generalized Hodge–de Rham Laplacian. Thus Yang–Mills connections are connections with harmonic curvature. The standard Yang–Mills equation arises in Physics when M is a Minkowski space, but if we pass to the imaginary time, M becomes the Euclidean 4-space $\mathbf{R^4}$. Moreover, since in dimension 4 the Yang–Mills equations are conformally invariant, $\mathbf{R^4}$ can be replaced by $\mathbf{S^4}$, provided we impose suitable conditions at infinity as a requirement for solutions ([6]). On a 4-dimensional Riemannian manifold (M, g), the Hodge map $*$ defines an endomorphism:

$$* : \Lambda^2 T^*M \to \Lambda^2 T^*M ,$$

such that $*^2 = I$. Then, denoting by Λ^+ , Λ^- the two eigenspaces corresponding to the eigenvalues $+1 , -1$ of the $*$-operator, we have:

$$\Lambda^2 T^*M = \Lambda^+ \oplus \Lambda^- . \tag{8.7}$$

We say that ∇ is a self-dual (anti-self-dual) connection if the curvature 2-form R^{∇} is an eigenvector of the Hodge operator, with eigenvalue $+1$ (-1), respectively. Alternatively, a self-dual (anti-self-dual) connection is called an *instanton* (*anti-instanton*). For a detailed study of the instantons on $SU(2)$-bundles over $\mathbf{S^4}$ we refer to ([8]). Using the Bianchi identity, it is easy to prove that any instanton (anti-instanton) is a solution of the Yang–Mills equation (8.6). Moreover, if E is an $SU(m)$-vector bundle over a compact oriented 4-manifold M, then any instanton (anti-instanton) yields an absolute minimum for the Yang–Mills functional (8.3) ([159]).

In [190], M. Mamone Capria and S. Salamon gave higher dimensional analogues of the self-dual (anti-self-dual) connections over a quaternionic Kähler manifold. A connection ∇ on a Riemannian vector bundle over a compact quaternionic Kähler manifold M is called a *quaternionic Yang–Mills connection* if:

$$\triangle^{\nabla}(R^{\nabla} \wedge F^{m-1}) = 0 ,$$

where F is the fundamental 4-form on M^{4m}. As proved in [283] if a connection ∇ is a quaternionic Yang–Mills connection, then it is a Yang–Mills connection.

A more geometric interpretation of Yang–Mills theory on $\mathbf{S^4}$ appears in the paper [328] of B. Watson, who studied the following commutative diagram of fibrations:

$$
\begin{array}{ccccc}
 & & SU(2) & \overset{\pi_4}{\longrightarrow} & \mathbf{P_1(C)} \\
 & & \downarrow & & \downarrow \\
U(1) \longrightarrow & & \mathbf{S^7} & \overset{\pi_2}{\longrightarrow} & \mathbf{P_3(C)} \\
 & \pi_1 & \downarrow & \swarrow \pi_3 & \\
 & & \mathbf{S^4} & &
\end{array}
\tag{8.8}
$$

where $SU(2) \simeq \mathbf{S^3}$ is the fibre type of the principal $SU(2)$-bundle $\mathbf{S^7}$ over $\mathbf{S^4}$ with canonical projection π_1 and $U(1) \simeq \mathbf{S^1}$ is the fibre type of the principal $\mathbf{S^1}$-bundle $\mathbf{S^7}$ over $\mathbf{P_3(C)}$ with projection π_2. All the projections π_i, $i \in \{1, \dots, 4\}$, are Riemannian submersions provided that all the manifolds are endowed with standard metrics. We remark that $\mathbf{S^7}$ is a 3-Sasakian manifold with three regular characteristic vector fields (ξ_1, ξ_2, ξ_3). Of course, π_2 is the Boothby–Wang fibration over $\mathbf{P_3(C)}$ and ξ_2, ξ_3 correspond to two regular vector fields on $\mathbf{P_3(C)}$. Then we obtain a Riemannian submersion π_3 which is a (not principal) fibre bundle with $\mathbf{S^2}$, identified with $\mathbf{P_1(C)}$, as fibre. On the other hand, ξ_1 on $\mathbf{S^7}$ fibers the fibres $\mathbf{S^3}$ of π_1 via the Hopf map $\pi_4 : \mathbf{S^3} \to \mathbf{S^2}$. Let η^1, η^2, η^3 be the 1-forms of the 3-Sasakian structure on $\mathbf{S^7}$, and consider the $\mathfrak{su}(2)$-valued 1-form $\omega = \sum_{i=1}^{3} \eta^i \otimes \overline{\xi}_i$ on $\mathbf{S^7}$, where $\overline{\xi}_i$ are the vectors of $\mathfrak{su}(2)$ corresponding to the vector fields ξ_i on the principal $SU(2)$-bundle $\mathbf{S^7} \to \mathbf{S^4}$, *i.e.* $\omega(\xi_i) = \overline{\xi}_i$. Then, ω defines a natural connection on this principal bundle. By a simple computation, the curvature 2-form Ω is given by:

$$\Omega(X, Y) = -\omega(A_X Y) \,,$$

for any horizontal vector fields X, Y on $\mathbf{S^7}$, A being the integrability tensor of π_1. We know that $A \neq 0$ for a 3-Sasakian submersion, so that the connection is not flat. Using the technique of Riemannian submersions, B. Watson proved that Ω is coclosed, *i.e.* $\delta\Omega = 0$; hence ω is a Yang–Mills connection.

Furthermore, in [292], A. Trautman also studied solutions to the $SU(2)$-Yang–Mills equations over $\mathbf{S}^4$ using the Hopf fibration $\pi_1 : \mathbf{S}^7 \to \mathbf{S}^4$.

We can generalize the diagram (8.8) using a Riemannian submersion from a 7-dimensional 3-Sasakian manifold over a quaternionic Kähler 4-manifold, $\pi : P^7 \to M^4$, where P^7 and M^4 are Einstein. Moreover if the ξ_i's are regular, then π is a principal $SU(2)$-bundle and the fibres are totally geodesic. For the existence of such submersions we refer to [54].

8.2 Einstein Equations and Kaluza–Klein Ansatz

An *electromagnetic field* ω is a 2-form over the space $\mathbf{R}^4$ viewed as space-time; it incorporates the electrical and magnetic fields. The *Maxwell equations* say that:

$$d\omega = 0; \qquad \delta\omega = \imath\ , \tag{8.9}$$

where d is the exterior differentiation operator, δ is its formal adjoint (using the Minkowski metric) and $\imath$ denotes a 4-vector having the charge density as time component and the current density as space components. In the vacuum we have $\imath = 0$, so that in this case the electromagnetic field ω is a harmonic 2-form. Unifying electricity and magnetism, Maxwell made a great step in glueing physical phenomena together.

In general relativity, a space-time (M, g) is a 4-dimensional manifold M with a Lorentz metric g (viewed as a gravitational potential). The metric g must be related, by a field equation, to the mass-energy distribution that generates the gravitational field. The field equation, called the *Einstein equation* (found by Einstein in 1915) is:

$$\rho - \frac{1}{2}\tau g = T\ , \tag{8.10}$$

where T is the *stress-energy tensor*, ρ the Ricci tensor and τ the scalar curvature. We observe that T is a symmetric 2-covariant tensor field on M. For details on the physical content of the stress-energy tensor T, we refer to [256].

The left-hand side of (8.10) is called the *Einstein tensor* of g. This equation must be viewed as a system of partial differential equations for the components of the metric g in local coordinates. We remark that such a system is hyperbolic, whereas in the Riemannian case it is an elliptic system. For $T = 0$, taking the trace of (8.10) we have the Einstein vacuum equation:

$$\rho = 0\,. \tag{8.11}$$

A current trend in modern physics is the search for a theory which provides a unification of the gravity with the other fundamental forces of the nature. One of the early possibilities for such a unification was suggested by Kaluza ([160]) and later expanded upon by Klein ([168]). It was shown within a 5-dimensional extension of Einstein's theory of general relativity how both gravity and electromagnetism could be treated on a similar footing. Both interactions were described as part of the five dimensional metric. The fifth coordinate was made invisible through a *cylindrical condition*: it was assumed that in the fifth direction, the world curled up into a cylinder of very small radius (10^{-33} cm, Planck's length). Kaluza considered $\mathbf{R^5}$ as a model for an extended space-time and a canonical projection $\pi : \mathbf{R^5} \to \mathbf{R^4}$, where $\mathbf{R^4}$ is the Minkowski space-time. The so-called *Kaluza ansatz* is to consider a metric $\tilde{g}$ on $\mathbf{R^5}$ of the form:

$$\tilde{g} = f^a[\pi^* g + f(\alpha + d\xi) \otimes (\alpha + d\xi)]\,, \tag{8.12}$$

where $\xi = x^5$ is the fifth coordinate on $\mathbf{R^5}$, $a \in \mathbf{R}$, g is a Lorentz metric on $\mathbf{R^4}$, α is a 1-form on $\mathbf{R^4}$ and f is a real function on $\mathbf{R^4}$. If $f \equiv 1$, then we have:

The Einstein vacuum equation for $\tilde{g}$ contains the Maxwell equations for the electromagnetic field $\omega = d\alpha$ and the Einstein equation for g with right-hand side T taken to be the interaction term coming from the electromagnetic field.

At a simple examination of (8.12) it appears necessary to put $f \equiv 1$. O. Klein ([168]) suggested that the extra-dimensional space has to be taken compact, *i.e.*, a circle $\mathbf{S^1}$. Then to a space-time M corresponds an extended space-time $\tilde{M} = M \times \mathbf{S^1}$.

The Kaluza–Klein model is obviously correlated to the Riemannian geometry of circle bundles. Let $\pi : \tilde{M} \to M$ be an $\mathbf{S^1}$-principal bundle over a Riemannian manifold (M, g) and consider an $\mathbf{S^1}$-connection with 1-form $\tilde{\omega}$ on $\tilde{M}$. One defines on $\tilde{M}$ the Riemannian metric $\tilde{g}$ putting:

$$\tilde{g} = \pi^* g + \tilde{\omega} \otimes \tilde{\omega}\,. \tag{8.13}$$

We have the following results due to J. P. Bourguignon (see also Sec. 5.1 and [26]):

(A) *The space of* $\mathbf{S^1}$*-invariant Riemannian metrics on the total space of an* $\mathbf{S^1}$*-principal bundle for which the fibres are totally geodesic is in 1-1 correspondence with the product of the space of Riemannian metrics on the base and the space of* $\mathbf{S^1}$*-connections over the bundle.*

(B) *On the total space* $\tilde{M}$ *of an* $\mathbf{S^1}$*-bundle* $\pi : \tilde{M} \to M$ *the Riemannian metric defined by* (8.13) *is Einstein with constant* $\tilde{\lambda}$ *if and only if:*

i) *the* 2*-form* Ω *on* M *which pulls back to the curvature* 2*-form of the connection* $\tilde{\omega}$ *is harmonic, with constant length* $2\sqrt{\tilde{\lambda}}$ *;*
ii) *the Ricci tensor of* M *is given by* $\rho = \tilde{\lambda} g - \frac{1}{2}\Omega \circ \Omega$,

where $\Omega \circ \Omega$ *is defined in a coordinate system by:*

$$(\Omega \circ \Omega)_{ij} = \sum_{kl} g^{kl} \Omega_{ik} \Omega_{lj} \,.$$

Remark 8.1 It is easy to see that the above equations also hold considering the Lorentzian metric. In the case of the vacuum Einstein equation, the bundle $\pi : \tilde{M} \to M$ is a trivial product and the electromagnetic field identically vanishes. This phenomenon is often referred to as the *Kaluza–Klein inconsistency.*

A natural generalization of the original Kaluza–Klein idea which incorporates non-Abelian gauge fields is to consider a higher than five dimensional theory in which the gauge fields become part of the metric in the same way as the electromagnetic field did in Kaluza's theory. In the modern Kaluza–Klein theories one starts with the hypothesis that the space-time has $(4+m)$ dimensions. The extra m spatial dimensions are static and curled up into a compact manifold of unobservable small size, typically of the order of the Planck length. It is not assumed that the space-time is of the form $M^4 \times M^m$ where M^4 is the Minkowski space and M^m is a compact space. Rather, such spaces should correspond to ground state solutions. Physical fields are then introduced as fluctuations around these ground state solutions.

In a large class of physically interesting models in $(4+m)$ dimensions, the compactification of the extra m dimensions is produced spontaneously by the non-trivial vacuum configuration of an antisymmetric tensor field. It is almost common to these models to get a huge cosmological constant for the space-time if the extra dimensions are Planck-sized.

Another mechanism for space-time compactification was proposed by Omero and Percacci ([225]) and Gell-Mann and Zwiebach ([112]). Their mechanism uses a scalar sector in the form of a nonlinear sigma model to trigger the compactification. The compactified space becomes isomorphic to the manifold in which the scalar fields take values and the four dimensional space has no cosmological term at the classical level. However, all symmetries in the extra dimensions are broken by the scalars so one would have to resort to solitons or bound states for the massless gauge bosons or put them in by hand.

There is also the possibility that the extra dimensions form a non-compact space of finite volume ([112]). The non-compact manifolds are interesting alternatives to the usual compact manifolds used in spontaneous compactifications. This mechanism can be used in several higher dimensional supergravity and superstring theories.

Many spontaneously compactifying schemes are solutions of higher dimensional gravity coupled with matter and the generalized nonlinear sigma model in curved space appears to be a convenient approach.

In few recent papers ([149; 150; 151; 148]) it was investigated the possibility to extend the model, assuming that the internal m-dimensional space can be larger than the manifold in which the fields take values. The general solution of the model can be expressed in terms of harmonic maps satisfying Einstein equations ([13; 80]). It was shown that a very general class of solutions is given by Riemannian submersions from the extra dimensional space onto the space in which the scalar fields take values. Investigating the isometries of the metric of the extra m-dimensional space one obtains that the gauge fields associated with the vertical Killing vectors are massless.

8.3 Generalized Nonlinear Sigma Model in Curved Space

We begin with some fundamental facts from the calculus of variations on Riemannian manifolds. As in Sec. 5.2 we consider the space $\mathcal{M}_1$ of Riemannian metrics of total volume 1 on a compact manifold M, *i.e.* $\int_M dv_g = 1$, for each $g \in \mathcal{M}_1$. Then, the integral

$$E(g) = \int_M \tau_g dv_g \ , \tag{8.14}$$

where τ_g is the scalar curvature of g, is called the *total scalar curvature* of (M, g) and the map $E : \mathcal{M}_1 \to \mathbf{R}$ is called the Einstein–Hilbert integral. It is known (by a Hilbert's theorem) that the metric g is a critical point of E restricted to $\mathcal{M}_1$ if and only if g is an Einstein metric ([26]).

Now, we consider a map $\psi : (M, g) \to (N, h)$ between two Riemannian manifolds (for the physicists ψ represents a scalar field on M). One defines its *energy density* at a point $x \in M$ by:

$$e(\psi)_x = \frac{1}{2} \parallel d\psi \parallel_x , \tag{8.15}$$

where $\parallel \quad \parallel_x$ denotes the Hilbert–Schmidt norm of $d\psi_x$. If M is compact, we define the *energy functional* E by:

$$E(\psi) = \int_M e(\psi) dv_g . \tag{8.16}$$

A critical point of the energy functional satisfies the Euler–Lagrange equations:

$$g^{ij} \frac{\partial^2 \psi^a}{\partial x^i \partial x^j} - g^{ij} \Gamma^k_{ij}(x) \frac{\partial \psi^a}{\partial x^k} + g^{ij} \Gamma^a_{bc}(\psi(x)) \frac{\partial \psi^b}{\partial x^i} \frac{\partial \psi^c}{\partial x^j} = 0 , \tag{8.17}$$

where (x^k) and $y^a = \psi^a(x)$ are local coordinates around points $x \in M$ and $y = \psi(x) \in N$, for $i, j, k \in \{1, \dots, m\}$ and $a \in \{1, \dots, n\}$ ([82; 80; 304]). We denote by $\tau^a(\psi)$ the functions of the left side in (8.17) which locally define the *tension field* $\tau(\psi)$ of ψ. If Δ^M is the Laplacian on (M, g), we have:

$$\tau^a(\psi) = \Delta^M(\psi^a) + g^{ij} \Gamma^a_{bc} \frac{\partial \psi^b}{\partial x^i} \frac{\partial \psi^c}{\partial x^j} . \tag{8.18}$$

This implies that if $N = \mathbf{R^n}$ and h is the Euclidean metric, then $\tau(\psi) \equiv 0$ if and only if $\{\psi^a\}$ are harmonic functions on (M, g).

We say that a map $\psi : (M, g) \to (N, h)$ between two Riemannian manifolds is a *harmonic map* if the Euler–Lagrange equations (8.17) are satisfied or, equivalently, $\tau(\psi)_x = 0$ for each $x \in M$. For the physicists a harmonic map is a *σ-model* ([109]).

Let $\psi^* h$ be the pull-back of the metric h by the map ψ. Then the symmetric tensor $S_\psi = e(\psi) g - \psi^* h$ is called the *stress-energy tensor* of ψ ([13]). It is easy to see that $\operatorname{tr} S_\psi = (m - 2) e(\psi)$. If ψ is harmonic, then $\operatorname{div} S_\psi = 0$; the converse is true if ψ is a Riemannian submersion ([13]).

The idea of coupling the Einstein's field equations to harmonic maps (nonlinear σ-models) seems to appear firstly in the papers of V. de Alfaro, S. Fubini, G. Furlan ([73]) and P. Baird and J. Eells ([13]). Now, we shall describe the model considered by M. Gell-Mann and B. Zwiebach ([112]) and C. Omero and R. Percacci ([225]). This model consists of Einstein's gravity in $(4+m)$ dimensions coupled to a nonlinear σ-model:

$$E(g,\psi) = \int_{\Omega}\left(-\frac{\tau}{4} + \frac{1}{\lambda^2}e(\psi)\right) dv_g \,, \tag{8.19}$$

where Ω is a compact domain of a $(4+m)$-dimensional space-time M^{4+m}. This integral action contains two parts. The first term is the usual gravitational action without a cosmological term. The second term is related to the energy integral of the map $\psi : (M^{4+m}, g) \to (B^n, h)$ where B^n is a compact Riemannian manifold. Varying the Lagrangian in (8.19) with respect to the metric g, the Euler–Lagrange equations describe the Einstein equation

$$\rho - \frac{1}{2}\tau g = 2T \,, \tag{8.20}$$

where the energy-momentum tensor T is, up to a coupling constant, the stress-energy tensor of ψ. Looking for the critical points of the integral (8.19) with respect to smooth variations of ψ, on any compact $\Omega \subset M$, one obtains the equations of the harmonic maps:

$$\tau(\psi) = 0 \,. \tag{8.21}$$

If ψ is a harmonic map, then $\mathrm{div} S_\psi = 0$ ([13]). On the other hand, S_ψ is obviously conservative, from (8.20), since the Einstein field tensor is divergence free. The system of equations (8.20) and (8.21) is usually termed as the *gravity system coupled to a harmonic map (nonlinear σ-model)*.

Now we assume that $\psi : (M^{4+m}, g) \to (B^n, h)$ is a (semi)-Riemannian submersion, $(m > n \geq 2)$. The stress-energy tensor S_ψ has divergence:

$$\mathrm{div} S_\psi = -h(\tau(\psi), d\psi) \,. \tag{8.22}$$

For a Riemannian submersion $\tau(\psi)$ vanishes if and only if S_ψ is conservative, *i.e.* $\mathrm{divS}_\psi = 0$ ([149]). We say that a (semi)-Riemannian submersion $\psi : (M, g) \to (B, h)$ is *coupled to gravity* if the following equation is satisfied:

$$\rho - \frac{1}{2}\tau g = \gamma S_\psi \,, \tag{8.23}$$

where ρ is the Ricci tensor, τ is the scalar curvature on (M, g), S_ψ is the stress-energy tensor of ψ and γ is *the coupling constant.* Of course in this case ψ is a harmonic map.

Spacetime with more than four dimensions has often been considered in attempt to find a unified theory including gravity. The supergravity and superstring theories strongly suggest the use of a higher-dimensional

background. In any case, there are general physical arguments in favor of this point of view. Following Kaluza and Klein, the extra dimensions are compactified to a small size, apparently beyond our present ability of experimental detection. Many papers on higher dimensional field theories assume that the universe (or its ground state) may be represented as:
i) a metric product $M = N^4 \times F$ *, where* N^4 *is a Lorentz manifold, and* F *is a (spacelike) Riemannian manifold,*
ii) F *is a compact manifold.*

There is a vast literature of physical motivations of i) and ii). The problem to providing such justification for i) may be called the *splitting problem* and for ii) the *spontaneous compactification* ([100; 200]). We must ask: what kind of splitting should we try to induce? There are many types: global products, local products, warped products, fibre bundles of various types and so on. Perhaps the simplest generalization of i) is obtained by taking a warped product ([16]).

For the models proposed by Omero–Percacci and Gell-Mann and Zwiebach, the compactification of the extra dimensions is induced by a harmonic map (nonlinear σ-model). The extra spatial dimensions are static and curled up into a compact manifold of unobservably small size. The harmonic map may be visualized as an identity map from the extra dimensional space to the space in which the scalar fields take values. In these models all symmetries in the extra dimensions are broken and there are no massless gauge bosons.

The second author and M.Visinescu investigated the possibility of extending the above models, assuming that the internal space can be larger than the manifold in which the fields take values. One takes $M^{4+m} = N^4 \times F^m$ as a Riemannian product and a parametrization by local coordinates (x^α, x^r), $\alpha = 1, \dots, 4$; $r = 5, 6, \dots, 4+m$. The scalar fields $\psi^a = \psi^a(x)$, $a = 5, \dots, 4+n$, are thought of as coordinates of an n-dimensional compact Riemannian space (B, h). From (8.23) we have:

$$\rho_{\alpha\beta} = 0, \quad \alpha, \beta = 1, \dots, 4 . \tag{8.24}$$

Therefore, the 4-dimensional spacetime N^4 can be taken Ricci-flat and has no cosmological term at the classical level.

For non-vanishing components of Ricci tensor, (8.23) was solved in [225; 112] when $m = n$, assuming that the extraspace is isometric to (B, h). In [150; 151], the authors study the case of a Riemannian submersion from (F^m, g) to $(B^n, h), m > n$. We remark the two following explicit examples:
1) F is a generalized Hopf manifold with Lee form η ([310; 231]). The dual

vector field ξ associated with η is parallel vector field on F, so that ξ is a Killing vector on M. Then we have the Riemannian submersion $\pi : F \to B$ where B is the quotient of F with respect to the regular distribution $\{\xi\}$ and h is defined in a canonical way ([310]). Let us remark that ξ is a vertical Killing vector field on the extra space which can generate massless gauge bosons.

2) The second example is provided by a cosymplectic manifold F ([33]) with a regular characteristic vector field ξ. Similarly, we have a Riemannian submersion $\pi : F \to B$, where $B = F/\xi$. The vector field ξ defines a massless gauge field.

8.4 Horizontally Conformal Submersions and Gravity

Let $\psi : M \to N$ be a smooth map between two semi-Riemannian manifolds. We consider

$$L(g, \psi) = \rho - \gamma e(\psi) , \tag{8.25}$$

a Lagrangian associated with the map ψ coupled to gravity, where ρ is the Ricci tensor field of (M, g) and $\gamma \neq 0$ is the coupling constant. The Euler–Lagrange equations associated with $L(g, \psi)$ are given by:

$$\rho - \frac{\tau}{2} g = \gamma S_\psi , \tag{8.26}$$

$$\tau(\psi) = 0 , \tag{8.27}$$

where τ is the scalar curvature of (M, g), S_ψ is the stress-energy tensor associated with ψ and $\tau(\psi)$ is the tension field of ψ ([268; 211]). If ψ is a constant map, then the map ψ satisfies (8.26) and (8.27) if and only if $\rho = 0$, *i.e.* (M, g) is Ricci flat. The identical map $\psi : (M, g) \to (M, h)$ fulfills (8.26) and (8.27) if and only if $\rho = \gamma h$.

Let $\psi : (M^m, g) \to (N^n, h)$ be a smooth map between semi-Riemannian manifolds and $C_\psi = \{x \in M \mid \text{rank } (d\psi)_x < n\}$. We denote by M^* the set of regular points of ψ, *i.e.* $M^* = M \setminus C_\psi$. For each regular point $x \in M$, the vertical and horizontal spaces are defined by $\mathcal{V}_x = \ker(d\psi)_x$ and $\mathcal{H}_x = (\mathcal{V}_x)^\perp$ respectively.

Definition 8.1 The map ψ is a *horizontally (weakly) conformal* submersion if at each point $x \in M$ with $(d\psi)_x \neq 0$ and $\mathcal{V}_x$ non-degenerate, the restriction of $(d\psi)_x$ to $\mathcal{H}_x$ is conformal in the sense that there exists $\lambda(x) \in \mathbf{R}^*$ such that:

$$h(d\psi(X), d\psi(Y)) = \lambda(x) g(X, Y), \quad \forall X, Y \in \mathcal{H}_x \,. \tag{8.28}$$

The extended function $\lambda : M \to \mathbf{R}$, obtained by putting $\lambda(x) = 0$ if $(d\psi)_x = 0$ or $\mathcal{V}_x$ is degenerate, is called the *dilation* of ψ. We remark that in the Riemannian case, λ is a non-negative function but in the semi-Riemannian case it can take negative values. Then, since $\lambda^2 = \frac{1}{n}||d\psi||^2$, λ is continuous on M and smooth on M^*; λ^2 is smooth on M ([102; 14]). A *horizontally conformal submersion* $\psi : (M^m, g) \to (N^n, h), m > n \geq 2$, *is said to be coupled to gravity* if ψ and g satisfy (8.26). From (8.28) and (8.22) it follows that S_ψ is divergence free. As in the case of Riemannian submersions, any solution of (8.26) also satisfies (8.27). We conclude that any horizontally conformal submersion coupled to gravity is a harmonic map.

M. Mustafa ([211]) gives some necessary conditions to construct a horizontally conformal submersion coupled to gravity on compact Riemannian manifolds and on certain semi-Riemannian manifolds. Finally, we remark that any horizontally conformal submersion coupled to gravity is a harmonic morphism ([14]).

8.5 The Dirac Monopole and the Hopf Map

In Physics the Hopf maps represent systems with non-trivial topological properties, e.g. Z_2 kink or sine-Gordon soliton, $U(1)$ magnetic monopole, $SU(2)$ instanton, etc. Other physical realizations of the Hopf maps are possible. Firstly, we outline the approach to the theory of magnetic monopoles initiated by Dirac. We examine the simplest possible situation assuming that a particle of mass m and electric charge q moves in the field of a magnetic monopole of strength g fixed at the origin of $\mathbf{R}^3$, so the magnetic field is:

$$\vec{B} = \frac{g}{r^3} \vec{r} \,, \tag{8.29}$$

r denoting the distance to the origin, $\vec{r}$ the position vector field of the particle ([93]). Denoting by F_{ij}, $i, j \in 1, 2, 3$, the components of the strength tensor field F, one has:

$$g = \frac{1}{2} \int_{\mathbf{S}^2(r)} F_{ij} dx^i \wedge dx^j \ . \tag{8.30}$$

This implies that F cannot be globally expressed in terms of a potential vector field $\vec{A}$ as:

$$F_{ij} = \partial_i A_j - \partial_j A_i \ , \tag{8.31}$$

since this would force the radius r of the 2-sphere to vanish. Dirac was able to circumvent this difficulty, introducing a fictitious string, namely a solenoid, joining the origin to infinity along which the magnetic field differs from that of the monopole by a singular magnetic flux. The Dirac quantization condition:

$$2qg = nc\hbar \ , \tag{8.32}$$

makes the quantum theory consistent and invariant under gauge transformations. Here n is an integer and we shall henceforth adhere to the convention of units $\hbar = c = 1$ where $\hbar$ is the Planck constant and c is the velocity of light. This is the famous Dirac quantization condition which implies *charge quantization*.

To make things more specific let us consider the magnetic field due to an infinitely long and thin solenoid placed along the negative z-axis with its positive pole g at the origin. The line occupied by the solenoid is called the *Dirac string*. The vector potential $\vec{A}$ of the solenoid can be written as:

$$A^x = g\frac{-y}{r(r+z)} \ , \quad A^y = g\frac{x}{r(r+z)} \ , \quad A^z = 0 \ ,$$

or, in polar coordinates:

$$A^r = A^\theta = 0 \ , \quad A^\phi = g\frac{1-\cos\theta}{r\sin\theta} \ , \tag{8.33}$$

which is singular on the negative z-axis. It is obvious that, by a suitable choice of coordinates, the Dirac string may be chosen along any direction and in fact we can choose a continuous curve from the origin to infinity. Choosing the Dirac string along the positive z-axis, we have

$$A^r = A^\theta = 0 \ , \quad A^\phi = -g\frac{1+\cos\theta}{r\sin\theta} \ . \tag{8.34}$$

Wu and Yang ([342]) have recast the theory of the Dirac monopole into a form which avoids the use of a singular vector potential. The sphere surrounding the monopole is divided into two overlapping regions R_a and R_b. R_a excludes the negative z-axis and in this region $\vec{A}$ is defined as in (8.33). The second region R_b excludes the positive z-axis and the vector potential $\vec{A}$ is defined as in (8.34). It is quite obvious that $\vec{A}_a$ and $\vec{A}_b$, the vector potentials in the regions R_a and R_b respectively, are both finite in their own domain. In the region of overlap they differ by a gauge transformation:

$$A_b^\mu = A_a^\mu - \frac{\mathrm{i}}{q} S \frac{\partial S^{-1}}{\partial x^\mu} .$$

In fact, we have

$$A_b^\phi = A_a^\phi - \frac{\mathrm{i}}{qr \sin\theta} S \frac{\partial S^{-1}}{\partial \phi} ,$$

where

$$S = e^{2\mathrm{i}gq\phi} .$$

If S is not single-valued, then the change in the phase of the wave function of a particle with the charge q, as the particle is transported around the equator, is ill defined. So we must demand

$$qg = \frac{n}{2} ,$$

which is precisely the Dirac quantization condition (8.32). From the above presentation of the physics of the Dirac monopole it is quite transparent that the presence of point magnetic monopoles necessitates a fibre bundle formulation of electrodynamics ([201; 255; 83]). The Hopf fibering of $\mathbf{S^3}$ over a base space $\mathbf{S^2}$ with fibre $\mathbf{S^1}$ yields the Wu–Yang potentials which describe the Dirac monopole.

For this purpose we start with the Hopf map, which maps $\mathbf{S^3}$ onto $\mathbf{S^2}$. $\mathbf{S^3}$ may be parametrized by x_1, x_2, x_3 and x_4, coordinates in $\mathbf{R^4}$, obeying:

$$x_1^2 + x_2^2 + x_3^2 + x_4^2 = 1 .$$

On the other hand $\mathbf{S^2}$ may be parametrized by ξ_1, ξ_2 and ξ_3, coordinates in $\mathbf{R^3}$, obeying:

$$\xi_1^2 + \xi_2^2 + \xi_3^2 = 1 .$$

The Hopf map f is described by:

$$\begin{aligned} \xi_1 &= 2(x_1x_3 + x_2x_4)\,, \\ \xi_2 &= 2(x_2x_3 - x_1x_4)\,, \\ \xi_3 &= x_1^2 + x_2^2 - x_3^2 - x_4^2\,. \end{aligned} \tag{8.35}$$

To see more neatly the relationship among Wu–Yang potentials we proceed to a parametrization of the sphere $\mathbf{S}^3$ by the Euler angles $\psi\,, \theta\,, \phi$:

$$\begin{aligned} x_1 &= \cos\left(\tfrac{\psi+\varphi}{2}\right)\cos\tfrac{\theta}{2}\,, \quad x_2 = \sin\left(\tfrac{\psi+\varphi}{2}\right)\cos\tfrac{\theta}{2}\,, \\ x_3 &= \cos\left(\tfrac{\psi-\varphi}{2}\right)\sin\tfrac{\theta}{2}\,, \quad x_4 = \sin\left(\tfrac{\psi-\varphi}{2}\right)\sin\tfrac{\theta}{2}\,, \end{aligned}$$

with $0 \le \theta \le \pi$ and $-\pi \le (\psi + \phi)/2 \le \pi$. The Hopf map (8.35) then gives

$$\xi_1 = \cos\phi\sin\theta\,,\ \xi_2 = \sin\phi\sin\theta\,,\ \xi_3 = \cos\theta\,.$$

The angles θ, ϕ may be identified with the polar angles on the sphere $\mathbf{S}^2$ and ψ is the angle on the fibre $\mathbf{S}^1$. The magnetic field of a monopole (8.29) is described by a 2-form σ_2 ([83]):

$$B = g\sigma_2 = g\sin\theta d\theta \wedge d\phi\,. \tag{8.36}$$

The 2-form σ_2 is closed but not exact on $\mathbf{S}^2$ and we have shown that there is no vector potential $\vec{A}$ on $\mathbf{S}^2$ such that $B = dA$. On the other hand, σ_2 as a 2-form on $\mathbf{S}^3$ is exact since the second cohomology group of $\mathbf{S}^3$ is trivial. Therefore on $\mathbf{S}^3$, B is exact and there exists a 1-form A on $\mathbf{S}^3$ such that

$$B = dA. \tag{8.37}$$

We can easily construct the following as a solution (not unique)

$$A = 2g(x_2dx_1 - x_1dx_2 + x_4dx_3 - x_3dx_4) = -g(d\psi + \cos\theta d\phi)\,,$$

verifying (8.37) with B given by (8.36). Now, taking the sections σ_a and σ_b of equations $\psi = -\phi$ and $\psi = \phi$, respectively and considering the pullback of A under such sections, we get:

$$\begin{aligned} &\sigma_a : \psi = -\phi \qquad A^a_\phi = g(1 - \cos\theta)\,, \quad A^a_\theta = 0\,, \\ &\sigma_b : \psi = \phi \qquad A^b_\phi = -g(1 + \cos\theta)\,, \quad A^b_\theta = 0\,, \end{aligned}$$

which together with $A^a_r = A^b_r = 0$ allow to recognize the Wu–Yang potentials (8.33) and (8.34), ([62; 255]). Finally let us remark that the monopole charge is connected with the first Chern number C_1 characterizing the bundle ([83]).

The extension of the Dirac monopole to non-Abelian gauge theories was realized by 't Hooft and Polyakov ([62]). In non-Abelian gauge theories of the Higgs type, magnetic monopoles can appear without any modification of the field equations. The 't Hooft-Polyakov monopole solution in non-Abelian gauge theory represents a synthesis of the Dirac monopole and the soliton solution.

Kaluza–Klein theories ([160; 168]) which try to unify Einstein's theory of gravitation with other gauge interactions also contain topological solitons that can be identified as magnetic monopoles.

8.6 Kaluza–Klein Monopole and Taub–NUT Instanton

In what follows we shall briefly present the Kaluza–Klein monopole ([135; 275]) which was obtained by embedding the Taub–NUT gravitational instanton into five-dimensional Kaluza–Klein theory ([138]). For this purpose it is necessary to construct soliton solutions in Kaluza–Klein theories. By solitons we mean non-singular solutions of the classical field equations which represent spatially localized lumps that are topologically stable. The construction of a soliton in Kaluza–Klein theories is analogous to the magnetic monopole of 't Hooft and Polyakov ([62]) that occurs in non-Abelian gauge theories.

The Kaluza–Klein monopole in a five-dimensional theory can be viewed as a principal fibre bundle with M^4 as the base manifold and $U(1)$ as the structure group. The vacuum is the trivial bundle $M^4 \times \mathbf{S^1}$ but of course there exist topologically inequivalent bundles. At spatial infinity a solution will describe an $\mathbf{S^1}$-bundle over $\mathbf{S^2}$ (the boundary of the 3-dimensional space) and there exist an infinite collection of such bundles, each characterized by an integer which can be identified with the magnetic charge of the soliton. The simplest and basic soliton is the magnetic monopole ([135; 275]). It is a generalization of the self-dual Taub–NUT solution ([203; 312]) and is described by the following metric:

$$\begin{aligned} ds^2 &= -dt^2 + V(r)[dx^4 + \vec{A}(\vec{r})d\vec{r}\,]^2 \\ &\quad + V^{-1}(r)(dr^2 + r^2 d\theta^2 + r^2 \sin^2\theta\, d\phi^2)\,, \end{aligned}$$

where $\vec{r}$ denotes a three-vector $\vec{r} = (r, \theta, \phi)$, the gauge field $\vec{A}$ is that of a monopole

$$\vec{B} = \vec{\nabla} \times \vec{A} = \frac{g}{r^3}\vec{r} \tag{8.38}$$

and the function $V(r)$ is

$$V(r) = \frac{r}{g + r}.$$

The trivial term $-dt^2$ corresponding to time was added, converting the four-dimensional Newman–Unti–Tamburino line element into a static solution of the five-dimensional vacuum Einstein's equations. There would appear to be a coordinate singularity when $r = 0$, where the Killing vector field has a fixed point. This is the so-called NUT singularity and it is absent if x^4 is periodic with period $4\pi g$ ([203]). From (8.38) it is clear that the gauge field A_μ is the one of a monopole and has a Dirac string singularity running from $r = 0$ to ∞. As usual this singularity is an artifact if and only if the period of x^4 is equal to $4\pi g$.

Bibliography

1 Alekseevskii, D. (1968). Riemannian spaces with exceptional holonomy groups. *Funct. Anal. Appl.*, **2**, 97–105.

2 Alekseevskii, D. (1987). Homogeneous Einstein metrics. In *Differential Geometry and its applications, Proc. Conf. Brno*, Czech., Commun., (ed. Krupka, D., Svec, A.), 1-21.

3 Alexandrov, B., Grantcharov, G., Ivanov, S. (1995). Curvature properties of twistor spaces of quaternionic Kähler manifolds. *J. Geom.*, **62**, 1–12.

4 Aloff, S., Wallach, N.R. (1975). An infinite family of distinct 7-manifolds admitting positively curved Riemannian structures. *Bull. Amer. Math. Soc.*, **81**, 93–97.

5 Ambrose, W., Singer, M. (1958). On homogeneous Riemannian manifolds. *Duke Math. J.*, **25**, 647–669.

6 Atiyah, M.F. (1978). *Geometry of Yang–Mills fields.* Lectures Notes in Physics, **80**, 216–221.

7 Atiyah, M.F. (1990). Hyper-Kähler manifolds. In *Complex Geometry and Analysis, Proceedings*, Pisa 1988, Lect. Notes Math.,(ed. Villani, V.), **1422**, Springer-Verlag Berlin, 1–13.

8 Atiyah, M.F., Hitchin, N.J., Singer, I.M. (1978). Self-duality in four-dimensional Riemannian Geometry. *Proc. R. Soc. London* Ser. A, **362**, 425–461.

9 Baditoiu, G. (2003). Semi-Riemannian submersions with totally geodesic fibres. *Tôhoku Math. J.*, (to appear).

10 Baditoiu, G., Buchner, K., Ianus, S. (1998). Some remarkable connections and semi-Riemannian submersions. *Bull. Math. Soc. Sci. Math. Roum.*, Nouv. Ser., **41**, 153–159.

11 Baditoiu, G., Ianus, S. (2002). Semi-Riemannian submersions with totally umbilic fibres. *Rend. Circ. Mat. Palermo*, Serie II, **51**, 249–276.

12 Baditoiu, G., Ianus, S. (2002). Semi-Riemannian submersions from real and complex pseudo-hyperbolic spaces. *Diff. Geom. and Appl.*, **16**, 79–84.

13 Baird, P., Eells, J. (1981). A conservation law for harmonic maps. In *Geometry, Proc. Symp.*, Utrecht 1980, Lect. Notes in Math., **894**, (ed. Looijenga, E., Siersma, D., Takens, F.), Springer-Verlag, Berlin, 1–25.

14 Baird, P., Wood, J.C. (2003). *Harmonic morphisms between Riemannian*

manifolds. London Math. Soc., Monographs, Oxford University Press.
15 Barletta, E., Dragomir, S. (1996). Submanifolds fibered in tori of a complex Hopf manifold. *5th Italian Conference on Integral Geometry, Geometric Probability and convex Bodies (Italian), Rend. Circ. Mat. Palermo*, II, suppl. **41**, 25–54.
16 Beem, J.K., Ehrlich, P.E., (1981). *Global Lorentzian Geometry*. Marcel Dekker, New York.
17 Bejancu, A., (1986). *Geometry of CR-submanifolds*. Mathematics and its applications, **23**, Dordrecht, D. Reidel Publishing Company, Kluwer Acad. Publ.
18 Bejancu, A., Duggal, K. (1996). *Lightlike submanifolds of Semi-Riemannian manifolds and applications*. **364**, Kluwer Academic Publishers, Dordrecht.
19 Bérard-Bergery, L. (1975). Sur certaines fibrations d'espaces homogenes Riemanniens. *Compositio Math.*, **30**, 43–61.
20 Bérard-Bergery, L. (1978). Sur la courbure des metriques Riemanniennes des groupes de Lie et des espaces homogenes. *Ann. Sci. Ec. Norm. Sup.*, IV Ser., **11**, no.4, 543–576.
21 Bérard-Bergery, L., Bourguignon, J.P. (1981). Laplacian and Riemannian submersions with totally geodesic fibres. In *Global Differential geometry and global analysis. Proc. Colloq.* (Berlin, 1979), 30–35, Lect. Notes Math., **838**, Springer-Verlag, Berlin.
22 Bérard-Bergery, L., Bourguignon, J.P. (1982). Laplacians and Riemannian submersions with totally geodesic fibres. *Ill. J. Math.*, **26**, 181–200.
23 Bérard-Bergery, L., Katz, M. (1994). On intersystolic inequalities in dimension 3. *Geom. Func. Anal.*, **4**, 621–632.
24 Berger, M. (1960). Sur quelques variétés Riemanniennes suffisament pincées. *Bull. Soc. Math. Fr.*, **88**, 57–71.
25 Berndt, J. (1997). Riemannian geometry of complex two-plane Grassmannians. *Rend. Semin. Mat. Torino*, **55**, no.1, 19–83.
26 Besse, A. (1987). *Einstein manifolds*. Springer-Verlag, Berlin.
27 Besson, G. (1986). A Kato type inequality for Riemannian submersions with totally geodesic fibers. *Ann. Global Anal. Geom.*, **4**, 273–289.
28 Besson, G., Bordoni, M. (1990). On the spectrum of Riemannian submersions with totally geodesic fibres. *Atti Accad. Naz. Lincei, Cl. Sci. Fis. Mat. Nat., IX Ser. Rend. Lincei, Mat. Appl.*, **1**, 335–340.
29 Bielawski, R. (1997). Betti numbers of 3-Sasakian quotients of spheres by tori. Bull. *London Math. Soc.*, **29**, No.6, 731–736.
30 Bishop, R.L. (1972). Clairaut submersions. In *Geometry in Honor of K.Yano, Kinokuniya*, Tokyo, 21–31.
31 Blair, D.E. (1967). The theory of quasi-Sasakian structures. *J. Differ. Geom.*, **1**, 331–345.
32 Blair, D.E. (1970). Geometry of manifolds with structural group $O(n) \times O(s)$. *J. Differ. Geom.*, **4**, 155–167.
33 Blair, D.E. (1976). *Contact manifolds in Riemannian Geometry*. Lecture Notes in Math., **509**, Springer-Verlag, Berlin.
34 Blair, D.E. (2002). *Riemannian Geometry of Contact and Symplectic Man-*

ifolds. Progress in Mathematics, **203**, Birkhäuser Boston.
35 Blair, D.E., Ludden, G., Yano, K. (1973). Differential geometric structures on principal toroidal bundles. *Trans. Am. Math. Soc.*, **181**, 175–184.
36 Blair, D.E., Ludden, G., Yano, K. (1976). Semi-invariant immersions. *Kodai Math. Sem. Rep.*, **27**, 313–319.
37 Blair, D.E, Vanhecke, L. (1987). Symmetries and φ-symmetric spaces. *Tôhoku Math. J.*, **39**, 373–383.
38 Blumenthal, R.A. (1984). Affine submersions and foliations of affinely connected manifolds. *C. R. Acad. Sci. Paris*, Ser. I, **299**, 1013–1015.
39 Blumenthal, R.A. (1987). Cartan submersions and Cartan foliations. *Ill. J. Math.*, **31**, 327–343.
40 Bonome Dopico, D., Castro Boleno, B. (1981). Quaternionic Kähler submersions. In *Proceedings of the Eighth Portuguese-Spanish Conference on Mathematics*, Vol. I, (Coimbra, 1981), 261–266.
41 Boothby, W. (1975). *An Introduction to Differentiable Manifolds and Riemannian Geometry.* Academic Press, New-York.
42 Boothby, W., Wang, H. (1958). On contact manifolds. *Ann. of Math.*, II Ser., **68**, 721–734.
43 Borel, A., Hirzebruch, F. (1958). Characteristic classes and homogeneous spaces, I. *Am. J. Math.*, **80**, 458–538.
44 Borisenko, A.A., Yampol'skij, A.L. (1991). Riemannian geometry of bundles. *Usp. Mat. Nauk.*, **46**, 51–95; translation in *Russ. Math. Surv.*, **46**, 55–106.
45 Bourbaki, N. (1968). *Groupes et algébre de Lie.* Fasc. **XXXIV**, Chap. 4, 5, 6, Hermann, Paris.
46 Bourbaki, N. (1975). *Groupes et algébre de Lie.* Fasc. **XXXVIII**, Chap. 7, 8, Hermann, Paris.
47 Bourguignon, J.P. (1980). Déformations des variétés d'Einstein. *Asterisque*, **80**, 21–31.
48 Bourguignon, J.P. (1989). A mathematician's visit to Kaluza-Klein theory. *Rend. Semin. Mat., Torino*, Fasc. Spec., 143–163.
49 Bourguignon, J.P., Lawson, H.B. (1981). Stability and isolation phenomena for Yang–Mills fields. *Commun. Math. Phys.*, **79**, 189–230.
50 Bourguignon, J.P. (1996). An introduction to geometric variational problems. In *Lectures on geometric variational problems. Sendai*, 1993, 1–41, Springer Tokyo.
51 Boyer, Ch., Galicki, K. (1999). 3-Sasakian Manifolds. In *Surveys in Differential Geometry: essays on Einstein Manifolds. Supplement to the Journal of Differential Geometry*, **6**, (ed. Le Brun, C., Wang, M.), International Press, Cambridge.
52 Boyer, Ch., Galicki, K., Mann, B. (1993). Quaternionic Reduction and Einstein manifolds. *Comm. Anal. Geom.*, **1**, 229–279.
53 Boyer, Ch., Galicki, K., Mann, B. (1994). The Geometry and Topology of 3-Sasakian manifolds. *J. Reine Angew. Math.*, **455**, 183–220.
54 Boyer, Ch., Galicki, K., Mann, B., Rees, E. (1998). Compact 3-Sasakian 7-manifolds with arbitrary second Betti number. *Invent. Math.*, **131**, 321–

344.
55 Cairns, G. (1986). A general description of totally geodesic foliations. *Tôhoku Math. J.* **II 38**, 37–55.
56 Capursi, M., Ianus, S. (1985). On the scalar curvature of a complex hypersurface of the product of two cosymplectic space-forms. *Tensor, New Ser.*, **42**, 131–136.
57 Castro,R., Hervella, L.M., Rozas, J.I. (1986). Almost contact submersions and $K_i - \varphi$-curvatures. *Riv. Mat. Univ. Parma*, IV Ser., **12**, 187–194.
58 Cheeger, J., Gromoll, D. (1972). On the structure of complete manifolds of nonnegative curvature. *Ann. of Math.*, II Ser., **96**, 413–443.
59 Chen, B.Y. (1973). *Geometry of submanifolds.* Marcel Dekker, New York.
60 Chen, B.Y. (1981). *Geometry of submanifolds and its applications.* Tokyo Science University of Tokyo.
61 Chen, B.Y., Vanhecke, L. (1989). Isometric, holomorphic and symplectic reflections. *Geom. Dedicata*, **29**, 259–277.
62 Cheng, T.P., Li, L.F. (1984). *Gauge theory of elementary particle physics.* Clarendon Press, Oxford.
63 Chinea, D. (1984). Quasi K-cosymplectic submersions. *Rend. Circ. Mat., Palermo*, II Ser., **33**, 319–330.
64 Chinea, D. (1985). On almost contact metric submersions. *An. Stiint. Univ. "Al.I. Cuza", Iasi*, Sect. I-a Mat. (N.S.), **31**, 77–88.
65 Chinea, D. (1985). Almost contact metric submersions. *Rend. Circ. Mat. Palermo*, II Ser., **34**, 89–104.
66 Chinea, D. (1987). Almost contact metric submersions and structure equations. *Publ. Math.*, **34**, 207–213.
67 Chinea, D. (1988). Transference of structures on almost complex contact metric submersions. *Houston J. Math.*, **14**, 9–22.
68 Chinea, D., Gonzales, C. (1990). A classification of almost contact metric manifolds. *Ann. Mat. Pura Appl.*, IV Ser., **CLVI**, 15–36.
69 Chinea, D., Marrero, J. (1992). Conformal changes of almost cosymplectic manifolds. *Rend. Mat. Appl.*, Ser. VII, **12**, 849–867.
70 Chinea, D., Marrero, J., Rocha, J. (1995). Almost contact submersions with total space a locally conformal cosymplectic manifold. *Ann. Fac. Sc. Toulouse*, IV Ser., Math., **4**, 473–517.
71 Cordero, L.A. (1989). Holomorphic principal torus bundles, curvature and compact complex submanifolds. In *Curvature Geometry. Proc. Workshop*, Lancaster/UK, (ed. Dodson, C.T.J.), 107–149.
72 Cordero, L.A., Dodson, C., de Leon, M. (1989). *Differential Geometry of frame Bundles.* Kluwer Acad. Publ., Dordrecht.
73 de Alfaro, V., Fubini, S., Furlan, G. (1978). Conformal invariance in field theory. *Differ. Geom. Math. math. Phys. II, Proc.* Bonn 1977, Lect. Notes Math., **676**, 255–293.
74 de Leon, M, Marrero, J.C. (1997). Compact cosymplectic manifolds with transversally positive definite Ricci tensor. *Rend. Mat. Appl.*, Ser. VII, **17**, 607–624.
75 Derdzinski, A. (2000). Einstein metrics in dimension four. In *Handbook of*

differential geometry, Vol. I, (ed. Dillen, F.), North-Holland, Amsterdam, 419–707.

76 Deshmukh, S., Ghazal, T., Hashem, H. (1992). Submersions of CR-submanifolds on an almost Hermitian manifold. *Yokohama Math. J.*, **40**, 45–57.

77 Dirac, P.A.M. (1931). Quantized singularities in the electromagnetic field. *Proc. R. Soc. London*, Ser. A, **133**, 60–72.

78 Dragomir, S., Ornea, L. (1998). *Locally conformal Kähler Geometry*. Birkhäuser, Boston.

79 Duggal, K., Ianus, S., Pastore, A.M. (2001). Maps interchanging f-structures and their harmonicity. *Acta Applicandae Mathematicae*, **67**, 91–115.

80 Eells, J., Lemaire, L. (1983). Selected topics in harmonic maps. *Reg. Conf. Ser. Math.*, **50**, Amer. Math. Soc., Providence, RI.

81 Eells, J., Ratto, A. (1993). *Harmonic maps and minimal immersions with symmetries. methods of ordinary differential equations applied to elliptic variational problems.* Annals of Mathematics Studies, **130**, Princeton University Press, Princeton.

82 Eells, J., Sampson, J.H. (1964). Harmonic mappings of Riemannian manifolds. *Am. J. Math.*, **86**, 109–160.

83 Eguchi, T., Gilkey, P.K., Hanson, A.J. (1980). Gravitation, gauge theories and Differential Geometry. *Phys. Rep.*, **66**, 213–393.

84 Eschenburg, J.H. (1992). Inhomogeneous spaces of positive curvature. *Diff. Geom. and Appl.*, **2**, 123–132.

85 Escobales, R.H.Jr. (1973). Submersions from spheres. *Bull. Am. Math. Soc.*, **79**, 71–74.

86 Escobales, R.H.Jr. (1975). Riemannian submersions with totally geodesic fibers. *J. Differ. Geom.*, **10**, 253–276.

87 Escobales, R.H.Jr. (1978). Riemannian submersions from complex projective space. *J. Differ. Geom.*, **13**, 93–107.

88 Escobales, R.H.Jr. (1982). Sufficient conditions for a bundle-like foliation to admit a Riemannian submersion onto its leaf space. *Proc. Am. Math. Soc.*, **84**, 280–284.

89 Escobales, R.H.Jr. (1982). The integrability tensor for bundle-like foliations. *Trans. Am. Math. Soc.*, **270**, 333–339.

90 Escobales, R.H.Jr. (1985). Riemannian foliations of the rank one symmetric spaces. *Proc. Am. Math. Soc.*, **95**, 495–498.

91 Escobales, R.H.Jr., Parker, P. (1988). Geometric consequences of the normal curvature cohomology class in umbilic foliations. *Indiana Univ. Math. J.*, **37**, 389–408.

92 Falcitelli, M. (2003). Some classes of almost contact metric manifolds and contact Riemannian submersions. *Acta Math. Hungar.*, to appear.

93 Falcitelli, M., Ianus, S., Pastore, A.M., Visinescu, M. (2002). Some applications of Riemannian submersions in physics. In *Proceeding of the First National Romanian Conference of Theoretical Physics*, (ed. Green, D., Visinescu, A.), to appear on *Romanian Journal of Physics*, vol. **48**, numbers 4-5, (2003). (elect. version www.theory.nipne.ro/ctp2002/proced/proced.html).

94 Falcitelli, M., Pastore, A.M. (2000). A note on almost Kähler and nearly Kähler submersions. *J. Geom.*, **69**, 79–87.

95 Farafanova, N.K. (1995). Riemannian metrics on fiber bundles, I. *Russian Mathematics*, (Iz.VUZ), **39**, 65–77.

96 Farafanova, N.K. (1996). Riemannian metrics on fiber bundles, II - The metric of the Grassmann bundle. *Russian Mathematics*, **40**, 59–68.

97 Feres, R. (1993). The center foliation of an affine diffeomorphism. *Geom. Dedicata*, **46**, 233–238.

98 Fernandez, M., Gray, A. (1982). Riemannian Manifolds with Structure Group G_2. *Ann. Mat. Pura Appl.*, IV Ser. **132**, 19–45.

99 Fischer, A.E. (1996). Riemannian submersions and the regular interval theorem of Morse theory. *Ann. Global Anal. Geom.*, **14**, 263–300.

100 Freund, P.G.C., Rubin, M.A. (1980). Dynamics of dimensional reduction. *Phys. Lett.*, **B97**, 233–235.

101 Fueki, S., Yamaguchi, S. (1997). Kählerian torse forming vector fields and Kählerian submersions. *SUT J. of Math.*, **33**, 257–275.

102 Fuglede, B. (1996). Harmonic morphisms between Semi-Riemannian manifolds. *Ann. Acad. Sci. Fenn. Math.*, **21**, n.1, 31–50.

103 Galicki, K., Salamon, S. (1996). Betti numbers of 3-Sasakian manifolds. *Geom. Dedicata*, **63**, 1, 45–68.

104 Garcia-Rio, E., Vasquez Abal, M.E. (1995). On the quaternionic sectional curvature of an indefinite quaternionic Kähler manifold. *Tsukuba J. Math.*, **19**, 2, 273–284.

105 Garcia-Rio, E., Kupeli, D. (1999). *Semi-Riemannian maps and their Applications.* Mathematics and its Applications, **475**, Kluwer Acad. Publ., Dordrecht.

106 Gauduchon, P. (1977). Le théorème de l'excentricitè nulle. *C.R.Acad. Sci. Paris*, Ser. A, **285**, 387–390.

107 Gauduchon, P. (1984). La 1-forme de torsion d'une variété hermitienne compacte. *Math. Ann.*, **267**, 495–518.

108 Gauduchon, P. (1995). Structures de Weyl-Einstein, espaces de twisteurs et variétés de type $S^1 \times S^3$. *J. Reine Angew. Math.*, **469**, 1–50.

109 Gauduchon, P. (ed) (1998). *Harmonic Mappings, Twistors and σ-models.* 9-13 June 1986, CIRM, Luming, France, Advanced Series in Math. Physics, World Sci. Publ. Singapore.

110 Geiges, H., Thomas, C.B. (1995). Hypercontact manifolds. *J. Lond. Math. Soc.*, II Ser., **51**, 342–352.

111 Gell-Mann, M., Zwiebach, B. (1984). Spacetime compactification induced by scalars. *Phys. Lett.*, **B, 141**, 333–336.

112 Gell-Mann, M., Zwiebach, B. (1985). Dimensional reduction of spacetime induced by nonlinear scalar dynamics and noncompact extra dimensions. *Nucl. Phys.*, **B260**, 569–592.

113 Ghika, G. (1986). Harmonic maps and submersions in local Euclidean gravity coupled to the σ-model. *Rev. Roumanie, Phys.*, **31**, no.7, 635–648.

114 Gilkey, P.B., Leahy, J.V., Park, J.H. (1997). The spectral geometry of Riemannian submersions. In *11th Yugoslav Geometrical Seminar*, (Divčibare,

1996), b. Rad. at. Inst. Beograd (N.S.), **6 (14)**, 36–54.

115 Gilkey, P.B., Leahy, J.V., Park,J.H. (1998). *Spinors, spectral geometry and Riemannian submersions.* Lecture Notes Series Seoul, **40**, Korea. Seoul National University, Seoul.

116 Gilkey, P.B., Leahy, J.V., Park,J.H. (1999). *Spectral geometry, Riemannian submersions and the Gromov-Lawson conjecture.* Studies in Advanced Mathematics, Boca Raton, Fl. Chapman Hall.

117 Gilkey, P.B., Park, J.H. (1995). Eigenvalues of the Laplacian and Riemannian submersions. *Yokohama Math. J.*, **43**, 7–11.

118 Gilkey, P.B., Park, J.H. (1996). Riemannian submersions which preserve the eigenforms of the Laplacian. *Ill. J. Math.*, **40**, 194–201.

119 Gluck, H., Warner, F., Ziller, W. (1986). The geometry of the Hopf fibrations. *Enseign. Math.*, II Sér., **32**, 173–198.

120 Gluck, H., Warner, F., Ziller, W. (1987). Fibrations of spheres by parallel great spheres and Berger's rigidity theorem. *Ann. Global Anal. Geom.*, **5**, 53–82.

121 Goldberg, S.I. (1972). A generalization of Kähler Geometry. *J. Differ. Geom.*, **6**, 343–355.

122 Goldberg, S.I., Ishihara, T. (1978). Riemannian submersions commuting with the Laplacian. *J. Differ. Geom.*, **13**, 139–144.

123 Goldberg, S.I., Yano, K. (1970). On normal globally framed f-manifolds. *Tôhoku Math. J.*, II Ser., **22**, 362–370.

124 Gónzalez-Dávila, J.C., Vanhecke, L. (1998). New examples of weakly symmetric spaces. *Monatsh. Math.*, **125**, 309–314.

125 Gray, A. (1965). Minimal varieties and almost Hermitian manifolds. *Mich. Math. J.*, **12**, 273–287.

126 Gray, A. (1967). Pseudo-Riemannian almost product manifolds and submersions. *J. Math. Mech.*, **16**, 715–737.

127 Gray, A. (1969). Vector cross products on manifolds. *Trans. Amer. Math. Soc.*, **141**, 465–504, (1970). Correction. **148**, 625.

128 Gray, A. (1976). The structure of Nearly Kähler manifolds. *Math. Ann.*, **223**, 233–248.

129 Gray, A. (1976). Curvature identities of Hermitian and almost Hermitian manifolds. *Tôhoku Math. J.*, II Ser., **28**, 601–612.

130 Gray, A., Hervella, L.M. (1980). The sixteen classes of almost Hermitian manifolds. *Ann. Mat. Pura Appl.*, **123**, 35–58.

131 Gray, J. (1959). Some global properties of contact structures. *Ann. of Math.*, II Ser., **69**, 421–450.

132 Green, M.B., Schwarz, J.H., Witten, E. (1987). *Superstring Theory.* Volume 1 *Introduction*, Volume 2 *Loop amplitudes, anomalies and phenomenology.* Cambridge Univ. Press, Cambridge.

133 Gromoll, D., Grove, K. (1988). The low-dimensional metric foliations of Euclidean spheres. *J. Differ. Geom.*, **28**, 143–156.

134 Gromoll, D., Meyer, W.T. (1974). An exotic sphere with nonnegative sectional curvature. *Ann. of Math.*, **100**, 401–406.

135 Gross, D.J., Perry, M.J. (1983). Magnetic monopoles in Kaluza-Klein the-

ory. *Nucl. Phys.*, **B 226**, 29–48.

136 Gudmundsson, S. (1990). On the geometry of harmonic morphisms. *Math. Proc. Camb. Philos. Soc.*, **108**, 461–466.

137 Gudmundsson, S., Wood, J. (1997). Harmonic morphisms between almost Hermitian manifolds. *Boll. U.M.I*, VII Ser., **11-B**, Suppl. fasc. 2, 185–197.

138 Hawking, S.W. (1977). Gravitational instantons. *Phys. Lett.*, **60A**, 81–85.

139 Helgason, S. (1978). *Differential Geometry, Lie groups and Symmetric spaces.* Academic Press, New York.

140 Hermann, R. (1960). A sufficient condition that a mapping of Riemannian manifolds be a fibre bundle. *Proc. Am. Math. Soc.*, **11**, 236–242.

141 Hernandez, G. (1996). On hyper f-structures. *Math. Ann.*, **306**, 205–230.

142 Higa, T. (1993). Weyl manifolds and Einstein-Weyl manifolds. *Comment. Math. Univ. St. Pauli*, **42**, 143–160.

143 Hitchin, N.J. (1992). Hyper-Kähler manifolds. *Séminaire Bourbaki*, Vol. 1991/92, *Astérisque* **206**, Exp. No. **748**, 137–166.

144 Hogan, P.A. (1984). Kaluza-Klein theory derived from a Riemannian submersion. *J. Math. Phys.*, **25**, 2301–2305.

145 Ianus, S. (1983). *Geometria diferentiala cu aplicatii in teoria relativitatii.* Editura Academiei RSR, Bucharest.

146 Ianus, S. (1994). Submanifolds of almost Hermitian manifolds. *Riv. Mat. Univ. Parma*, V Ser., **3**, 123–142.

147 Ianus, S., Ornea, L., Vutulescu, V. (1995). Holomorphic and harmonic maps of locally conformal Kähler manifolds. *Boll. U.M.I.*, VII Ser., **9-A**, 569–579.

148 Ianus, S., Pastore, A.M., Visinescu, M. (1998). Riemannian Submersions. Recent results relevant to Mathematical Physics. *Analele Univ. Vest. din Timisoara, Theoretical Mathematical and Computational Physics*, **1**, 1–14.

149 Ianus, S., Visinescu, M. (1986). Spontaneous compactification induced by non-linear scalar dynamics, gauge fields and submersions. *Class. Quantum Grav.*, **3**, 889–896.

150 Ianus, S., Visinescu, M. (1987). Kaluza-Klein theory with scalar fields and generalized Hopf manifolds. *Class. Quantum Grav.*, **4**, 1317–1325.

151 Ianus, S., Visinescu, M. (1991). Space-time compactification and Riemannian submersions. In *The mathematical heritage of C.F. Gauss*, (ed. Rassias, G.), World Sci. Publishing, River Edge NJ, 358–371.

152 Ishihara, S. (1973). Quaternion Kählerian manifolds and fibered Riemannian spaces with Sasakian 3-structure. *Kodai Math. Sem. Rep.*, **25**, 321–329.

153 Ishihara, S. (1974). Quaternion Kählerian manifolds. *J. Diff. Geom.*, **9**, 483–500.

154 Ishihara, S., Konishi, M. (1972). Fibred Riemannian spaces with Sasakian 3-structure. *Collection Diff. Geometry in honor of K. Yano*, Kinokuniya - Tokyo, 179–194.

155 Jelonek, W, (2001). Positive and negative $3-K$-contact structures. *Proc. Am. Math. Soc.*, **129**, 247–256.

156 Jensen, G.R. (1973). Einstein metrics on principal fibre bundles. *J. Diff. Geom.*, **8**, 599–614.

157 Johnson, D.L. (1980). Kähler submersions and holomorphic connections. *J. Diff. Geom.*, **15**, 71–79.
158 Johnson, D.L., Whitt, L.B. (1980). Totally geodesic foliations. *J. Diff. Geom.*, **15**, 225–235.
159 Jost, J. (1995). *Riemannian geometry and geometric analysis.* Springer-Verlag, Berlin.
160 Kaluza, T. (1921). Zum Unitätsproblem der Physic. *Sitzber. Preuss. Akad. Wiss. Kl. Berlin, Math. Phys.*, **2**, 966–970.
161 Kamishima, Y. (1994). Uniformization of Kähler manifolds with vanishing Bochner tensor. *Acta Math.*, **172**, 299–308.
162 Kamishima, Y. (1999). Geometric rigidity of spherical hypersurfaces in quaternionic manifolds. *Asian J. Math.*, **3**, 519–559.
163 Karcher, H. (1999). Submersions via Projections. *Geom. Dedicata*, **74**, No.3, 249–260.
164 Kashiwada, T. (2001). On a contact 3-structure. *Math. Zeit.* **238**, 829–832.
165 Kasue, A., Mendori, A. (1998). Riemannian submersions and isoperimetric inequalities. *Geom. Dedicata*, **70**, 27–47.
166 Kenmotsu, K. (1972). A class of almost contact Riemannian manifolds. *Tôhoku Math. J.*, **24**, 93–103.
167 Kerr, M. (1996). Some new homogeneous Einstein metrics on symmetric spaces. *Trans. Am. Math. Soc.*, **348**, 153–171.
168 Klein, O. (1929). Quantentheorie und fünf-dimensionale Relativitätstheorie. *Z. Phys.*, **37**, 895–901.
169 Klingenberg, W. (1982). *Riemannian Geometry.* de Gruyter Studies in Mathematics, **1**, Walter de Gruyter, Berlin-New York.
170 Kobayashi, S. (1963). Topology of positively pinched Kähler manifolds. *Tôhoku Math. J.*, **15**, 121–139.
171 Kobayashi, S. (1987). Submersions of CR submanifolds. *Tôhoku Math. J.*, II Ser., **39**, 95–100.
172 Kobayashi, S., Nomizu, K. (1963), (1969). *Foundations of Differential Geometry.* Vol. I, II, Interscience Publishers, New-York.
173 Koda, T., Watanabe, Y. (1997). Homogeneous almost contact Riemannian manifolds and infinitesimal models. *Boll. Un. Mat. Ital.*, B (7) **11**, no.2 Suppl., 11–24.
174 Konishi, M. (1975). On manifolds with Sasakian 3-structure over quaternionic Kählerian manifolds. *Kodai Math. Sem. Rep.*, **26**, 194–200.
175 Kowalski, O. (1971). Curvature of the induced Riemannian metric on the tangent bundle of a Riemannian manifold. *J. Reine Angew. Math.*, **250**, 24–129.
176 Kowalski, O. (1980). *Generalized Symmetric Spaces.* Lect. Notes Math., **805**, Springer-Verlag Berlin.
177 Kowalski, O., Tricerri, F. (1987). Riemannian manifolds of dimension $n \leq 4$ admitting a homogeneous structure of class T_2. *Conf. Sem. Mat. Univ. Bari*, **222**.
178 Kuo, Y. (1970). On almost contact 3-structure. *Tôhoku Math. J.*, **22**, 325–332.

179 Kupeli, D.M. (1995). The eikonal equation of an indefinite metric. *Acta Appl. Math.*, **40**, 245–253

180 Kupeli, D.M. (1999). On semi-Riemannian submersions. In *New Developments in Differential Geometry, Proceedings of the Conference Budapest*, Hungary, 1996, (ed. Szenthe J.), Kluwer Academic Publishers, 203–211.

181 Kwon, J.H., Kim, B.H., Kim, N.G. (1994). Fibred Riemannian spaces with critical Riemannian metrics. *J. Korean Math. Soc.*, **31**, 205–211.

182 Kwon, J.H., Suh, Y.J. (1997). On semi-Riemannian submersions of codimension one. *Math. J. Toyama Univ.*, **20**, 15–35.

183 Kwon, J.H., Suh, Y.J. (1997). On sectional and Ricci curvatures of semi-Riemannian submersions. *Kodai Math. J.*, **20**, 53–66.

184 Lawson, H.B., Michelson, M.L. (1989). *Spin Geometry.* Mathematical Series, **38**, Princeton University Press, Princeton.

185 Le Brun, C., Wang, M. (ed) (1999). *Surveys in Differential Geometry: essays on Einstein Manifolds.* Supplement to the Journal of Differential Geometry, **6**, International Press, Cambridge.

186 Libermann, P. (1959). Sur les automorphismes infinitèsimaux des structures symplectiques et des structures de contact. In *Colloq. Geometrie Differentielle Globale*, (Bruxelles, 1958), Louvain, 37–59.

187 Ludden, G. (1970). Submanifolds of cosymplectic manifolds. *J. Diff. Geom.*, **4**, 237–244.

188 Lutz, R. (1974). Structures de contact en codimension quelconque. In *Geom. Differ. Colloqu., Santiago de Compostela.* Lect. Notes Math., (ed. Vidal, E.), **392**, 23–36, Springer Berlin.

189 Magid, M.A. (1981). Submersions from anti-de Sitter space with totally geodesic fibers. *J. Diff. Geom.*, **16**, 323–331.

190 Mamone Capria, M., Salamon, S. (1988). Yang–Mills fields on quaternionic spaces. *Nonlinearity*, **1**, 517–530.

191 Marathe, K.B., Martucci, G. (1992). *The mathematical foundations of gauge theories.* North-Holland, Amsterdam.

192 Marchiafava, S. (1991). On the local geometry of a quaternionic Kähler manifold. *Boll. Un. Mat. Ital.*, Ser.VII, **B5**, 417–447.

193 Marchiafava, S., Pontecorvo, M., Piccinni, P. (ed) (2001). *Proceedings of the 2nd meeting on quaternionic structures in mathematics and physics*, Roma 1999. Dedicated to the memory of A. Lichnerowicz and E. Martinelli, Dip. Matematica "G. Castelnuovo", Roma "La Sapienza", Dip. Matematica Roma tre, 469 (electronic).

194 Marrero, J.C. (1992). The local structure of trans-sasakian manifolds. *Ann. Mat. Pura Appl.*, IV Ser., **CLXII**, 77–86.

195 Marrero, J.C., Rocha, J. (1994). Locally conformal Kähler submersions. *Geom. Dedicata*, **52**, 271–289.

196 Marrero, J.C., Rocha, J. (1995). Complex conformal submersions with total space a locally conformal Kähler manifold. *Publ. Math. Debrecen*, **47**, 335–348.

197 Matsuzawa, T. (1983). Einstein metrics and fibred Riemannian structures. *Kodai Math. J.*, **6**, 340–345.

198 Matzeu, P. (2002). Almost contact Einstein–Weyl structures. *Manuscr. Math.*, **108**, 275–288.
199 Matzeu, P., Ornea, L. (2000). Local almost contact metric 3-structures. *Publ. Math.*, **57**, 3-4, 499–508.
200 Mc Innes, B. (1987). On the splitting problem in higher dimensional field theories. *Class. Quantum Grav.*, **4**, 329–342.
201 Minami, M. (1979). Dirac's monopole and the Hopf map. *Prog. Theor. Phys.*, **62**, 1128–1142.
202 Miron, R., Anastasiei, M. (1997). *Vector bundles and Lagrange spaces with applications to relativity.* Balkan Society of Geometers. Monographs and Textbooks, vol 1.
203 Misner, C.W. (1980). The flatter regions of Newman, Unti and Tamburino's generalized Schwarzschild space. *J. Math. Phys.*, **4**, 924–937.
204 Molino, P. (1988). *Riemannian foliations.* Progress in Math., **73**, Birkhäuser, Boston.
205 Monar, D. (1987). 3-estructuras casi contacto. Doctoral Thesis, Ser. de Pubblic. Univ. de La Laguna.
206 Morimoto, A. (1964). On normal almost contact structure with a regularity. *Tôhoku Math. J.*, **16**, 90–104.
207 Moroianu, A. (1998). Spin^c manifolds and complex contact structures. *Comm. Math. Phys.*, **193**, 661–674.
208 Murakami, S. (1959). Sur certain espaces fibrés principaux différentiable et holomorphes. *Nagoya Math. J.*, **15**, 171–199.
209 Musso, E. (1989). Submersioni localmente conformemente Kähleriane. *Boll. Un. Mat. Ital.*, (7), **3-A**, 171–176.
210 Musso, E., Tricerri, F. (1988). Riemannian metrics on tangent bundles. *Ann. Mat. Pura Appl.*, IV Ser., **150**, 1–20.
211 Mustafa, M.T. (2000). Applications of harmonic morphisms to gravity. *J. Math. Phys.*, **41**, n. 10, 6918–6929.
212 Muto, Y. (1952). On some properties of a fibred Riemannian manifold. *Semin. Rep.Yokohama*, Yokohama Nat. Univ., **1**, 1–14.
213 Nagano, T. (1960). On fibred Riemannian manifolds. *Sci. Pap. Coll. Gen. Educ. Univ. Tokyo*, **10**, 17–27.
214 Nagy, P.T. (1983). Non-horizontal geodesics of a Riemannian submersion. *Acta Sci. Math.*, **45**, 347–355.
215 Narita, F. (1992). Riemannian submersions and isometric reflections with respect to submanifolds. *Math. J. Toyama Univ.*, **15**, 83–94.
216 Narita, F. (1993). Riemannian submersions with isometric reflections with respect to the fibers. *Kodai Math. J.*, **16**, 416–427.
217 Narita, F. (1996). CR-submanifolds of locally conformal Kaehler manifolds and Riemannian submersions. *Colloq. Math.*, **70**, 165–179.
218 Narita, F. (1997). Riemannian submersions from k-generalized Hopf manifolds. *Kyungpook Math. J.*, **37**, 355–364.
219 Narita, F. (1997). Riemannian submersions and Riemannian manifolds with Einstein Weyl structures. *Geom. Dedicata*, **65**, 103–116.
220 Narita, F. (1998). Einstein–Weyl structures on almost contact metric man-

ifolds. *Tsukuba J. Math.*, **22**, 87–98.

221 Nash, J.C. (1979). Positive Ricci curvature on fibre bundles. *J. Diff. Geom.*, **14**, 241–254.

222 Nishiyama, M. (1984). Classification of invariant complex structures on irreducible compact simply connected coset spaces. *Osaka J. Math.*, **21**, 39–58.

223 Nishikawa, S., Tondeur, Ph., Vanhecke, L. (1992). Spectral geometry for Riemannian foliations. *Ann. Global Anal. Geom.*, **10**, No.3, 291–304.

224 Ogiue, K. (1965). On fiberings of almost contact manifolds. *Kodai Math. Sem. Rep.*, **17**, 53–62.

225 Omero, C., Percacci, R. (1980). Generalized nonlinear sigma models in curved space and spontaneus compactification. *Nuclear Phys.*, **B, 165**, 351–364.

226 O'Neill, B. (1966). The fundamental equations of a submersion. *Mich. Math. J.*, **13**, 459–469.

227 O'Neill, B. (1967). Submersions and geodesics. *Duke Math. J.*, **34**, 363–373.

228 O'Neill, B. (1983). *Semi-Riemannian Geometry with Applications to Relativity.* Pure and Applied Math., **103**, Academic Press, New-York.

229 Oproiu, V. (1999). A generalization of natural almost Hermitian structures on the tangent bundles. *Math. J. Toyama Univ.*, **22**, 1–14.

230 Ornea, L., Piccinni, P. (1997). Locally conformal Kähler structures in quaternionic Geometry. *Trans. Am. Math. Soc.*, **349**, 641–655.

231 Ornea, L., Piccinni, P. (1999). Induced Hopf bundles and Einstein metrics. In *New developments in differential geometry. Proceedings of the Conference Budapest*, Hungary, 1996 295–305, (ed. Zsenthe, J.), Kluwer Acad. Publ., Dordrecht.

232 Ornea, L., Romani, G. (1993). The fundamental equations of conformal submersions. *Beiträge. Algebra Geom.*, **34**, n.2, 233–243.

233 Oubiña, J.A. (1985). New classes of almost contact metric structures. *Publ. Math. Debrecen*, **32**, 187–193.

234 Paiva, J.C.Alvarez, Duran, C.E. (2001). Isometric submersions of Finsler manifolds. *Proc. Am. Math. Soc.*, **129**, No.8, 2409–2417.

235 Palais, R. (1957). A global formulation of the Lie theory of transformation groups. *Mem. Am. Math. Soc.*, **22**, 1–123.

236 Pantilie, R. (1998). On the fundamental equations of a Riemannian submersion. *Rev. Roum. Math. Pures Appl.*, **43**, No. 7-8, 779–783.

237 Papaghiuc, N. (1989). Submersions of semi-invariant submanifolds of a Sasakian manifold. *An. Stiint. Univ. "Al.I. Cuza" Iasi*, Sect. I-a, Mat., **35**, 281–288.

238 Pastore, A.M. (1989). On the homogeneous Riemannian structures of type $T_1 \oplus T_3$. *Geom. Dedicata*, **30**, 235–246.

239 Pastore, A.M., Verroca, F. (1991). Some results on the homogeneous Riemannian structures of class $T_1 \oplus T_2$. *Rend. Mat. Appl.*, VII Ser., Roma, **11**, 105–121.

240 Pastore, A.M., Verroca, F. (1993). Alcuni risultati sulle strutture Riemanniane omogenee di classe $T_1 \oplus T_2 \oplus T_3$ con 1-forma fondamentale chiusa. *Rapporto*, Dip. di Matem. Bari.

241 Pedersen, H., Poon, Y., Swann, A. (1993). The Einstein–Weyl equations in complex and quaternionic Geometry. *Differential Geom. Appl.*, **3**, 309–321.
242 Pedersen, H., Swann, A. (1993). Riemannian submersions, four-manifolds and Einstein-Weyl geometry. *Proc. Lond. Math. Soc.*, III Ser., **66**, 381–399.
243 Perelman, G. (1994). Proof of the soul conjecture of Cheeger and Gromoll. *J. Differential Geom.*, **40**, 209–212.
244 Poor, W.A. (1981). *Differential geometric structures.* McGraw-Hill, New York.
245 Popovici, I., Turtoi, A. (1976). Sur une classe d'espaces de Einstein. *Ann. Mat. Pura Appl.*, **109**, 117–133.
246 Ranjan, A. (1985). Riemannian submersions of spheres with totally geodesic fibres. *Osaka J. Math.*, **22**, 243–260.
247 Ranjan, A. (1986). On a remark of O'Neill. *Duke Math. J.*, **53**, 113–115.
248 Ranjan, A. (1986). Riemannian submersions of compact simple Lie groups with connected totally geodesic fibres. *Math. Z.*, **191**, 239–246.
249 Ranjan, A. (1986). Structural equations and an integral formula for foliated manifolds. *Geom. Dedicata*, **20**, 85–91.
250 Reckziegel, H. (1985). Horizontal lifts of isometric immersions into the bundle space of a pseudo-Riemannian submersion. In *Global diff. geom. and global analysis*, (Berlin, 1984), 264–279, (ed. Ferus, P., Gardner, R.B., Helgason, S., Simon, U.), Lect. Notes Math., **1156**, Springer, Berlin - New York.
251 Reckziegel, H. (1988). A correspondence between horizontal submanifolds of Sasakian manifolds and totally real submanifolds of Kähler manifolds. In *Topics in Differential Geometry*, Debrecen, (Hungary), 1984, (ed. Szenthe, J., Tamassy, I.), *Colloq. Math. Soc. János Bolyai*, **46**, 1063–1081.
252 Rizza, G.B. (1974). Varietà parakähleriane. *Ann. Mat. Pura Appl.*, IV Ser. **98**, 47–61.
253 Rovenskii, V.Y. (1994). Classes of submersions of Riemannian manifolds with compact fibres. *Siberian Math. J.*, **35**, no.5, 1027–1035, translated from *Sibirsk. Mat. Zh.*, **35**, no.5, 1154–1164.
254 Rovenskii, V.Y. (1998). *Foliations on Riemannian Manifolds and Submanifolds.* Birkhäuser Boston.
255 Ryder, L.H. (1980). Dirac monopoles and the Hopf Map $S^3 \to S^2$. *J. Phys.* A, **13**, 437–447.
256 Sachs, R.K., Wu, H. (1977). *General relativity for Mathematicians*, Springer-Verlag, New York.
257 Sakamoto, K. (1974). On the topology of quaternionic Kähler manifolds. *Tôhoku Math. J.*, **26**, 389–405.
258 Sakane, Y. (1993). Homogeneous Einstein metrics on principal circle bundle. In *Complex Geometry, Proceedings of the Osaka Mathematical Conference* (Osaka, 1990), (ed. Komatsu, G., Sakane, Y.), Lect. Notes Pure Appl. Math., **143**, Marcel Dekker, New York, 161–178.
259 Sakane, Y. (1993). Homogeneous Einstein metrics on a principal circle bundle, II. In *Differential geometry, Proceedings of the Symposium in honour of Professor Su Buchin.* (Shangai, China, 1991), (ed. Gu, C.H., Hu, H.S.,

Xin, Y.L.), World Sc., Singapore, 177–186.
260 Salam, A., Strathdee, J. (1982). On Kaluza-Klein Theory. *Annals of Physics*, **141**, 316–351.
261 Salamon, S. (1989). *Riemannian Geometry and holonomy Groups.* Pitman Research Notes in Math., **201**, Longman Sc. Techn., New York.
262 San Martin, L., Negreiros, C. (2003). Invariant almost Hermitian structures on flag manifolds. *Adv. Math.*, **178**, 277–310.
263 Sasaki, S. (1958). On the differential geometry of tangent bundles of Riemannian manifolds. *Tôhoku Math. J.*, II Ser., **10**, 338–354.
264 Sasaki, S. (1965). *Almost Contact Manifolds.* Lecture Notes, Mathematical Institute, Tôhoku University, Vol.1.
265 Sasaki, S. (1966). On almost contact manifolds. In *Proc. United States - Japan, Semin. Diff. Geom.*, Kyoto 1965, 128–136.
266 Sasaki, S. (1985). *Selected Papers.* Shun-ichi Tachibana (ed) Kinokuniya Co. Ltd., Tokyo.
267 Sasaki, S., Hatakeyama, Y. (1961). On differentiable manifolds with certain structures which are closely related to almost contact structures. *Tôhoku Math. J.*, II Ser., **13**, 281–294.
268 Schimming, R., Hirschmann, T. (1988). Harmonic maps from spacetime and their coupling to gravitation. *Astronom. Nachr.*, **309**, 311-321.
269 Sekigawa, K. (1987). On some compact Einstein almost Kähler manifolds. *J. Math. Soc. Japan*, **39**, 677–684.
270 Sekizawa, M. (1991). Curvatures of tangent bundles with Cheeger-Gromoll metric. *Tokyo J. Math.*, **14**, 407–417.
271 Shahid, A., Husain, S.I. (1991). Submersions of CR-submanifolds of a nearly Kähler manifold I. *Rad. Mat.*, **7**, 197–205.
272 Smith, J.W. (1968). An exact sequence for submersions. *Bull. Am. Math. Soc.*, **74**, 233–236.
273 Smith, J.W. (1969). Submersions of codimension 1. *J. Math. Mech.*, **18**, 437–443.
274 Solovév, A.F. (1982). Second fundamental form of a distribution. *Math. Notes*, **31**, 71–75, transl. from *Mat. Zametki*, **31**, 139–146.
275 Sorkin, R.D. (1983). Kaluza-Klein monopole. *Phys. Rev. Lett.*, **51**, 87–90.
276 Steenrod, M. (1951). *The topology of fibre bundles.* Princeton University Press, Princeton.
277 Stepanov, S.E. (1995). On the global theory of projective mappings. *Math. Notes*, **58**, 752–756, transl. from *Mat. Zametki*, **58**, 1995, 111–118.
278 Stepanov, S.E. (1996). $O(n) \times O(m-n)$-structures on m-dimensional manifolds and submersions of Riemannian manifolds. *St. Petersb. Math. J.*, **7**, 1005–1016, transl. from *Algebra i Analiz*, **7**, 1995, 188–204.
279 Strake, M., Walschap, G. (1989). Σ-flat manifolds and Riemannian submersions. *Manuscr. Math.*, **64**, 213–226.
280 Suh, Y.J., Kwon, J.H. (1998). On geodesics and semi-Riemannian submersions with totally geodesic fibers. *Math. J. Toyama Univ.*, **21**, 67–86.
281 Takahashi, T. (1969). Sasakian manifold with pseudo-Riemannian metric. *Tôhoku Math. J.*, **21**, 271–290.

282 Takahashi, T. (1977). Sasakian φ-symmetric space. *Tôhoku Math. J.*, **29**, 97–113.

283 Taniguchi, T. (1998). Isolation phenomena for quaternionic Yang–Mills connections. *Osaka J. Math.*, **35**, n.1, 147–164.

284 Tanno, S. (1965). Almost complex structures in bundle spaces over almost contact manifolds. *J. Math. Soc. Japan*, **17**, 167–186.

285 Tanno, S. (1971). Killing vectors on contact Riemannian manifolds and fiberings related to the Hopf fibrations. *Tôhoku Math. J.*, **23**, 313–333.

286 Tanno, S. (1996). Remarks on a triple of K-contact structures. *Tôhoku Math. J.*, **48**, 519–531.

287 Tapp, K. (2002). Finitess theorems for submersions and souls. *Proc. Am. Math. Soc.*, **130**, 1809–1817.

288 Tondeur, P. (1988). *Foliations on Riemannian Manifolds.* Springer-Verlag, New York.

289 Tondeur, P. (1994). *Geometry of foliations.* Monographs in Mathematics, **90**, Birkhäuser Basel.

290 Toth, G. (1982). Harmonic submersions onto nonegatively curved manifolds. *Acta Math. Acad. Sci. Hung.*, **39**, 49–53.

291 Toth, G. (1983). One-parameter families of harmonic maps into spaces of constant curvature. *Stud. Sci. Math. Hung.*, **18**, 183–193.

292 Trautman, A. (1977). Solutions of the Maxwell and Yang–Mills equations associated with Hopf fiberings. *Int. J. Theor. Phys.*, **16**, 561–565.

293 Trautman, A. (1979). The geometry of gauge fields. *Czechoslovak J. Phys.*, **B, 29**, 107–116.

294 Tricerri, F., Vanhecke, L. (1981). Curvature tensors on almost Hermitian manifolds. *Trans. Am. Math. Soc.*, **267**, 365–398.

295 Tricerri, F., Vanhecke, L. (1983). *Homogeneous structures on Riemannian manifolds.* London Math. Soc., Lecture Note Series, **83**, Cambridge Univ. Press, Cambridge.

296 Tshikuna-Matamba, T. (1993). On the φ-linearity of the configuration tensor of an almost contact metric submersion. *Bull. Math. Soc. Sci. Math. Roum.*, **37**, 165–174.

297 Tshikuna-Matamba, T. (1994). Submersions metriques presque de contact sur le nouvelles variétés de Kenmotsu. *An. Stiint. Univ. "Al.I. Cuza" Iasi*, Ser. Noua Mat., **40**, 117–126.

298 Tshikuna-Matamba, T. (1995). A classification of almost contact metric submersions based on the fibres. *An. Stiint. Univ. "Al.I. Cuza" Iasi*, Ser. noua Mat., **41**, 135–144.

299 Tshikuna-Matamba, T. (1997). On the structure of the base space and the fibres of an almost contact metric submersion. *Houston J. Math.*, **23**, 291–305.

300 Tsukada, K. (1997). Holomorphic maps of compact generalized Hopf manifolds. *Geom. Dedicata*, **68**, 61–71.

301 Tsukada, K. (1999). The canonical foliation of a compact generalized Hopf manifold. *Diff. Geom. Appl.*, **11**, 13–28.

302 Ucci, J. (1983). On the nonexistence of Riemannian submersions from CP(7)

and QP(3). *Proc. Am. Math. Soc.*, **88**, 698–700.

303 Udriste, C. (1969). Structures presque coquaternioniennes. *Bull. Math. Soc. Sci. Math., R. S. Roumanie*, **13**, 487–507.

304 Urakawa, H. (1987). Stability of harmonic maps and eigenvalues of the Laplacian. *Trans. Am. Math. Soc.*, **301**, 557–589.

305 Vaisman, I. (1976). On locally conformal almost Kähler manifolds. *Israel J. Math.*, **24**, 338–351.

306 Vaisman, I. (1979). Locally conformal Kähler manifolds with parallel Lee form. *Rend. Mat.*, Roma, V Ser., **12**, 263–284.

307 Vaisman, I. (1980). Some curvature properties of locally conformal Kaehler manifolds. *Trans. Am. Math. Soc.*, **259**, 439–447.

308 Vaisman, I. (1980). Conformal changes of almost contact metric structures. In *Geometry and Differential Geometry, Proc. Conf. Haifa* 1979, (ed. Artzy, R., Vaisman, I.), Lect. Notes Math., **792**, 435–443.

309 Vaisman, I. (1981). A geometric condition for a locally conformal Kähler manifold to be Kähler. *Geom. Dedicata*, **10**, 129–134.

310 Vaisman, I. (1982). Generalized Hopf manifolds. *Geom. Dedicata*, **13**, 231–255.

311 Vaisman, I. (1983). A survey of generalized Hopf manifolds. In *Differential Geometry of Homogeneous Spaces, Conf.*, Torino-Italy, *Rend. Semin. Mat. Torino*, Fasc. spec., 205–221.

312 Vaman, D., Visinescu, M. (1998). Spinning particles in Taub-NUT space. *Phys. Rev.*, **D57**, 3790–3793.

313 Vilms, J. (1970). Totally geodesic maps. *J. Differ. Geom.*, **4**, 73–79.

314 Walczak, P.G. (1992). A finiteness theorem for Riemannian submersions. *Ann. Polon. Math.*, **57**, 283–290.

315 Walczak, P.G. (1993). Erratum to the paper. A finiteness theorem for Riemannian submersions. *Ann. Polon. Math.*, **58**, 319.

316 Walschap, G. (1994). Soul-preserving submersions. *Michigan. Math. J.*, **41**, 609–617.

317 Walschap, G. (1995). Causality relations on a class of spacetimes. *Gen. Relativity-Gravitation*, **27**, 721–733.

318 Walschap, G. (2002). On the geometry of the base space of a metric fibration. *Differential Geometry Appl.*, **16**, 141–147.

319 Wang, M. (1986). Einstein metrics with positive scalar curvature. In *Curvature and Topology of Riemannian manifolds (Proc. 17th. Int. Taniguchi Symp.* Katata, Jap., 1985), 319–336, Lect. Notes Math., **1201**, Springer, Berlin-New York.

320 Wang, M., Ziller, W. (1985). On normal homogeneous Einstein manifolds. *Ann. Sci. Ec. Norm. Sup.*, **18**, 563–633.

321 Wang, M., Ziller, W. (1986). Existence and non-existence of homogeneous Einstein metrics. *Invent. Math.*, **84**, 177–194.

322 Wang, M., Ziller, W. (1990). Einstein metrics on principal torus bundles. *J. Differ. Geom.*, **31**, 215–248.

323 Watanabe, Y., Mori, H. (1998). From Sasakian 3-structures to quaternionic geometry. *Arch. Math.*, (Brno), **34**, no.3 379–386.

324 Watson, B. (1973). Manifold maps commuting with the Laplacian. *J. Diff. Geom.*, **8**, 85–94.
325 Watson, B. (1975). δ-commuting mappings and Betti numbers. *Tôhoku Math. J.*, II Ser., **27**, 135–152.
326 Watson, B. (1976). Almost Hermitian submersions. *J. Diff. Geom.*, **11**, 147–165.
327 Watson, B. (1980). Riemannian submersions and instantons. *Math. Modelling*, **1**, 381–393.
328 Watson, B. (1983). G, G'-Riemannian submersions and nonlinear gauge field equations of general relativity. In *Global Analysis-Analysis on Manifolds*, dedic. M. Morse, (ed. Rassias, T.), Teubner-Texte Math., Leipzig, **57**, 324–349.
329 Watson, B. (1984). Almost contact metric 3-submersions. *Int. J. Math. Math. Sci.*, **7**, 667–688.
330 Watson, B. (1991). The differential geometry of two types of almost contact metric submersions. In *The mathematical heritage of C.F. Gauss*, (ed. Rassias, G.), World Scientific Publ., 827–861.
331 Watson, B. (2000). The four-dimensional Goldberg conjecture is true for almost Kähler submersions. *J. Geom.*, **69**, 215–226.
332 Watson, B. (2000). Superminimal fibres in an almost Hermitian submersion. *Boll. Un. Mat. Ital.*, (**8**), I-B, 159–172.
333 Watson, B., Vanhecke, L. (1977). J-symmetries and J-linearities of the configuration tensors of an almost Hermitian submersion. *Simon Stevin*, **51**, 139–156.
334 Watson, B., Vanhecke, L. (1978). K_i-curvatures and almost Hermitian submersions. *Rend. Sem. Mat. Torino*, **36**, 205–224.
335 Watson, B., Vanhecke, L. (1979). The structure equation of an almost semi-Kähler submersion. *Houston Math. J.*, **5**, 295–305.
336 Weinberg, S. (1967). A model of Leptons. *Phys. Rev. Lett.*, **19**, 1264–1266.
337 Weinstein, A. (1980). Fat bundles and symplectic manifolds. *Adv. Math.*, **37**, 239–250.
338 Wilking, B. (2001). Index parity of closed geodesics and rigidity of Hopf fibrations. *Invent. Math.*, **144**, No.2, 281–295.
339 Wolf, J.A. (1963). Elliptic spaces in Grassmann manifolds. *Ill. J. Math.*, **7**, 447–462.
340 Wolf, J.A. (1967). *Spaces of constant curvature.* Mc Graw-Hill.
341 Wong, Y.C. (1961). Isoclinic n-planes in Euclidean $2n$-space, Clifford parallels in elliptic $(2n-1)$-space, and the Hurwitz matrix equations. *Mem. Amer. Math. Soc.*, **41**, 1–112.
342 Wu, T.T., Yang, C.N. (1975). Concept of non-integrable phase factors and global formulations of gauge fields. *Phys. Rev.*, **D12**, 3845–3852.
343 Yamaguchi, S., Chuman, G. (1983). Critical Riemannian metrics on Sasakian manifolds. *Kodai Math. J.*, **6**, 1–13.
344 Yanamoto, H. (1972). Vector cross Product on 7-dimensional Riemannian manifold. *Res. Pep. Nagaoka Tech. Coll.*, **8**, 145–147.
345 Yang, C.N., Mills, R.L. (1954). Conservation of Isotopic Spin and Isotopic

Gauge Invariance. *Phys. Rev.*, **96**, 191–195.

346 Yano, K. (1963). On a structure defined by a tensor field of type $(1, 1)$ satisfying $f^3 + f = 0$. *Tensor*, New Ser., **14**, 99–109.

347 Yano, K., Ishihara, S. (1973). *Tangent and cotangent bundles.* Marcel Dekker, New York.

348 Yano, K., Kon, M. (1976). *Antiinvariant submanifolds.* Lect. Notes Pure Applied Math., **21**, Marcel Dekker, New York.

349 Yano, K., Kon, M. (1980). Generic submanifolds of Sasakian manifolds. *Kodai Math. J.*, **3**, 163–196.

350 Yano, K., Kon, M. (1984). *Structures on manifolds.* Series in Pure Mathematics, **3**, World Scientific, Singapore.

351 T.Yasui, T. (1992). Vector bundle epimorphisms and submersions. *Bol. Soc. Mat. Mexicana*, II Ser., **37**, 579–585.

352 Xin, Y.L. (1993). Riemannian submersions and equivariant harmonic maps. In *Differential geometry, Proceedings of the Symposium in honour of Prof. Su Buchin*, (Shangai, China, 1991), (ed. Gu, C.H., Hu, H.S., Xin, Y.L.), World Sc., Singapore, 272–287.

353 Zhukova, N. (1988). Submersions with an Ehresmann connection. *Sov. Math.*, **32**, 27–38, transl. from *Iz. VUZ, Mat.*, **312**.

354 Ziller, W. (1982). Homogeneous Einstein metrics on Spheres and Projective spaces. *Math. Ann.*, **259**, 351–358.

Index

Printed in the United States
By Bookmasters